Edward L. Wolf

Nanophysics and Nanotechnology

Edward L. Wolf

Nanophysics and Nanotechnology

An Introduction to Modern Concepts in Nanoscience

Second, Updated and Enlarged Edition

Wiley India Pvt. Ltd.

Nanophysics and Nanotechnology
An Introduction to Modern Concepts in Nanoscience
Second, Updated and Enlarged Edition

Authorized reprint by Wiley India Pvt. Ltd., 4435-36/7, Ansari Road, Daryaganj, New Delhi – 110002.

Reprint: 2013

Printed at: *Chaudhary Press, Delhi*

ISBN: 978-81-265-3843-0

To

Carol, Doug, Dave, Ben

And

Phill, Ned, Dan, Mehdi, Michael

Preface

Nanophysics, in this non-specialist book, deals with physical effects at the nanometer and sub-nanometer scales; particularly aspects of importance to the *smallest size scales of any possible technology.*

"Nanophysics" thus includes physical laws applicable from the 100 nm scale down to the sub-atomic, sub-0.1 nm, scale. This includes "quantum mechanics" as advanced by the theoretical physicist Erwin Schrodinger, ca. 1925; "mesocale physics", with more diverse and recent origins; and the physics of the atomic nucleus, on the 10^{-15} m (fm) scale. From a pedagogical point of view, the 1 nm scale requires the concepts of "quantum mechanics" (sometimes here described as "nanophysics") which, once introduced, are key to understanding behavior down to the femtometer scale of the atomic nucleus.

New material in the 2nd Edition centers on "nanoelectronics", from magnetic and quantum points of view, and also relating to the possibilities for "quantum computing" as an extension of the existing successful silicon technology. The new Chapter 8 is called "Quantum technologies based on magnetism, electron spin, superconductivity", and is followed by the new Chapter 9 titled "Silicon nanoelectronics and beyond". New electronics-related applications of carbon nanotubes are included. Sections have been added on superconductivity: a concrete example of quantum coherence, and to help understand devices of the "rapid single flux quantum" (RSFQ) computer logic (already mentioned in the original Chapter 7), notable for low power dissipation and fast operation. The old Chapter 8 ("Looking into the Future") becomes the new Chapter 10.

Additional material has been added (in Chapters 4 and 5, primarily), giving concepts needed for the most important new areas, including the absolutely most recent advances in nanotechnology. The basic ideas of ferromagnetic interactions and quantum computing, now included, are central to any quantum- or magnetic-based technology. The new edition is more self-contained, with the addition of a short list of useful constants and a glossary.

A criterion in choice of new material (many astonishing developments have occurred since the 2004 publication of the 1st Edition of this book) has been the author's view of what may be important in the development of nanotechnology. For this reason, nuclear physics is now touched on (briefly), in connection with proposals to use the "nuclear spin ½" as the "qubit" of information in a "quantum com-

Nanophysics and Nanotechnology: An Introduction to Modern Concepts in Nanoscience. Second Edition.
Edward L. Wolf

ISBN: 3-527-40651-4

puter"; and with a recent small-scale experiment demonstrating neutron generation (via a standard "nuclear fusion reaction") which exploits nanotechnology for its success.

Another essential and relevant aspect of fundamental physics, the "exchange interaction of identical particles", has already been incorporated, as essential to a basic understanding of covalent bonds, ferromagnetism (essential to computer disk drive nanotechnology), and, more recently, to proposals for a "charge-qubit" for a quantum computer. This topic (the exchange interaction) is of importance beyond being the basis for covalent bonds in organic chemistry.

From the beginning, this book was intended as an introduction to the phenomena and laws of nature applicable on such tiny size scales (without excluding the nuclear, fm, size scale) for those who have taken college mathematics and physics, but who have not necessarily studied atomic physics or nuclear physics. Primarily, the reader will need facility with numbers, and an interest in new ideas.

The Exercises have been conceived more as self-learning aids for the interested reader, than as formal problems. Some new material, especially in regard to field-ionization by tips, and aspects of the collapse of ultrasonically induced bubbles in dense liquids, appears now in the Exercises, not to clutter the text for the more general reader.

It is hoped that the interested reader can find stimulating, even profitable, new ideas in this (still rather slim) book. For details, the reader can use the copious and absolutely current references that are included.

E. L. Wolf
New York
February, 2006

Preface to 1st Edition

This book originated with an elective sequence of two upper level undergraduate Physics courses, which I initiated at Polytechnic University. "Concepts of Nanotechnology" and "Techniques and Applications of Nanotechnology" are taken in the spring of the junior year and the following fall, and the students have a number of such sequences to choose from. I have been pleased with the quality, diversity (of major discipline), interest, and enthusiasm of the students who have taken the "Nano" sequence of courses, now midway in the second cycle of offering. Electrical engineering, computer engineering, computer science, mechanical engineering and chemical engineering are typical majors for these students, which facilitates breaking the class into interdisciplinary working groups who then prepare term papers and presentations that explore more deeply topics of their choice within the wealth of interesting topics in the area of nanotechnology. The Physics prerequisite for the course is 8 hours of calculus-based physics. The students have also had introductory Chemistry and an exposure to undergraduate mathematics and computer science.

I am grateful to my colleagues in the Interdisciplinary Physics Group for helping formulate the course, and in particular to Lorcan Folan and Harold Sjursen for help in getting the course approved for the undergraduate curriculum. Iwao Teraoka suggested, since I told him I had trouble finding a suitable textbook, that I should write such a book, and then introduced me to Ed Immergut, a wise and experienced consulting editor, who in turn helped me transform the course outlines into a book proposal. I am grateful to Rajinder Khosla for useful suggestions on the outline of the book. At Wiley-VCH I have benefited from the advice and technical support of Vera Palmer, Ron Schultz, Ulrike Werner and Anja Tschortner. At Polytechnic I have also been helped by DeShane Lyew and appreciate discussions and support from Stephen Arnold and Jovan Mijovic. My wife Carol has been a constant help in this project.

I hope this modest book, in addition to use as a textbook at the upper undergraduate or masters level, may more broadly be of interest to professionals who have had a basic background in physics and related subjects, and who have an interest in the developing fields of nanoscience and nanotechnology. I hope the book may play a career enhancing role for some readers. I have included some exercises to go with each chapter, and have also set off some tutorial material in half-tone sections of text, which many readers can pass over.

At the beginning of the 21st century, with a wealth of knowledge in scientific and engineering disciplines, and really rapid ongoing advances, especially in areas of nanotechnology, robotics, and biotechnology, there may be a need also to look more broadly at the capabilities, opportunities, and possible pitfalls thus enabled. If there is to be a "posthuman era", a wide awareness of issues will doubtless be beneficial in making the best of it.

Edward L. Wolf
New York
July, 2004

Contents

Nanophysics and Nanotechnology: An Introduction to Modern Concepts in Nanoscience. Second Edition.
Edward L. Wolf

ISBN: 3-527-40651-4

1
Introduction

Technology has to do with the application of scientific knowledge to the economic (profitable) production of goods and services. This book is concerned with the size or scale of working machines and devices in different forms of technology. It is particularly concerned with the smallest devices that are possible, and equally with the appropriate laws of nanometer-scale physics: "nanophysics", which are available to accurately predict behavior of matter on this invisible scale. Physical behavior at the nanometer scale is predicted accurately by quantum mechanics, represented by Schrodinger's equation. Schrodinger's equation provides a quantitative understanding of the structure and properties of atoms. Chemical matter, molecules, and even the cells of biology, being made of atoms, are therefore, in principle, accurately described (given enough computing power) by this well tested formulation of nanophysics.

There are often advantages in making devices smaller, as in modern semiconductor electronics. What are the limits to miniaturization, how small a device can be made? Any device must be composed of atoms, whose sizes are the order of 0.1 nanometer. Here the word "nanotechnology" will be associated with human-designed working devices in which some essential element or elements, produced in a controlled fashion, have sizes of 0.1 nm to thousands of nanometers, or, one Angstrom to one micron. There is thus an overlap with "microtechnology" at the micrometer size scale. Microelectronics is the most advanced present technology, apart from biology, whose complex operating units are on a scale as small as micrometers.

Although the literature of nanotechnology may refer to nanoscale machines, even "self-replicating machines built at the atomic level" [1], it is admitted that an "assembler breakthrough" [2] will be required for this to happen, and no nanoscale machines presently exist. In fact, scarcely any micrometer μm scale machines exist either, and it seems that the smallest mechanical machines readily available in a wide variety of forms are really on the millimeter scale, as in conventional wristwatches. (To avoid confusion, note that the prefix "micro" is sometimes applied, but never in this book, to larger scale techniques guided by optical microscopy, such as "microsurgery".)

The reader may correctly infer that Nanotechnology is presently more concept than fact, although it is certainly a media and funding reality. That the concept has

Nanophysics and Nanotechnology: An Introduction to Modern Concepts in Nanoscience. Second Edition.
Edward L. Wolf

ISBN: 3-527-40651-4

great potential for technology, is the message to read from the funding and media attention to this topic.

The idea of the limiting size scale of a miniaturized technology is fundamentally interesting for several reasons. As sizes approach the atomic scale, the relevant physical laws change from the classical to the quantum-mechanical laws of nanophysics. The changes in behavior from classical, to "mesoscopic", to atomic scale, are broadly understood in contemporary physics, but the details in specific cases are complex and need to be worked out. While the changes from classical physics to nanophysics may mean that some existing devices will fail, the same changes open up possibilities for new devices.

A primary interest in the concept of nanotechnology comes from its connections with biology. The smallest forms of life, bacteria, cells, and the active components of living cells of biology, have sizes in the nanometer range. In fact, it may turn out that the only possibility for a viable complex nanotechnology is that represented by biology. Certainly the present understanding of molecular biology has been seen as an existence proof for "nanotechnology" by its pioneers and enthusiasts. In molecular biology, the "self replicating machines at the atomic level" are guided by DNA, replicated by RNA, specific molecules are "assembled" by enzymes and cells are replete with molecular scale motors, of which kinesin is one example. Ion channels, which allow (or block) specific ions (e.g., potassium or calcium) to enter a cell through its lipid wall, seem to be exquisitely engineered molecular scale devices where distinct conformations of protein molecules define an open channel vs. a closed channel.

Biological sensors such as the rods and cones of the retina and the nanoscale magnets found in magnetotactic bacteria appear to operate at the quantum limit of sensitivity. Understanding the operation of these sensors doubtless requires application of nanophysics. One might say that Darwinian evolution, a matter of odds of survival, has mastered the laws of quantum nanophysics, which are famously probabilistic in their nature. Understanding the role of quantum nanophysics entailed in the molecular building blocks of nature may inform the design of man-made sensors, motors, and perhaps much more, with expected advances in experimental and engineering techniques for nanotechnology.

In the improbable event that engineering, in the traditional sense, of molecular scale machines becomes possible, the most optimistic observers note that these invisible machines could be engineered to match the size scale of the molecules of biology. Medical nanomachines might then be possible, which could be directed to correct defects in cells, to kill dangerous cells, such as cancer cells, or even, most fancifully, to repair cell damage present after thawing of biological tissue, frozen as a means of preservation [3].

This book is intended to provide a guide to the ideas and physical concepts that allow an understanding of the changes that occur as the size scale shrinks toward the atomic scale. Our point of view is that a general introduction to the concepts of nanophysics will add greatly to the ability of students and professionals whose undergraduate training has been in engineering or applied science, to contribute in the various areas of nanotechnology. The broadly applicable concepts of nanophysics are worth study, as they do not become obsolete with inevitable changes in the forefront of technology.

1.1 Nanometers, Micrometers, Millimeters

A nanometer, 10^{-9} m, is about ten times the size of the smallest atoms, such as hydrogen and carbon, while a micron is barely larger than the wavelength of visible light, thus invisible to the human eye. A millimeter, the size of a pinhead, is roughly the smallest size available in present day machines. The range of scales from millimeters to nanometers is one million, which is also about the range of scales in present day mechanical technology, from the largest skyscrapers to the smallest conventional mechanical machine parts. The vast opportunity for making new machines, spanning almost six orders of magnitude from 1 mm to 1nm, is one take on Richard Feynman's famous statement [4]: "there is plenty of room at the bottom". If L is taken as a typical length, 0.1 nm for an atom, perhaps 2 m for a human, this scale range in L would be 2×10^{10}. If the same scale range were to apply to an area, 0.1 nm by 0.1 nm vs 2 m × 2 m, the scale range for area L^2 is 4×10^{20}. Since a volume L^3 is enclosed by sides L, we can see that the number of atoms of size 0.1 nm in a $(2\text{ m})^3$ volume is about 8×10^{30}. Recalling that Avogadro's number $N_A = 6.022\times10^{23}$ is the number of atoms in a gram-mole, supposing that the atoms were ^{12}C, molar mass 12 g; then the mass enclosed in the $(2\text{ m})^3$ volume would be 15.9×10^4 kg, corresponding to a density $1.99\times10^4\text{ kg/m}^3$ (19.9 g/cc). (This is about 20 times the density of water, and higher than the densities of elemental carbon in its diamond and graphitic forms (which have densities 3.51 and 2.25 g/cc, respectively) because the equivalent size L of a carbon atom in these elemental forms slightly exceeds 0.1 nm.)

A primary working tool of the nanotechnologist is facility in scaling the magnitudes of various properties of interest, as the length scale L shrinks, e.g., from 1 mm to 1 nm.

Clearly, the number of atoms in a device scales as L^3. If a transistor on a micron scale contains 10^{12} atoms, then on a nanometer scale, $L'/L = 10^{-3}$ it will contain 1000 atoms, likely too few to preserve its function!

Normally, we will think of scaling as an isotropic scale reduction in three dimensions. However, scaling can be thought of usefully when applied only to one or two dimensions, scaling a cube to a two-dimensional (2D) sheet of thickness a or to a one-dimensional (1D) tube or "nanowire" of cross sectional area a^2. The term "zero-dimensional" is used to describe an object small in all three dimensions, having volume a^3. In electronics, a zero-dimensional object (a nanometer sized cube a^3 of semiconductor) is called a "quantum dot" (QD) or "artificial atom" because its electron states are few, sharply separated in energy, and thus resemble the electronic states of an atom.

As we will see, a quantum dot also typically has so small a radius a, with correspondingly small electrical capacitance $C = 4\pi\varepsilon\varepsilon_o a$ (where $\varepsilon\varepsilon_o$ is the dielectric constant of the medium in which the QD is immersed), that the electrical charging energy $U = Q^2/2C$ is "large". (In many situations, a "large" energy is one that exceeds the thermal excitation energy, $k_B T$, for $T = 300$ K, basically room temperature. Here T is the absolute Kelvin temperature, and k_B is Boltzmann's constant, 1.38×10^{-23} J/K.) In this situation, a change in the charge Q on the QD by even one electron charge e, may effectively, by the "large" change in U, switch off the possibility of the QD being part of the path of flow for an external current.

This is the basic idea of the "single electron transistor". The role of the quantum dot or QD in this application resembles the role of the grid in the vacuum triode, but only one extra electron change of charge on the "grid" turns the device off. To make a device of this sort work at room temperature requires that the QD be tiny, only a few nm in size.

Plenty of room at the bottom

Think of reducing the scale of working devices and machines from 1mm to 1nm, six orders of magnitude! Over most of this scaling range, perhaps the first five orders of magnitude, down to 10 nm (100 Angstroms), the laws of classical Newtonian physics may well suffice to describe changes in behavior. This classical range of scaling is so large, and the changes in magnitudes of important physical properties, such as resonant frequencies, are so great, that completely different applications may appear.

Scaling the xylophone

The familiar xylophone produces musical sounds when its keys (a linear array of rectangular bars of dimensions $a \times b \times c$, with progressively longer key lengths c producing lower audio frequencies) are struck by a mallet and go into transverse vibration perpendicular to the smallest, a, dimension. The traditional "middle C" in music corresponds to 256 Hz. If the size scale of the xylophone key is reduced to the micrometer scale, as has recently been achieved, using the semiconductor technology, and the mallet is replaced by electromagnetic excitation, the same transverse mechanical oscillations occur, and are measured to approach the Gigahertz (10^9 Hz) range [5]!

The measured frequencies of the micrometer scale xylophone keys are still accurately described by the laws of classical physics. (Actually the oscillators that have been successfully miniaturized, see Figure 1.1, differ slightly from xylophone keys, in that they are clamped at both ends, rather than being loosely suspended. Very similar equations are known to apply in this case.) Oscillators whose frequencies approach the GHz range have completely different applications than those in the musical audio range!

Could such elements be used in new devices to replace Klystrons and Gunn oscillators, conventional sources of GHz radiation? If means could be found to fabricate "xylophone keys" scaling down from the micrometer range to the nanometer range, classical physics would presumably apply almost down to the molecular scale. The limiting vibration frequencies would be those of diatomic molecules, which lie in the range 10^{13} – 10^{14} Hz. For comparison, the frequency of light used in fiberoptic communication is about 2×10^{14} Hz.

Figure 1.1 Silicon nanowires in a harp-like array. Due to the clamping of the single-crystal silicon bars at each end, and the lack of applied tension, the situation is more like an array of xylophone keys. The resonant frequency of the wire of 2 micrometer length is about 400 MHz. After Reference [5].

Reliability of concepts and approximate parameter values down to about $L = 10$ nm (100 atoms)

The large extent of the "classical" range of scaling, from 1 mm down to perhaps 10 nm, is related to the stability (constancy) of the basic microscopic properties of condensed matter (conventional building and engineering materials) almost down to the scale L of 10 nm or 100 atoms in line, or a million atoms per cube.

Typical microscopic properties of condensed matter are the interatomic spacing, the mass density, the bulk speed of sound v_s, Young's modulus Y, the bulk modulus B, the cohesive energy U_o, the electrical resistivity ρ, thermal conductivity K, the relative magnetic and dielectric susceptibilities κ and ε, the Fermi energy E_F and work function φ of a metal, and the bandgap of a semiconductor or insulator, E_g. A timely example in which bulk properties are retained down to nanometer sample sizes is afforded by the CdSe "quantum dot" fluorescent markers, which are described below.

Nanophysics built into the properties of bulk matter

Even if we can describe the size scale of 1 mm – 10 nm as one of "classical scaling", before distinctly size-related anomalies are strongly apparent, a nanotechnologist must appreciate that many properties of bulk condensed matter already require concepts of nanophysics for their understanding. This might seem obvious, in that atoms themselves are completely nanophysical in their structure and behavior!

Beyond this, however, the basic modern understanding of semiconductors, involving energy bands, forbidden gaps, and effective masses m^* for free electrons and free holes, is based on nanophysics in the form of Schrodinger's equation as applied to a periodic structure.

Periodicity, a repeated unit cell of dimensions a,b,c (in three dimensions) profoundly alters the way an electron (or a "hole", which is the inherently positively charged absence of an electron) moves in a solid. As we will discuss more completely below, ranges (bands) of energy of the free carrier exist for which the carrier will pass through the periodic solid with no scattering at all, much in the same way that an electromagnetic wave will propagate without attenuation in the passband of a transmission line. In energy ranges between the allowed bands, gaps appear, where no moving carriers are possible, in analogy to the lack of signal transmission in the stopband frequency range of a transmission line.

So, the "classical" range of scaling as mentioned above is one in which the consequences of periodicity for the motions of electrons and holes (wildly "non-classical", if referred to Newton's Laws, for example) are unchanged. In practice, the properties of a regular array of 100 atoms on a side, a nanocrystal containing only a million atoms, is still large enough to be accurately described by the methods of solid state physics. If the material is crystalline, the properties of a sample of 10^6 atoms are likely to be an approximate guide to the properties of a bulk sample. To extrapolate the bulk properties from a 100-atom-per-side simulation may not be too far off.

It is probably clear that a basic understanding of the ideas, and also the fabrication methods, of semiconductor physics is likely to be a useful tool for the scientist or engineer who will work in nanotechnology. Almost all devices in the Micro-electromechanical Systems (MEMS) category, including accelerometers, related angular rotation sensors, and more, are presently fabricated using the semiconductor microtechnology.

The second, and more challenging question, for the nanotechnologist, is to understand and hopefully to exploit those changes in physical behavior that occur at the end of the classical scaling range. The "end of the scaling" is the size scale of atoms and molecules, where nanophysics is the proven conceptual replacement of the laws of classical physics. Modern physics, which includes quantum mechanics as a description of matter on a nanometer scale, is a fully developed and proven subject whose application to real situations is limited only by modeling and computational competence.

In the modern era, simulations and approximate solutions increasingly facilitate the application of nanophysics to almost any problem of interest. Many central problems are already (adequately, or more than adequately) solved in the extensive literatures of theoretical chemistry, biophysics, condensed matter physics and semiconductor device physics. The practical problem is to find the relevant work, and, frequently, to convert the notation and units systems to apply the results to the problem at hand.

It is worth saying that information has no inherent (i.e., zero) size. The density of information that can be stored is limited only by the coding element, be it a bead on an abacus, a magnetized region on a hard disk, a charge on a CMOS capacitor, a nanoscale indentation on a plastic recording surface, the presence or absence of a particular atom at a specified location, or the presence of an "up" or "down" electronic or nuclear spin (magnetic moment) on a density of atoms in condensed matter, $(0.1\,\text{nm})^{-3} = 10^{30}/\text{m}^3 = 10^{24}/\text{cm}^3$. If these coding elements are on a surface, then the limiting density is $(0.1\,\text{nm})^{-2} = 10^{20}/\text{m}^2$, or $6.45 \times 10^{16}/\text{in}^2$.

The principal limitation may be the physical size of the reading element, which historically would be a coil of wire (solenoid) in the case of the magnetic bit. The limiting density of information in the presently advancing technology of magnetic computer hard disk drives is about 100 Gb/in^2, or 10^{11}/in^2. It appears that non-magnetic technologies, perhaps based on arrays of AFM tips writing onto a plastic film such as polymethylmethacrylate (PMMA), may eventually overtake the magnetic technology.

1.2
Moore's Law

The computer chip is certainly one of the preeminent accomplishments of 20th century technology, making vastly expanded computational speed available in smaller size and at lower cost. Computers and email communication are almost universally available in modern society. Perhaps the most revolutionary results of computer technology are the universal availability of email to the informed and at least minimally endowed, and magnificent search engines such as Google. Without an unexpected return to the past, which might roll back this major human progress it seems rationally that computers have ushered in a new era of information, connectedness, and enlightenment in human existence.

Moore's empirical law summarizes the "economy of scale" in getting the same function by making the working elements ever smaller. (It turns out, as we will see, that smaller means faster, characteristically enhancing the advantage in miniaturization). In the ancient abacus, bead positions represent binary numbers, with infor-

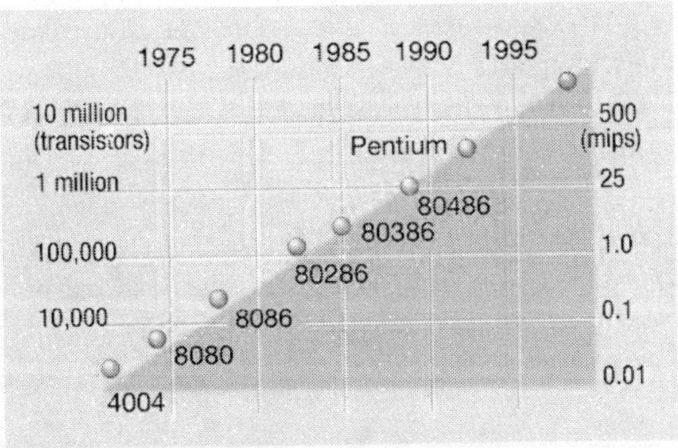

Figure 1.2 Moore's Law. [6]. The number of transistors in successive generations of computer chips has risen exponentially, doubling every 1.5 years or so. The notation "mips" on right ordinate is "million instructions per second". Gordon Moore, co-founder of Intel, Inc. predicted this growth pattern in 1965, when a silicon chip contained only 30 transistors! The number of Dynamic Random Access Memory (DRAM) cells follows a similar growth pattern. The growth is largely due to continuing reduction in the size of key elements in the devices, to about 100 nm, with improvements in optical photolithography. Clock speeds have similarly increased, presently around 2 GHz. For a summary, see [7].

mation recorded on a scale of perhaps 1 bit [(0,1) or (yes/no)] per cm^2. In silicon microelectronic technology an easily produced memory cell size of one micron corresponds to 10^{12} bits/cm^2 (one Tb/cm^2). Equally important is the continually reducing size of the magnetic disk memory element (and of the corresponding read/write sensor head) making possible the ~100 Gb disk memories of contemporary laptop computers. The continuing improvements in performance (reductions in size of the performing elements), empirically summarized by Moore's Law (a doubling of performance every 1.5 years, or so), arise from corresponding reductions in the size scale of the computer chip, aided by the advertising-related market demand.

The vast improvements from the abacus to the Pentium chip exemplify the promise of nanotechnology. Please note that this is all still in the range of "classical scaling"! The computer experts are absolutely sure that nanophysical effects are so far negligible.

The semiconductor industry, having produced a blockbuster performance over decades, transforming advanced society and suitably enriching its players and stockholders, is concerned about its next act!

The next act in the semiconductor industry, if a second act indeed shows up, must deal with the nanophysical rules. Any new technology, if such is feasible, will have to compete with a base of universally available applied computation, at unimaginably low costs. If Terahertz speed computers with 100 Mb randomly accessible memories and 100 Gb hard drives, indeed become a commodity, what can compete with that? Silicon technology is a hard act to follow.

Nanotechnology, taken literally, also represents the physically possible limit of such improvements. The limit of technology is also evident, since the smallest possible interconnecting wire on the chip must be at least 100 atoms across! Moore's law empirically has characterized the semiconductor industry's success in providing faster and faster computers of increasing sophistication and continually falling price. Success has been obtained with a larger number of transistors per chip made possible by finer and finer scales of the wiring and active components on the silicon chips. There is a challenge to the continuation of this trend (Moore's Law) from the economic reality of steeply increasing plant cost (to realize reduced linewidths and smaller transistors).

The fundamental challenge to the continuation of this trend (Moore's Law) from the change of physical behavior as the atomic size limit is approached, is a central topic in this book.

1.3 Esaki's Quantum Tunneling Diode

The tunneling effect is basic in quantum mechanics, a fundamental consequence of the probabilistic wave function as a measure of the location of a particle. Unlike a tennis ball, a tiny electron may penetrate a barrier. This effect was first exploited in semiconductor technology by Leo Esaki, who discovered that the current–voltage (I/V) curves of semiconductor p–n junction rectifier diodes (when the barrier was made very thin, by increasing the dopant concentrations) became anomalous, and in fact double-valued. The forward bias I vs. V plot, normally a rising exponential $\exp(eV/kT)$, was preceded by a distinct "current hump" starting at zero bias and extending to $V = 50\,\text{mV}$ or so. Between the region of the "hump" and the onset of the conventional exponential current rise there was a region of negative slope, $dI/dV < 0$!

The planar junction between an N-type region and a P-type region in a semiconductor such as Si contains a "depletion region" separating conductive regions filled with free electrons on the N-side and free holes on the P-side. It is a useful non-trivial exercise in semiconductor physics to show that the width W of the depletion region is

$$W = [2\varepsilon\varepsilon_o V_B(N_D + N_A)/e(N_D N_A)]^{1/2}\,. \tag{1.1}$$

Here $\varepsilon\varepsilon_o$ is the dielectric constant, e the electron charge, V_B is the energy shift in the bands across the junction, and N_D and N_A, respectively, are the concentrations of donor and acceptor impurities.

The change in electrical behavior (the negative resistance range) resulting from the electron tunneling (in the thin depletion region limit) made possible an entirely new effect, an oscillation, at an extremely high frequency! (As often happens with the continuing advance of technology, this pioneering device has been largely supplanted as an oscillator by the Gunn diode, which is easier to manufacture.)

The Esaki tunnel diode is perhaps the first example in which the appearance of quantum physics at the limit of a small size led to a new device. In our terminology the depletion layer tunneling barrier is two-dimensional, with only one small dimension, the depletion layer thickness W. The Esaki diode falls into our classification as an element of nanotechnology, since the controlled small barrier W is only a few nanometers in thickness.

1.4 Quantum Dots of Many Colors

"Quantum dots" (QDs) of CdSe and similar semiconductors are grown in carefully controlled solution precipitation with controlled sizes in the range $L = 4$ or 5 nm. It is found that the wavelength (color) of strong fluorescent light emitted by these quantum dots under ultraviolet (uv) light illumination depends sensitively on the size L.

There are enough atoms in this particle to effectively validate the concepts of solid state physics, which include electron bands, forbidden energy band gaps, electron and hole effective masses, and more.

Still, the particle is small enough to be called an "artificial atom", characterized by discrete sharp electron energy states, and discrete sharp absorption and emission wavelengths for photons.

Transmission electron microscope (TEM) images of such nanocrystals, which may contain only 50 000 atoms, reveal perfectly ordered crystals having the bulk

Figure 1.3 Transmission Electron Micrograph (TEM) Image of one 5 nm CdSe quantum dot particle, courtesy Andreas Kornowski, University of Hamburg, Germany.

crystal structure and nearly the bulk lattice constant. Quantitative analysis of the light emission process in QDs suggests that the bandgap, effective masses of electrons and holes, and other microscopic material properties are very close to their values in large crystals of the same material. The light emission in all cases comes from radiative recombination of an electron and a hole, created initially by the shorter wavelength illumination.

The energy E_R released in the recombination is given entirely to a photon (the quantum unit of light), according to the relation $E_R = h\nu = hc/\lambda$. Here ν and λ are, respectively, the frequency and wavelength of the emitted light, c is the speed of light 3×10^8 m/s, and h is Planck's constant $h = 6.63 \times 10^{-34}$ Js $= 4.136 \times 10^{-15}$ eVs. The color of the emitted light is controlled by the choice of L, since $E_R = E_G + E_e + E_h$, where E_G is the semiconductor bandgap, and the electron and hole confinement energies, E_e and E_h, respectively, become larger with decreasing L.

It is an elementary exercise in nanophysics, which will be demonstrated in Chapter 4, to show that these confinement (blue-shift) energies are proportional to $1/L^2$. Since these terms increase the energy of the emitted photon, they act to shorten the wavelength of the light relative to that emitted by the bulk semiconductor, an effect referred to as the "blue shift" of light from the quantum dot.

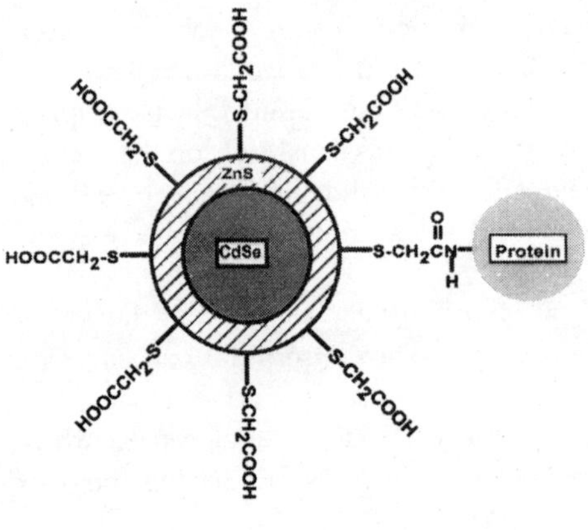

Figure 1.4 Schematic of quantum dot with coatings suitable to assure water solubility, for application in biological tissue. This ZnS-capped CdSe quantum dot is covalently coupled to a protein by mercaptoacetic acid. The typical QD core size is 4.2 nm. [8]

These nanocrystals are used in biological research as markers for particular kinds of cells, as observed under an optical microscope with background ultraviolet light (uv) illumination.

In these applications, the basic semiconductor QD crystal is typically coated with a thin layer to make it impervious to (and soluble in) an aqueous biological environment. A further coating may then be applied which allows the QD to preferentially bond to a specific biological cell or cell component of interest. Such a coated quantum dot is shown in Figure 1.4 [8]. These authors say that the quantum dots they use as luminescent labels are 20 times as bright, 100 times as stable against photobleaching, and have emission spectra three times sharper than conventional organic dyes such as rhodamine.

The biological researcher may, for example, see the outer cell surface illuminated in green while the surface of the inner cell nucleus may be illuminated in red, all under illumination of the whole field with light of a single shorter wavelength.

1.5 GMR 100 Gb Hard Drive "Read" Heads

In modern computers, the hard disk encodes information in the form of a linear array of planar ferromagnetic regions, or bits. The performance has recently improved with the discovery of the Giant Magnetoresistance (GMR) effect allowing a smaller assembly (read head), to scan the magnetic data. The ferromagnetic bits are written into (and read from) the disk surface, which is uniformly coated with a ferromagnetic film having a small coercive field. This is a "soft" ferromagnet, so that a small imposed magnetic field B can easily establish the ferromagnetic magnetization M along the direction of the applied B field. Both writing and reading operations are accomplished by the "read head".

The density of information that can be stored in a magnetic disk is fundamentally limited by the minimum size of a ferromagnetic "domain". Ferromagnetism is a cooperative nanophysical effect requiring a minimum number of atoms: below this number the individual atomic magnetic moments remain independent of each other. It is estimated that this "super-paramagnetic limit" is on the order of 100 Gb/in^2.

The practical limit, however, has historically been the size of the "read head", as sketched in Figure 1.5, which on the one hand impresses a local magnetic field **B** on the local surface region to create the magnetized domain, and then also senses the magnetic field of the magnetic domain so produced. In present technology, the linear bits are about 100 nm in length (**M** along the track) and have widths in the range 0.3–1.0 µm. The ferromagnetic domain magnetization **M** is parallel or anti-parallel to the linear track.

The localized, perpendicular, magnetic fields **B** that appear at the junctions between parallel and anti-parallel bits are sensed by the read head. The width of the transition region between adjacent bits, in which the localized magnetic field is present, is between 10 and 100 nm. The localized **B** fields extend linearly across the track and point upward (or down) from the disk surface, as shown in Figure 1.5.

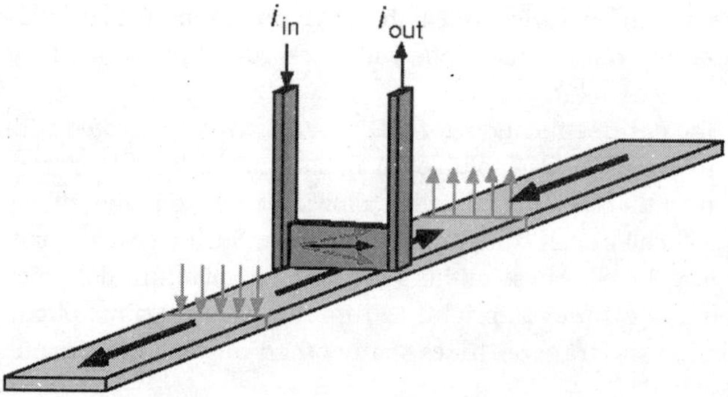

Figure 1.5 Schematic diagram of the GMR read head, showing two current leads connected by the sensing element, which itself is a conducting copper sheet sandwiched between a hard and a soft magnet [9].

The state-of-the-art magnetic field sensor is an exquisitely thin sandwich of magnetic and non-magnetic metals oriented vertically to intercept the fringe **B** field between adjacent bits. The total thickness of the sensor sandwich, along the direction of the track, is presently about 80 nm, but this thickness may soon fall to 20 nm. The GMR detector sandwich is comprised of a sensing soft ferromagnetic layer of NiFe alloy, a Cu spacer, and a "magnetically hard" Co ferromagnetic film. A sensing current is directed along the sandwich in the direction transverse to the track, and the voltage across the sandwich, which is sensitive to the magnetic field in the plane of the sandwich, is measured. In this read-head sandwich the Cu layer is about 15 atoms in thickness! The sensitivity of this GMR magnetoresistive sensor is presently in the vicinity of 1% per Oersted.

Writing the magnetic bits is accomplished by an integrated component of the "read head" (not shown in Figure 1.5) which generates a surface localized magnetic field **B** parallel or antiparallel to the track. The local surface magnetic field is produced inductively in a closed planar loop, resembling an open-ended box, of thin magnetic film, which is interrupted by a linear gap facing the disk, transverse to the track.

These "read head" units are fabricated in mass, using methods of the silicon lithographic microtechnology. It is expected that even smaller sensor devices and higher storage densities may be possible with advances in the silicon fabrication technology.

These units, which have had a large economic impact, are a demonstration of nanotechnology in that their closely controlled dimensions are in the nm range. The mechanism of greatly enhanced magnetic field sensitivity in the Giant Magnetoresistance Effect (GMR) is also fully nanophysical in nature, an example of (probably unexpectedly) better results at the quantum limit of the scaling process.

1.6
Accelerometers in your Car

Modern cars have airbags which inflate in crashes to protect drivers and passengers from sudden accelerations. Micro-electro-mechanical semiconductor acceleration sensors (accelerometers) are located in bumpers which quickly inflate the airbags. The basic accelerometer is a mass m attached by a spring of constant k to the frame of the sensor device, itself secured to the automobile frame. If the car frame (and thus the frame of the sensor device) undergo a strong acceleration, the spring connecting the mass to the sensor frame will extend or contract, leading to a motion of the mass m relative to the frame of the sensor device. This deflection is measured, for example, by change in a capacitance, which then triggers expansion of the airbag. This microelectromechanical (MEM) device is mass-produced in an integrated package including relevant electronics, using the methods of silicon microelectronics.

Newton's laws of motion describe the position, x, the velocity, $v = \mathrm{d}x/\mathrm{d}t$, and the acceleration $a = \mathrm{d}^2x/\mathrm{d}t^2$ of a mass m which may be acted upon by a force F, according to

$$m\mathrm{d}^2x/\mathrm{d}t^2 = \mathrm{F}. \tag{1.2}$$

Kinematics describes the relations among x, v, a, and t. As an example, the time-dependent position x under uniform acceleration a is

$$x = x_o + v_o t + at^2, \tag{1.3}$$

where x_o and v_o, respectively, are the position and velocity at $t = 0$. Also, if a time-varying acceleration $a(t)$ is known, and $x_o = 0$ and $v_o = 0$, then

$$x(t) = \iint a(t)\mathrm{d}t^2. \tag{1.4}$$

Newton's First Law states that in the absence of a force, the mass m remains at rest if initially so, and if initially in motion continues unchanged in that motion. (These laws are valid only if the coordinate system in which the observations are made is one of uniform motion, and certainly do not apply in an accelerated frame of reference such as a carrousel or a merry-go-round. For most purposes the earth's surface, although accelerated toward the earth's rotation axis, can be regarded as an "inertial frame of reference", i.e., Newton's Laws are useful.)

The Second Law is $F = ma$, (1.2).

The Third Law states that for two masses in contact, the force exerted by the first on the second is equal and opposite to the force exerted by the second on the first.

A more sophisticated version of such an accelerometer, arranged to record accelerations in x,y,z directions, and equipped with integrating electronics, can be used to record the three-dimensional displacement over time.

These devices are not presently built on a nanometer scale, of course, but are one example of a wide class of microelectronic sensors that could be made on smaller scales as semiconductor technology advances, and if smaller devices are useful.

1.7 Nanopore Filters

The original nanopore (Nuclepore) filters [10,11] are sheets of polycarbonate of 6 – 11 μm thickness with closely spaced arrays of parallel holes running through the sheet. The filters are available with pore sizes rated from 0.015 μm – 12.0 μm (15 nm – 12 000 nm). The holes are made by exposing the polycarbonate sheets to perpendicular flux of ionizing α particles, which produce linear paths of atomic scale damage in the polycarbonate. Controlled chemical etching is then employed to establish and enlarge the parallel holes to the desired diameter. This scheme is an example of nanotechnology.

The filters are robust and can have a very substantial throughput, with up to 12% of the area being open. The smallest filters will block passage of bacteria and perhaps even some viruses, and are used in many applications including water filters for hikers.

A second class of filters (Anapore) was later established, formed of alumina grown by anodic oxidation of aluminum metal. These filters are more porous, up to 40%, and are stronger and more temperature resistant than the polycarbonate filters. The Anapore filters have been used, for example, to produce dense arrays of nanowires. Nanowires are obtained using a hot press to force a ductile metal into the pores of the nanopore alumina filter.

1.8 Nanoscale Elements in Traditional Technologies

From the present knowledge of materials it is understood that the beautiful colors of stained glass windows originate in nanometer scale metal particles present in the glass. These metal particles have scattering resonances for light of specific wavelengths, depending on the particle size L. The particle size distribution, in turn, will depend upon the choice of metal impurity, its concentration, and the heat treatment of the glass. When the metallic particles in the glass are illuminated, they preferentially scatter light of particular colors. Neutral density filters marketed for photographic application also have distributions of small particles embedded in glass.

Carbon black, commonly known as soot, which contains nanometer-sized particles of carbon, was used very early as an additive to the rubber in automobile tires.

(As we now know, carbon black contains small amounts of ^{60}C (Buckminsterfullerene), other fullerenes, and graphitic nanotubes of various types.)

The AgBr and AgI crystals of conventional photography are nanometer-sized single crystals embedded in a thin gelatin matrix. It appears that the fundamental light absorption in these crystals is close to the quantum sensitivity limit. It further appears that the nanoscopic changes in these tiny crystals, which occur upon absorption of one or more light photons, enable them to be turned into larger metallic silver particles in the conventional photographic development process. The conventional photographic negative image is an array of solid silver grains embedded in a gelatin matrix. As such it is remarkably stable as a record, over decades or more.

The drugs that are so important in everyday life (and are also of huge economic importance), including caffeine, aspirin and many more, are specific molecules of nanometer size, typically containing fewer than 100 atoms.

Controlled precipitation chemistry for example is employed to produce uniform nanometer spheres of polystyrene, which have long been marketed as calibration markers for transmission electron microscopy.

References

[1] R. Kurzweil, *The Age of Spiritual Machines,* (Penguin Books, New York, 1999), page 140.

[2] K. E. Drexler, *Engines of Creation,* (Anchor Books, New York, 1986), page 49.

[3] K. E. Drexler, op. cit., p. 268.

[4] R. Feynman, "There's plenty of room at the bottom", in *Miniaturization*, edited by H.D. Gilbert (Reinhold, New York, 1961).

[5] Reprinted with permission from D.W. Carr *et al.*, Appl. Phy. Lett. **75**, 920 (1999). Copyright 1999. American Institute of Physics.

[6] Reprinted with permission from Nature: P. Ball, Nature **406**, 118–120 (2000). Copyright 2000. Macmillan Publishers Ltd.

[7] M. Lundstrom, Science **299**, 210 (2003).

[8] Reprinted with permission from C.W. Warren & N. Shumig, Science **281**, 2016–2018 (1998). Copyright 1998 AAAS.

[9] Courtesy G.A. Prinz, U.S. Naval Research Laboratory, Washington, DC.

[10] These filters are manufactured by Nuclepore Corporation, 7035 Commerce Circle, Pleasanton, CA 94566.

[11] G.P.Crawford, L.M. Steele, R. Ondris-Crawford, G. S. Iannocchione, C. J. Yeager, J. W. Doane, and D. Finotello, J. Chem. Phys. **96**, 7788 (1992).

2
Systematics of Making Things Smaller, Pre-quantum

2.1
Mechanical Frequencies Increase in Small Systems

Mechanical resonance frequencies depend on the dimensions of the system at hand. For the simple pendulum, $\omega = (g/l)^{1/2}$, where l is the length of the pendulum rod and g is the acceleration of gravity. The period $T = 2\pi/\omega$ of the pendulum of a grandfather clock is exactly one second, and corresponds to a length l about one meter (depending on the exact local value of the gravitational acceleration, g, which is approximately 9.8 m/s^2). The relation $\omega = (g/l)^{1/2}$ indicates that the period of the pendulum scales as $\sqrt{l}$, so that for l one micron T is one ms; a one-micron sized grandfather clock (oscillator) would generate a 1000 Hz tone. Either clock could be used (in conjunction with a separate frequency measurement device or counter), to measure the local vertical acceleration $a = d^2y/dt^2$ according to $\omega = [(g+a)/l]^{1/2}$.

If used in this way as an accelerometer, note that the miniature version has a much faster response time, 1ms, than the original grandfather clock, which would have a time resolution of about a second.

A mass m attached to a rigid support by a spring of constant k has a resonance frequency $\omega = (k/m)^{1/2}$. This oscillator, and the pendulum of the grandfather clock, are examples of simple harmonic oscillation (SHO).

Simple harmonic oscillation occurs when a displacement of a mass m in a given direction, x, produces an (oppositely directed) force $F = -kx$. The effective spring constant k has units of N/m in SI units. According to Newton's Second Law (1.2) $F = ma = md^2x/dt^2$. Applied to the mass on the spring, this gives the differential equation

$$d^2x/dt^2 + (k/m)x = 0. \tag{2.1}$$

$$x = x_{max}\cos(\omega t + \delta) \tag{2.2}$$

is a solution of the equation for arbitrary amplitude x_{max} and arbitrary initial phase angle δ, but only when

$$\omega = (k/m)^{1/2}. \tag{2.3}$$

Nanophysics and Nanotechnology: An Introduction to Modern Concepts in Nanoscience. Second Edition.
Edward L. Wolf

ISBN: 3-527-40651-4

The period of the motion is therefore $T = 2\pi(m/k)^{1/2}$. The maximum values of the speed $v = dx/dt$ and the acceleration d^2x/dt^2 are seen to be $x_{max}\omega$ and $x_{max}\omega^2$, respectively. The total energy $E = U + K$ in the motion is constant and equal to $1/2kx_{max}^2$. (In nanophysics, which is needed when the mass m is on an atomic scale, the same frequency $\omega = (k/m)^{1/2}$ is found, but the energies are restricted to $E_n = (n + 1/2)\,\hbar\omega$, where the quantum number n can take zero or positive integer values, $\hbar = h/2\pi$, and Planck's constant is $h = 6.67 \times 10^{-34}$ Joule · s.)

Simple harmonic oscillation is a more widely useful concept than one might think at first, because it is applicable to any system near a minimum, say x_o, in the system potential energy $U(x)$. Near x_o the potential energy $U(x)$ can be closely approximated as a constant plus $k(x-x_o)^2/2$, leading to the same resonance frequency for oscillations of amplitude A in $x-x_o$. An important example is in molecular bonding, where x_o is the interatomic spacing.

More generally, the behavior applies whenever the differential equation appears, and the resonant frequency will be the square root of the coefficient of x in the equation. In the case of the pendulum, if $x \cong L$ is the horizontal displacement of the mass m., then $F \cong -g\,x/L$ and $\omega = (g/L)^{1/2}$.

Considering the mass and spring to be three-dimensional, mass m will vary as L^3 and k will vary as L, leading to $\omega \propto a\,L^{-1}$. Frequency inversely proportional to length scale is typical of mechanical oscillators such as a violin or piano string and the frequency generated by a solid rod of length L struck on the end. In these cases the period of the oscillation T is the time for the wave to travel $2L$, hence $T = 2L/v$. (This is the same as $L = \lambda/2$, where $\lambda = vT$ is the wavelength. If the boundary conditions are different at the two ends, as in a clarinet, then the condition will be $L = \lambda/4$ with half the frequency.) Hence $\omega = 2\pi(v/L)$ where $v = (F/\rho)^{1/2}$ for the stretched string, where F is the tension and ρ the mass per unit length.

The speed of sound in a solid material is $v = (Y/\rho)^{1/2}$, with Y the Young's modulus. Young's modulus represents force per unit area (pressure stress) per fractional deformation (strain). Young's modulus is therefore a fundamental rigidity parameter of a solid, related to the bonding of its atoms. For brass, $Y = 90\,\text{GPa} = 90 \times 10^9\,\text{N/m}^2$. (This means, e.g., that a pressure F/A of 101 kPa applied to one end of a brass bar of length $L = 0.1$ m would compress its length by $\Delta l = LF/YA = 11\,\mu\text{m}$.) Note that $Y = 90$ GPa and $\rho = 10^4\,\text{kg/m}^3$, values similar to brass, correspond to a speed of sound $v = 3000$ m/s. On this basis the longitudinal resonant frequency of a 0.1 m brass rod is $f = v/2L = 15$ kHz. This frequency is in the ultrasonic range.

If one could shorten a brass rod to 0.1 micron in length, the corresponding frequency would be 15 GHz, which corresponds to an electromagnetic wave with 2 cm wavelength. This huge change in frequency will allow completely different applications to be addressed, achieved simply by changing the size of the device!

A connection between macroscopic and nanometer scale descriptions can be made by considering a linear chain of N masses m spaced by springs of constants K, of length a. The total length of the linear chain is thus $L = Na$.

Vibrations on a Linear Atomic Chain of length $L = Na$

On a chain of N masses of length L, and connected by springs of constant K, denote the longitudinal displacement of the nth mass from its equilibrium position by u_n. The differential equation (Newton's Second Law) $F = ma$ for the nth mass is

$$md^2u_n/dt^2 + K(u_{n+1} - 2u_n - u_{n-1}) = 0. \qquad (2.4)$$

A traveling wave solution to this equation is

$$u_n = u_o\cos(\omega t + kna). \qquad (2.5)$$

Here kna denotes $kx = 2\pi x/\lambda$, where k is referred to as the wave number. Substitution of this solution into the difference equation reveals the auxiliary condition $m\omega^2 = 4K\sin^2(ka/2)$. This "dispersion relation" is the central result for this problem. One sees that the allowed frequencies depend upon the wavenumber $k = 2\pi/\lambda$, as

$$\omega = 2(K/m)^{1/2}\,|\sin(ka/2)|\,. \qquad (2.6)$$

The highest frequency, $2(K/m)^{1/2}$ occurs for $ka/2 = \pi/2$ or $k = \pi/a$; where the wavelength $\lambda = 2a$, and nearest neighbors move in opposite directions. The smallest frequency is at $k = \pi/Na = \pi/L$, which corresponds to $L = \lambda/2$. Here one can use the expansion of $\sin(x) \cong x$ for small x. This gives $\omega = 2(K/m)^{1/2}ka/2 = a(K/m)^{1/2}k$, representing a wave velocity $\omega/k = v = a(K/m)^{1/2}$.

Comparing this speed with $v = (Y/\rho)^{1/2}$ for a thin rod of Young's modulus Y, and mass density ρ, we deduce that $Y/\rho = Ka^2/m$. Here K and a are, respectively, the spring constant, and the spacing a of the masses. [1]

Young's modulus can thus be expressed in microscopic quantities as $Y = \rho Ka^2/m$ if the atoms have spacing a, mass m, and the interactions can be described by a spring constant K.

Table 2.1 Mechanical properties of some strong solids

Material	Young's modulus Y (GPa)	Strength (GPa)	Melting point (K)	Density ρ (kg/m^3)
Diamond	1050	50	1800	3500
Graphite	686	20	3300	2200
SiC	700	21	2570	3200
Si	182	7	1720	2300
Boron	440	13	2570	2300
Al_2O_3	532	15	2345	4000
Si_3N_4	385	14	2200	3100
Tungsten	350	4	3660	19300

A cantilever of length L clamped at one end and free at the other, such as a diving board, resists transverse displacement y (at its free end, $x = L$) with a force $-Ky$. The effective spring constant K for the cantilever is of interest to designers of scanning tunneling microscopes and atomic force microscopes, as well as to divers. The resonant frequency of the cantilever varies as L^{-2} according to the relation $\omega = 2\pi(0.56/L^2)(YI/\rho A)^{1/2}$. Here ρ is the mass density, A the cross section area and I is the moment of the area in the direction of the bending motion. If t is the thickness of the cantilever in the y direction, then $I_A = \int A(y)y^2 dy = wt^3/12$ where w is the width of the cantilever. It can be shown that $K = 3YI/L^3$. For cantilevers used in scanning tunneling microscopes the resonant frequencies are typically 10 kHz – 200 kHz and the force constant K is in the range 0.01–100 Newtons/m. It is possible to detect forces of a small fraction of a nanoNewton (nN).

Cantilevers can be fabricated from Silicon using photolithographic methods. Shown in Figure 1.1 is a "nanoharp" having silicon "wires" of thickness 50 nm and lengths L varying from 1000 nm to 8000 nm. The silicon rods are not under tension, as they would be in a musical harp, but function as doubly clamped beams. The resonant frequency for a doubly clamped beam differs from that of the cantilever, but has the same characteristic L^{-2} dependence upon length: $\omega = (4.73/L)^2(YI/\rho A)^{1/2}$. The measured resonant frequencies in the nanoharp structure range from 15 MHz to 380 MHz.

The largest possible vibration frequencies are those of molecules, for example, the fundamental vibration frequency of the CO molecule is 6.42×10^{13} Hz (64.2 THz). Analyzing this vibration as two masses connected by a spring, the effective spring constant is 1860 N/m.

2.2
Scaling Relations Illustrated by a Simple Harmonic Oscillator

Consider a simple harmonic oscillator (SHO) such as a mass on a spring, as described above, and imagine shrinking the system in three dimensions. As stated, $m\alpha L^3$ and $K\alpha L$, so $\omega = (K/m)^{1/2}\,\alpha L^{-1}$.

It is easy to see that the spring constant scales as L if the "spring" is taken a (massless) rod of cross section A and length L described by Young's modulus $Y = (F/A)/(\Delta L/L)$. Under a compressive force F, $\Delta L = -(LY/A)F$, so that the spring constant $K = LY/A$, αL.

A more detailed analysis of the familiar coiled spring gives a spring constant $K = (\pi/32R^2)\mu_S d^4/\ell$, where R and d, respectively, are the radii of the coil and of the wire, μ_S is the shear modulus and ℓ the total length of the wire. [1] This spring constant K scales in three dimensions as L^1.

Insight into the typical scaling of other kinetic parameters such as velocity, acceleration, energy density, and power density can be understood by further consideration of a SHO, in operation, as it is scaled to smaller size. A reasonable quantity to hold constant under scaling is the strain, x_{max}/L, where x_{max} is the amplitude of the motion and L is length of the spring. So the peak velocity of the mass $v_{max} = \omega x_{max}$

which is then constant under scaling: $v\alpha L^0$, since $\omega\alpha L^{-1}$. Similarly, the maximum acceleration is $a_{max}=\omega^2 x_{max}$, which then scales as $a\alpha L^{-1}$. (The same conclusion can be reached by thinking of a mass in circular motion. The centripetal acceleration is $a = v^2/r$, where r is the radius of the circular motion of constant speed v.) Thus for the oscillator under isotropic scaling the total energy $U = 1/2\ Kx_{max}{}^2$ scales as L^3.

In simple harmonic motion, the energy resides entirely in the spring when $x = x_{max}$, but has completely turned into kinetic energy at $x = 0$, a time $T/4$ later. The spring then has done work U in a time $1/\omega$, so the power $P = dU/dt$ produced by the spring is $\alpha\ \omega U$, which thus scales as L^2. Finally, the power per unit volume (power density) scales as L^{-1}. The power density strongly increases at small sizes. These conclusions are generally valid as scaling relations.

2.3 Scaling Relations Illustrated by Simple Circuit Elements

A parallel plate capacitor of area A and spacing t gives $C = \varepsilon_o A/t$, which under isotropic scaling varies as L. The electric field in a charged capacitor is $E = \sigma/\varepsilon_o$, where σ is the charge density. This quantity is taken as constant under scaling, so E is also constant. The energy stored in the charged capacitor $U = Q^2/2C = (1/2)\ \varepsilon_o E^2 At$, where At is the volume of the capacitor. Thus U scales as L^3. If a capacitor is discharged through a resistor R, the time constant is $\tau = RC$. Since the resistance $R = \rho\ell/A$, where ρ is the resistivity, ℓ the length, and A the constant cross section of the device, we see that R scales as L^{-1}. Thus the resistive time constant RC is constant (scales as L^0). The resistive electrical power produced in the discharge is $dU/dt = U/RC$, and thus scales as L^3. The corresponding resistive power density is therefore constant under scale changes.

Consider a long wire of cross section A carrying a current I. Ampere's Law gives $B = \mu_o I/2\pi R$ as the (encircling) magnetic field B at a radius R from the wire. Consider scaling this system isotropically. If we express $I = AE/\rho$, where E is the electric field in the wire, assumed constant in the scaling, and ρ is the resistivity, then B scales as L. The assumption of a scale-independent current density driven by a scale-independent electric field implies that current I scales as L^2. The energy density represented by the magnetic field is $\mu_o B^2/2$. Therefore the magnetic energy U scales as L^5. The time constant for discharge of a current from an inductor L' through a resistor R is L'/R. The inductance L' of a long solenoid is $L' = \mu_o n^2 A\ell$, where n is the number of turns per unit length, A is the cross section and ℓ the length. Thus inductance L' scales as length L, and the inductive time constant L'/R thus scales as L^2.

For an LC circuit the charge on the capacitor $Q = Q(0)\cos[(C/L)^{1/2}t]$. The radian resonant frequency $\omega_{LC} = (C/L)^{1/2}$ thus scales as L^0.

2.4
Thermal Time Constants and Temperature Differences Decrease

Consider a body of heat capacity C (per unit volume) at temperature T connected to a large mass of temperature $T = 0$ by a thermal link of cross section A, length L and thermal conductivity k_T. The heat energy flow dQ/dt is $k_T AT/L$ and equals the loss rate of thermal energy from the warm mass, $dQ/dt = CV dT/dt$. The resulting equation $dT/T = -(k_T A/LCV)dt$ leads to a solution $T = T(0)\exp(-t/\tau_{th})$, where $\tau_{th} = LCV/k_T A$. Under isotropic scaling τ_{th} varies as $L^2\, C/k_T$. Thermal time constants thus strongly decrease as the size is reduced. As an example, a thermal time constant of a few μs is stated [2] for a tip heater incorporated into a 200 kHz frequency AFM cantilever designed for the IBM "Millipede" 1024 tip AFM high density thermomechanical memory device. The heater located just above the tip, mounted at the vertex of two cantilever legs each having dimensions $50 \times 10 \times 0.5$ micrometers.

In steady state with heat flow dQ/dt, we see that the temperature difference T is $T = (dQ/dt)(L/k_T A)$. Since the mechanical power dQ/dt scales as L^2, this result implies that the typical temperature difference T scales, in three dimensions, as L. Temperature differences are reduced as the size scale is reduced.

2.5
Viscous Forces Become Dominant for Small Particles in Fluid Media

The motion of a mass in a fluid, such as air or water, eventually changes from inertial to diffusive as the mass of the moving object is reduced. Newton's Laws (inertial) are a good starting point for the motions of artillery shells and baseballs, even though these masses move through a viscous medium, the atmosphere. The first corrections for air resistance are usually velocity-dependent drag forces. A completely different approach has to taken for the motion of a falling leaf or for the motion of a microscopic mass in air or in water.

The most relevant property of the medium is the viscosity η, defined in terms of the force $F = \eta v A/z$ necessary to move a flat surface of area A parallel to an extended surface at a spacing z and relative velocity v in the medium in question. The unit of viscosity η is the Pascal-second (one Pascal is a pressure of 1 N/m^2). The viscosity of air is about 0.018×10^{-3} Pa·s, while the value for water is about 1.8×10^{-3} Pa·s. The traditional unit of viscosity, the Poise, is 0.1 Pa·s in magnitude.

The force needed to move a sphere of radius R at a velocity v through a viscous medium is given by Stokes' Law,

$$F = 6\pi\eta Rv. \tag{2.7}$$

This is valid only for very small particles and small velocities, under conditions of streamline flow such that the Reynolds number $N_{Reynolds}$ is less than approximately 2000. $N_{Reynolds}$, which is dimensionless, is defined as $N_{Reynolds} = 2R\rho v/\eta$, where R is the radius, ρ the mass density, v the velocity and η the viscosity.

The fall, under the acceleration of gravity g, of a tiny particle of mass m in this regime is described, following Stokes' Law, by a limiting velocity obtained by setting F (from equation 2.6) equal to mg. This gives

$$v = mg/6\pi\eta R. \tag{2.8}$$

As an example, a particle of 10 μm radius and density 2000 kg/m^3 falls in air at about 23 mm/s, while a 15 nm particle of density 500 kg/m^3 will fall in air at about 13 nm/s. In the latter case one would expect random jostling forces $f(t)$ on the particle by impacts with individual air molecules (Brownian motion) to be present as well as the slow average motion. Newton's laws of motion as applied to the motion of artillery shells are not useful in such cases, nor for any cases of cells or bacteria in fluid media.

An appropriate modification of Newton's Second Law for such cases is the Langevin equation [3]

$$F_{ext} + f(t) = [4\pi\rho R^3/3]\mathrm{d}^2x/\mathrm{d}t^2 + 6\pi\eta R\,\mathrm{d}x/\mathrm{d}t. \tag{2.9}$$

This equation gives a motion $x(t)$ which is a superposition of drift at the terminal velocity (resulting, as above, from the first and last terms in the equation) and the stochastic diffusive (Brownian) motion represented by $f(t)$.

In the absence of the external force, the diffusive motion can be described by

$$P(x,t) = (4\pi Dt)^{-3/2}\exp(-x^2/4Dt), \tag{2.10}$$

where

$$D = kT/6\pi\eta R \tag{2.11}$$

is the diffusivity of the particle of radius R in a fluid of viscosity η at temperature T. Consideration of the exponential term allows one to define the "diffusion length" as

$$x_{rms} = (4Dt)^{1/2}. \tag{2.12}$$

Returning to the fall of the 15 nm particle in air, which exhibits a drift motion of 13 nm in one second, the corresponding diffusion length for 300 K is $x_{rms} = 2D^{1/2} = 56$ μm. It is seen that the diffusive motion is dominant in this example.

The methods described here apply to slow motions of small objects where the related motion of the viscous medium is smooth and not turbulent. The analysis of diffusion is more broadly applicable, for example, to the motion of electrons in a conductor, to the spreading of chemical dopants into the surface of a silicon crystal at elevated temperatures, and to the motion of perfume molecules through still air.

In the broader but related topic of flying in air, a qualitative transition in behavior is observed in the vicinity of 1 mm wingspan. Lift forces from smooth flow over airfoil surfaces, which derive from Bernoulli's principle, become small as the scale is reduced. The flight of the bumblebee is not aerodynamically possible, we are told, and the same conclusion applies to smaller flying insects such as mosquitoes and gnats. In these cases the action of the wing is more like the action of an oar as it is forced against the relatively immovable water. The reaction force against moving the viscous and massive medium is the force that moves the rowboat and also the force that lifts the bumblebee.

No tiny airplane can glide, another consequence of classical scaling. A tiny airplane will simply fall, reaching a terminal velocity that becomes smaller as its size is reduced.

These considerations only apply when one is dealing with small particles in a liquid or a gas. They do not apply in the prospect of making smaller electronic devices in silicon, for example.

2.6
Frictional Forces can Disappear in Symmetric Molecular Scale Systems

Viscous and frictional forces are essentially zero in a nanotechnology envisioned by Drexler [4] in which moving elements such as bearings and gears of high symmetry are fashioned from diamond-like "diamondoid" covalently bonded materials. (To be sure, these are only computer models, no such structures have been fabricated.) The envisioned nanometer scaled wheels and axles are precisely self-aligned in vacuum by balances of attractive and repulsive forces, with no space for any fluid. Such moving parts basically encounter frictional forces only at a much lower level than fluid viscous forces.

There are natural examples of high symmetry nested systems. One example is provided by nested carbon nanotubes, which are shown in Figures 2.1 and 2.2 (after Cumings and Zettl, [5]). Nanotubes are essentially rolled sheets of graphite, which

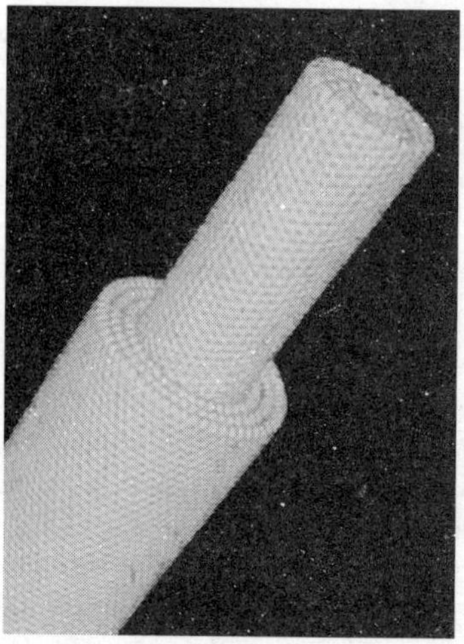

Figure 2.1 Nested carbon nanotubes [5]. This is a computer generated image. Zettl [5] has experimentally demonstrated relative rotation and translation of nested nanotubes, a situation very similar to that in this image. The carbon–carbon bonding is similar to that in graphite. The spacer between the tubes is simply vacuum.

have no dangling bonds perpendicular to their surfaces. Graphite is well known for its lubricating properties, which arise from the easy translation of one sheet against the next sheet. It is clear that there are no molecules at all between the layers of graphite, and the same is true of the nanotubes. The whole structure is simply made of carbon atoms. The medium between the very closely spaced moving elements is vacuum.

A second example in nature of friction-free motion may be provided by the molecular bearings in biological rotary motors. Such motors, for example, rotate flagella (propellers) to move cells in liquid. The flagellum is attached to a shaft which rotates freely in a molecular bearing structure. The rotation transmits torque and power and the motors operate continuously over the life of the cell, which suggests a friction-free molecular bearing. This topic will be taken up again in Chapter 3.

The double nanotube structure is maintained in its concentric relation by forces between the carbon atoms in the inner and outer tubes. These forces are not easy to fully characterize, but one can say that there is negligible covalent bonding between individual atoms on the adjacent tubes. Presumably there are repulsive overlap forces between atoms on adjacent tubes, such that a minimum energy (stable configuration) occurs when the tubes are parallel and coaxial. Attractive forces are presumably of the van der Waals type, and again the symmetry would likely favor the concentric arrangement. In the elegant experiments of Zettl [5], it was found that a configuration as shown in Figure 2.2 would quickly revert to a fully nested configuration when the displaced tube was released. This indicates a net negative energy of interaction between the two tubes, which will pull the inner tube back to full nesting. This is in accordance with the fact that the attractive van der Waals force is of longer range than the repulsive overlap force, which would be expected to have a negative exponential dependence on the spacing of the two tubes.

The structures of Figures 2.1 and 2.2 make clear that there will be a corrugated potential energy function with respect to relative translation and relative rotation, with periodicity originating in the finite size of the carbon atoms. The barriers to rotation and translation must be small compared to the thermal energy kT, for free rotation and translation to occur, and also for apparently friction-free motions to occur without damage to the structures. Incommensurability of the two structures will reduce such locking tendencies. An analysis of incommensurability in the

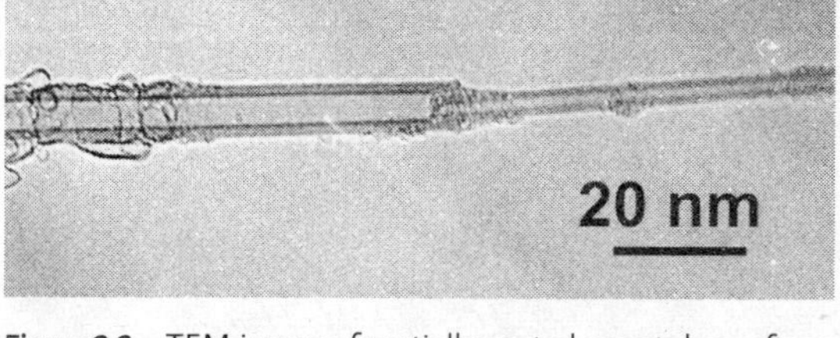

Figure 2.2 TEM image of partially nested nanotubes, after relative translation (Cumings and Zettl [6]). It was found that the inner tube could repeatedly be slid and rotated within the outer tube, with no evidence of wear or friction. Attractive forces very rapidly pulled a freed tube back into its original full nesting.

design of molecular bearings has been given by Merkel [7]. Again, there are no known methods by which any such structures can be fabricated, nor are there immediate applications.

Rotational and translational relative motions of nested carbon nanotubes, essentially free of any friction, are perhaps prototypes for the motions envisioned in the projected diamondoid nanotechnology. The main question is whether elements of such a nanotechnology can ever be fabricated in an error-free fashion so that the unhindered free motions can occur.

References

[1] For more details on this topic, see A. Guinier and R. Jullien, *The Solid State*, (Oxford, New York, 1989).

[2] P. Vettiger, M. Despont, U. Drechsler, U. Durig, W. Haberle, M.I. Lutwyche, H.E. Rothuizen, R. Stutz, R. Widmer, and G.K. Binnig, IBM J. Res. Develop. **44**, 323 (2000).

[3] E. A. Rietman, *Molecular Engineering of Nanosystems*, (Springer, New York, 2001), p. 55.

[4] K. Eric Drexler, *Nanosystems*, (Wiley, New York, 1992).

[5] Image courtesy of Zettl Research Group, University of California at Berkeley and Lawrence Berkeley National Laboratory.

[6] Reprinted with permission from J. Cummings and A. Zettl, Science **289**, 602–604 (2000). Copyright 2000 AAAS.

[7] R. C. Merkle, Nanotechnology **8**, 149 (1997).

3
What are Limits to Smallness?

3.1
Particle (Quantum) Nature of Matter: Photons, Electrons, Atoms, Molecules

The granular nature of matter is the fundamental limit to making anything arbitrarily small. No transistor smaller than an atom, about 0.1 nm, is possible. That chemical matter is composed of atoms is well known! In practice, of course, there are all sorts of limits on assembling small things to an engineering specification. At present there is hardly any systematic approach to making arbitrarily designed devices or machines whose parts are much smaller than a millimeter! A notable exception is the photolithographic technology of the semiconductor electronics industry which make very complex electronic circuits with internal elements on a much smaller scale, down to about 100 nm. However, this approach is essentially limited to forming two-dimensional planar structures.

It is not hard to manufacture Avogadro's number of H_2O molecules, which are individually less than one nm in size. One can react appropriate masses of hydrogen and oxygen, and the H_2O molecules will "self-assemble" (but stand back!). But to assemble even 1000 of those H_2O molecules (below 0 °C) in the form, e.g., of the letters "IBM", is presently impossible. Perhaps this will not always be so.

The most surprising early recognition of the granularity of nature was forced by the discovery that light is composed of particles, called photons, whose precise energy is $h\nu$. Here h is Planck's constant, 6.6×10^{-34} J·s, and ν is the light frequency in Hz. The value of the fundamental constant h was established by quantitative fits to the measurements of the classically anomalous wavelength distribution of light intensity emitted by a body in equilibrium at a temperature T, the so-called "black body spectrum" [1].

The energy of a particle of light in terms of its wavelength, λ, is

$$E = h\nu = hc/\lambda. \tag{3.1}$$

A convenient approach to calculating E, in eV, giving λ in nm, involves remembering that the product $hc = 1240\,\text{eV}\cdot\text{nm}$.

Nanophysics and Nanotechnology: An Introduction to Modern Concepts in Nanoscience. Second Edition.
Edward L. Wolf

ISBN: 3-527-40651-4

Later, it was found that electrons are emitted in a particular discrete manner from a metal surface illuminated by light of wavelength $\lambda = c/\nu$. The maximum kinetic energy of the photoelectrons, K, was measured to be

$$K = (hc/\lambda) - \phi, \tag{3.2}$$

where ϕ, the work function, is characteristic of each metal but usually in the range of a few electron volts (eV) [2]. The important thing was that the same value of h was obtained from this experiment as from the light spectrum analysis of Planck. It was found that the energy of the electrons (as distinct from their number) was not affected by the intensity of the light but only by its frequency. So it was clear that the light was coming in discrete chunks of energy, which were completely absorbed in releasing the electron from the metal (the binding energy of the electron to the metal is ϕ).

As far as electrical charge Q is concerned, the limit of smallness is the charge $-e$ of the single electron, where $e = 1.6 \times 10^{-19}$ C. This is an exceedingly small value, so the granularity of electrical charge was not easily observed. For most purposes electrical charge can be considered to be a continuously distributed quantity, described by volume density ρ or surface density σ.

The first measurement of the electron charge e was made by the American physicist Robert Millikan, who carefully observed [3] the fall of electrically charged microscopic droplets of oil in air under the influence of gravity and a static electric field. Following the application of Stokes' Law, (see equations 2.7 and 2.8) the velocity of fall in air is given by

$$v = (mg + neE)/(6\pi\nu R), \tag{3.3}$$

where n is the number of electron charges on the droplet. By making measurements of the electric field E needed to make the velocity v of a single drop come to zero, as its charge number n changed in his apparatus, Millikan was able to deduce the value of the electron charge e.

These historical developments, seminal and still highly relevant to the origins of nanophysics, are summarized in introductory sections of [6] and [7] which are quickly available and inexpensive.

3.2 Biological Examples of Nanomotors and Nanodevices

Biology provides examples of nanometer scale motors and electrical devices, which can be seen as limits of smallness. If nature can make these (only recently perceived) nanoscale machines, why, some ask, cannot human technology meet and eventually exceed these results? It is certainly a challenge.

The contraction of muscle occurs through the concerted action of large numbers of muscle myosin molecules, which "walk" along actin filaments in animal tissue.

Other sorts of motors rotate flagella, providing motion of bacteria through liquid media. Myosin, Kinesin, and rotary motors for flagella appear in extremely primitive life forms. It is believed that original forms of these proteins were present in single cell organisms appearing about a billion years ago. Simpler motors (similarly ancient) resembling springs are exemplified by the spasmoneme in *Vorticella*.

Electrically controlled valves in biology are exemplified by ion channels. One of the most studied is the voltage-gated potassium channel. This protein assembly controllably allows potassium ions to cross the lipid membrane of a neuron, generating nerve impulses.

These ion channels can be compared to transistors, in that a voltage controls a current flow. Ion channels, embedded in lipid cell walls, are truly nanometer scale electrically controlled gate devices.

3.2.1 Linear Spring Motors

An example of a biological motor, the spasmoneme spring, is shown in Figure 3.1 [6,7].

This system, first observed by the famous microscopist Leeuwenhoek in 1676, has a large literature. When extended, the spring may be mm in length. When exposed to calcium, which neutralizes the net negative charge in the extended state, the stalk contracts in a few ms to 40% of its length, at velocities approaching 8 cm/s.

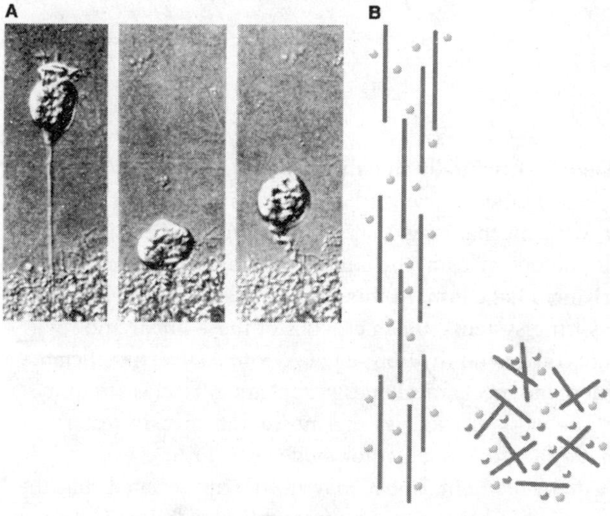

Figure 3.1 Spasmoneme spring. (A). The spasmoneme in *Vorticella* in its fully extended (left), fully contracted (middle), and partly extended (right) states [6]. (B). The extended spring state (left) consists of aligned filaments held apart by negative charges (dots). On the right, with plus and minus charges (dots) equally likely, the stalk collapses to a rubber-like state.

In consideration of the viscous drag forces that are involved, as described in Chapter 2, noting that the diameter of the head is as much as 50 μm, the force of contraction is estimated to be on the order of 10 nN, and the power is of order 100 pW.

As sketched in Figure 3.1 (B), the stalk is modeled [6] to consist of a bundle of filaments, each about 2 nm in diameter, which are roughly aligned but only weakly cross-linked. The extended structure is net negatively charged, forced into this extended state simply by Coulombic repulsion within the linear constraint of the filament bundle. Many further subtleties of this situation are known, including a helical aspect, so that the spasmoneme rotates as it contracts.

A somewhat similar spring motor is believed to operate in sperm cells exemplified by the horseshoe crab *Limulus polyphemus* in which a 60 μm finger, called the "acrosomal process", extends quickly across a jelly-like barrier to accomplish fertilization of egg plasma. Again, the extension motion seems clearly to be the result of Coulombic repulsion, as the ionic charge in the system is changed. There seems to be an opportunity to reanalyze some of this careful and extensive literature in more simple electrostatic terms.

The microbiological literature scarcely contains mention of electrostatic forces, yet these must be the essential origin of motions in bio-molecular motors, from a perhaps naïve, but unavoidably basic point of view. (Magnetic forces are certainly to be excluded, in the context of motors, but not in the context of sensors, as we will see below.)

Interdisciplinary approaches involving applied physicists familiar with semiconductor charge layers, screened electrostatic forces, etc., working with biologists, are likely to be very productive.

3.2.2
Linear Engines on Tracks

The spring and ratchet systems, structurally ill-defined but clearly electrostatic in mechanism, contrast with well defined linear and rotary engines, which move in stepwise fashions, however, with mechanisms that remain unclear. In this discussion we mention the linear motors myosin and kinesin [8]; and an example of a rotary motor, suitable for driving a flagellum, in this case F_1-adenosine triphosphate synthase. In contrast to the spring systems, the energetics of these linear and rotary motors is definitely tied to the conversion (hydrolysis) of ATP to ADP. The efficiency of conversion of chemical energy to mechanical energy appears to be close to unity.

Footnote 33 to [8] states "The work efficiencies of kinesin and myosin (each 50–60% efficient) are much greater than that of an automobile (10–15%)".

There is no quarrel with this statement, but it may not be appreciated that the biological and internal combustion engines are of entirely different types. The automobile engine is a heat engine, as described by the laws of thermodynamics, with efficiency that depends essentially on the ratio of the input and exhaust gas temperatures on the Kelvin scale, where room temperature is about 300 K. The *compression ratio* is a quantity to be maximized, for example, for an efficient diesel or gasoline internal combustion engine, and the mathematics for the ideal or limiting case is based on the Carnot cycle. The compression ratio has nothing to do with biological engines.

The biological motors, in contrast, provide direct conversion of chemical to mechanical energy. Temperature plays no essential role, and the system is more like a chemical battery except that the output is realized in mechanical rather than electrical terms.

The linear engines, myosin and kinesin, move along protein polymers, (actin and microtubules, respectively), in quantized steps, using up a fixed number of ATP molecules per step [8]. The actin filaments, for example, are about 5nm in diameter and 1 – 4 micrometers in length. The essence of the step-wise motion seems to be a change in *conformation* (shape or configuration) of the large motor molecule, energetically enabled by the hydrolysis of ATP, which allows the motor to move one well-defined step along its substrate. The rotary motor is less transparent in its operation. Yet it seems likely that these cases may eventually all be understood in essentially electrostatic terms.

A summary of various types of biological cellular engines is provided in Table 3.1 [6].

Table 3.1 Cellular engines of biology [6]. The performance of various cellular engines is compared with thermal energy (kT) and a typical automobile engine. Calculations for the specific power are based on the molecular weight of the smallest unit of the engine. Thus, molecular motors and polymerization-based engines are more powerful than the cellular structures in which they are found; for example, compare myosin to striated muscle

Engine	Velocity (μm s^{-1})	Force (dynes)	Specific Power (erg s^{-1} g^{-1})
kT (thermal energy)	–	4×10^{-7} dyne nm	–
Actin polymerization [9]	1.0	1×10^{-6}	1×10^{9}
Microtubule polymerization [10]	0.02	4×10^{-7}	5×10^{8}
Myosin II [11]	4.0	1×10^{-6}	2×10^{8}
Kinesin [12]	1.0	6×10^{-7}	7×10^{7}
Vorticellid spasmoneme [13]	8×10^{4}	1×10^{-3}	4×10^{7}
Automobile engine. typical [14]			3×10^{6}
Striated muscle [14]			2×10^{6}
Bacterial flagellar motor [15,16]	100 Hz	4.5×10^{-11} dyne cm	1×10^{6}
Thyone acrosomal reaction [17]	6–9	5×10^{-4}	1×10^{5}
Limulus acrosome reaction	10	1×10^{-6}	1×10^{4}
Eukaryotic flagellum [14]	–		3×10^{2}
Mitotic spindle [14]	2.0	1×10^{-5}	

Models for the movement of muscle myosin and kinesin are given in Figure 3.2 [8]. In both cases the motor moves along a strong filament which is part of the internal structure of the cell. The function of kinesin is to move material within the cell, the upward extended filament in Figure 3.2 B is attached to such a load.

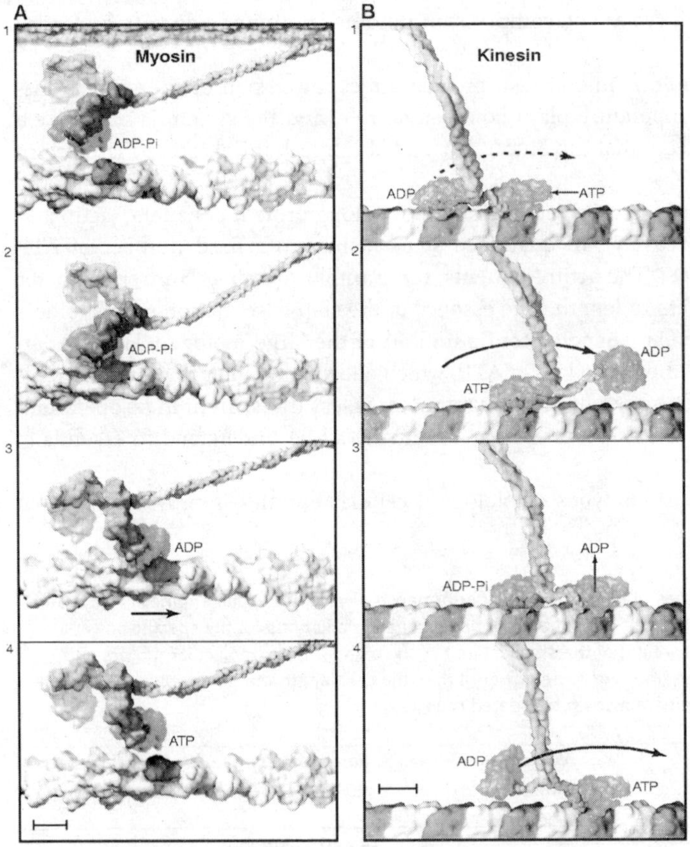

Figure 3.2 Models of the motion of muscle myosin and conventional kinesin [8]. (A). Frame 1: Muscle myosin is a dimer of two identical motor heads, anchored to a thick filament (top). Frames 2–4 show docking of heads on actin filament (lower), which serves to move the actin. The motion is about 10 nm per ATP hydrolyzed. (B). The two heads of the kinesin dimer move along a tubulin filament as indicated in Frames 1–4. In these frames the coiled coil extends to the top and to the attached cargo. The step length is about 8 nm. The head motion is associated with hydrolysis of ATP to ADP. The scale bar in A is 6 nm; that in B is 4 nm. (Taken from [8]).

The details of the muscle myosin movement have been recently confirmed [18,19]. It is now definitely known that the trailing muscle myosin head leaps ahead of the forward fixed head, in a motion similar to bipedal walking, or hand-over-hand rope climbing.

3.2.3
Rotary Motors

Two examples of biomolecular rotary motors are F_1-adenosine triphosphate synthase (F_1-ATPase) and ATP Synthase (F_0F_1) [20,21]. These motors depicted in Figures 3.3 and 3.4, are similar to motors that rotate flagella in bacteria and appear to have originated in cells about a billion years ago. They have evolved and serve many different purposes. These motors occur mounted in the wall of a cell and transmit rotational torque and power across the cell wall. The cell wall is a lipid hydrophobic membrane impervious to flows of liquids and ions and generally sealing the interior of the cell from its environment. The cell is internally structured with rigid actin filaments and tubulins, which act as tracks for the linear engines mentioned above.

The F_1-adenosine triphosphate synthase (F_1-ATPase) motor is shown in Figure 3.3 [20]. This rotary engine, which is found in mitochondria and bacterial membranes, "couples with an electrochemical proton gradient and also reversibly hydrolyzes ATP to form the gradient" [20]. It is shown in these experiments that this motor does work by rotating an attached actin filament through a viscous fluid. The work done by this motor in some biological contexts may be understood as $W = e\Delta V$, where e is the charge of the proton (hydrogen ion H^+) and ΔV is electrostatic potential difference. In such a case we have a "proton pump".

ATP Synthase (F_0F_1) motors from *e. coli* were attached to a glass coverslip, using the His tag linked to the lower end of the α subunit (lower portion of Figure 3.3). In

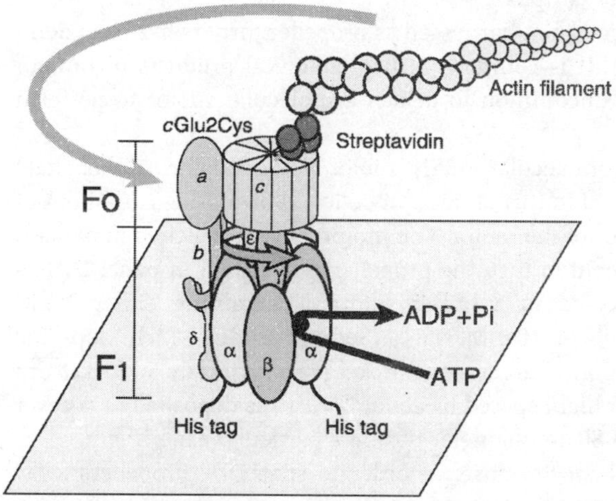

Figure 3.3 Fluorescently labeled actin filament permits observation of rotation of the *c* subunit in ATP Synthase (F_0F_1) [20]. A *c* subunit of Glu^2 was replaced by cysteine and then biotinylated to bind streptavidin and the actin filament. The γ, ε, and *c* units are thus shown to be a rotor, while the α, β, δ, *a*, and *b* complex is the stator. The rotation rate of the actin filament in the viscous medium was found to depend upon its length. Rotational rates in the range 0.5 Hz – 10 Hz were measured, consistent with a torque τ of 40 pN·nm.

this motor, the γ, ε, and c units rotate together, as shown in this work. This was demonstrated by attaching the actin filament to the top of the motor, the c unit. Attachment was accomplished in an elegant fashion making use of Streptavidin as shown. The actin filament was tagged with molecules which give characteristic fluorescence when illuminated, in the fashion described above in connection with quantum dots. Video microscope pictures of the fluorescing actin filament revealed counter clockwise rotation (see arrows), when the ambient solution contained 5 mM Mg ATP. The rotation rate of the actin filament in the viscous medium was found to depend upon its length. Rotational rates in the range 0.2 Hz – 3.5 Hz were measured, consistent with a torque τ of 40 pN · nm.

Analysis of the rotational torque exerted in the viscous fluid was based on the expression

$$\tau = (4\pi/3)\omega\eta L^3[\ln(L/2r) - 0.447]^{-1}. \tag{3.4}$$

Here ω is the rotation rate, η is the viscosity 10^{-3} Pa · s; L and r, respectively, are the length and radius of the actin filament. The actin filament radius r is taken as 5 nm. The measurements indicate a constant torque τ of about 40 pN · nm.

These experiments demonstrate an accumulated virtuosity in integrating biological, chemical, and microphysical insight, experimental design, and measurements. To form the experimental structure, alone, is a great accomplishment. The "fluorescently labeled actin filament-biotin-streptavidin complex" by which the rotation was observed in [20] is an excellent example.

The durability of such motors when harnessed as propeller-turners has been demonstrated by Soong *et al.* [21]. It is common to use biological products in human endeavors, but certainly very uncommon to detach a molecular motor to perform work in the inanimate world.

Figure 3.4 [21] shows a biomolecular rotary motor, F_1-adenosine triphosphate synthase (F_1-ATPase), harnessed to turn a nanopropeller, allowing measurement of the power output and efficiency of the motor. The motor itself is sketched in panel B of Figure 3.4, and its deployment to turn the propeller is sketched in panel D. The assembled motors are immersed in a buffer solution containing either 2 mM Na_2ATP (which fuels rotation), or 10 mM NaN_3 (Sodium azide) which stops the motion. Panel A shows a portion of an array of nickel posts, typically 50 to 120 nm in diameter and about 200 nm high, spaced by about 2.5 μm, as deposited on a cover glass. The inset to A shows an 80 nm diameter post in a particular experiment.

The motors were mounted on the posts in order to space the propellers away from the base plane. This allows calculations of the work required to turn the propeller in the viscous fluid to be simplified.

The Ni propellers were fabricated, coated suitably to bind to the top of the rotary engines, and suspended in a buffer solution which flowed past the array of engines mounted on Ni posts. In the reported experiment [21] about 400 propellers led to observation of five that rotated continuously in a counter clockwise direction. The attachment points along the propeller length of the rotor axis showed a random distribution, but the lengths L_1 and L_2 of the two sides from the rotor axis were mea-

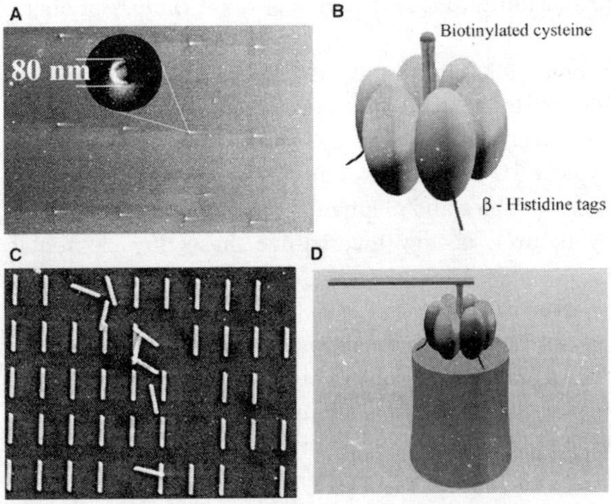

Figure 3.4 Bio-molecular rotary motor powered propellers. [21] (A) 80 nm Ni post from array. (B) Schematic view of F_1-ATPase molecular motor. (C) Array of Ni propellers, 750 nm – 1400 nm in length, 150 nm in diameter. (D) Schematic view of one assembled device from array. Rotation in fluid of propeller (0.8 – 8.3 rps) fueled with ATP is 50% efficient.

sured. The observed speeds of rotation depended upon the values of L_1 and L_2, ranging from 0.8 Hz to 8.3 Hz. It was found that some of the propellers ran for almost 2.5 hours while ATP was present in the cell. The rotation could be stopped by adding sodium azide to the cell.

An analysis of the work done by the rotating propellers was based on an expression for the force per unit length dF/dR exerted by the viscous medium on an element dR of propeller at radius R from the axis, on the assumption that the height h above a flat surface is small:

$$dF/dR = 4\pi\eta\omega R[\cosh^{-1}(h/r)]^{-1}. \tag{3.5}$$

In this expression η is the viscosity of the medium, 10^{-3} Pa·s, ω is the angular velocity of rotation, R the radius from the axis of rotation, h the pillar height (200 nm), and r half the width of the propeller (75 nm). Integration of this expression on R led to the expression for the torque τ

$$\tau = 4\pi\eta\omega({L_1}^3 + {L_2}^3)\,[3\cosh^{-1}(h/r)]^{-1}. \tag{3.6}$$

Here L_1 and L_2 are the lengths of the propeller sections extending from the rotational axis.

The determined values of the torque τ were 20 pN·nm for the 750 nm propellers (8 Hz, attached 200 nm from the end) and 19 pN·nm for the 1400 nm propellers (1.1 Hz, attached 350 nm from the end). The energy for one turn of these propellers is then 119 – 125 pN·nm. The energy released by hydrolysis of three ATP molecules is reported as about 240 pN · nm, leading to an efficiency of these motors of about 50% [21].

The experiments depicted in Figures 3.3 and 3.4 prove that engines of biology can act as stand-alone machines which function in suitably buffered environments, living or dead, if given ATP. The question: "How many pounds of muscle myosin would be needed to power a typical SUV?", if irreverent, is not entirely irrelevant. After all, large mammals, including whales and elephants, move by muscle myosin. The basic question, (solved by nature), is how to organize these tiny elemental engines to pull together.

It has been suggested that these tiny motors (available, and abundantly tested, proven over millions of years) might fill a gap in the development of nanoelectromechanical systems (NEMS), namely, for suitable power sources. As is pointed out in Reference [21], the typical biological rotary motor is about 8 nm in diameter, 14 nm in length and is capable of producing 20 – 100 pN · nm of rotary torque. Its operation depends on the availability of ATP in a suitable aqueous environment.

3.2.4
Ion Channels, the Nanotransistors of Biology

The smallest forms of life are bacteria, which are single cells of micrometer size. Cells are enclosed by an impermeable lipid bilayer membrane, the cell wall. This hydrophobic layer is akin to a soap bubble. Lipid cell walls are ubiquitous in all forms of life. Communication from the cell to the extracellular environment is accomplished in part by ion channels, which allow specific ion species to enter or leave the cell.

Two specific types of transmembrane protein ion channels are the Ca^{++} gated potassium channel, and the voltage-gated potassium ion channel, which is essential to the generation of nerve impulses. The dimensions of these transmembrane proteins are on the same order as their close relatives, the rotary engines, which were characterized as 8 nm in diameter and 14 nm in length.

These ion channel structures are "highly conserved", meaning that the essential units which appeared about 1 billion years ago in single cells, have been elaborated upon, but not essentially changed, in the many different cellular applications that have since evolved.

Ca^{++} gated potassium channel

A model for the Ca^{++} gated potassium channel is shown in Figure 3.5, after Schumacher and Adelman [22] based on [23] and [24]. Here the shaded horizontal slabs represent are the upper (outside) and lower (inside) cell walls. The envisioned ion channel, as closed, is shown on the left (a). The structure, which extends across the cell membrane, has an upper "selectivity filter" (which passes only potassium ions), a central cavity, and a Ca^{++}-controlled gate (lower) which opens the channel (right view, b). In this condition the channel will pass potassium ions, but not other ions. The nanometer scale dimensions of the ancient molecular structure that is modeled in Figure 3.5 are similar to those of the rotary motors shown above.

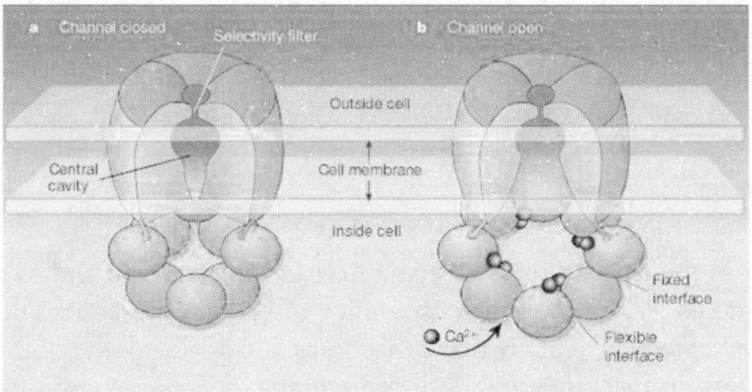

Figure 3.5 Model for Ca^{++}-gated K channel, after [22].

Voltage-gated potassium channel

The long-standing interest in the voltage-gated K^+ channel, stems in part from the discovery of Hodgkin and Huxley [25], that nerve impulses are generated by the flux of ions across the lipid membrane of a neuron. These ion channels are responsible for bringing a nerve impulse to an end so a neuron can prepare to fire again. This unit is key to the operation of the central nervous system. It is clear that the voltage difference across the cell membrane serves to open and shut the channel to the flow of potassium ions.

Two competing contemporary models for this channel are depicted in Figure 3.6 [26,27]. In the currently accepted conventional model (upper) the open pore configuration (left) is closed by the lifting of positively cylindrical (helical) units upward by the electric potential difference. In the competing model (lower), closing of the pore to potassium flow results from rotation of "paddle" structures in response to the potential difference.

In both cases the dimensions are nanometer scale and flows of electrical currents are controlled by potential differences. These are the functions of a transistor in elec-

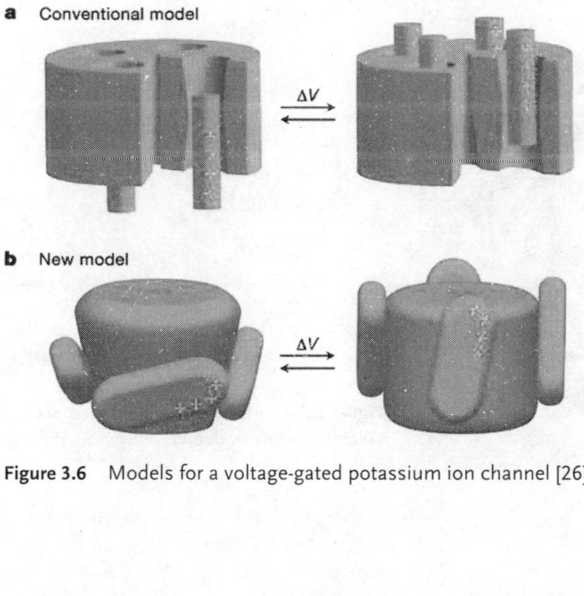

Figure 3.6 Models for a voltage-gated potassium ion channel [26].

tronics. These ancient biological devices have dimension typically 10 nm, and are thus nanoscale natural versions of transistors.

3.3
How Small can you Make it?

The fundamental limits to the sizes of machines and devices, presented by the size of atoms and as represented by the molecular machines of biology, are clear. The ability within present technology to actually make machines and devices is very limited, so that practical limits on machines and devices are many orders of magnitude larger in size. (There are some notable exceptions: really atomic sized assemblies have been made using the scanning tunneling microscope.) The conventional machine tools that make small mechanical parts scarcely work below millimeter size at present.

Making many identical small molecules, or even many very large molecules is easy for today's chemist and chemical engineer. The challenge of nanotechnology is much harder, to engineer (design and make to order) a complex structure out of molecular sized components.

3.3.1
What are the Methods for Making Small Objects?

The microelectronics process is able to make complex planar structures containing millions of working components in a square centimeter. This process is adaptable to making essentially planar mechanical machines with components on a micrometer scale.

An example of a turbine wheel is shown in Figure 3.7 [28]. This radial inflow turbine is made from silicon, using deep reactive ion etching. It is projected that a hydrocarbon burning gas turbine-electric generator under 1 cm^3 in volume could be

Figure 3.7 Turbine wheel produced on silicon wafer with deep reactive ion etching [28].

built elaborating on this approach, generating up to 50 watts, at an energy density 10 – 30 times that of the most advanced battery materials [28].

3.3.2 How Can you See What you Want to Make?

The motors of biology and the transistors of microelectronics are invisible. Using the tip of a Scanning Tunneling Microscope (STM) as an assembly tool, allows the product to be imaged at the same time. In principle this is a way of making some kinds of nanostructures.

Figure 3.8, after [29] shows steps in the assembly of a ring of iron atoms on a 111 atomically clean surface of copper. In this case the iron atoms are not chemically bonded to the copper, but remain in fixed positions because the equilibrium thermal energy is extremely low in the ultra-high-vacuum cryogenic environment. The Fe atoms actually rest in fixed positions between nearest-neighbor Cu atoms on the 111 Cu surface. The tip of the STM is used to nudge the atoms gently from one of these depressions to a neighboring one, thus assembling the ring. The tip is not used to carry an Fe atom: if it were, the property of the tip for also providing an image would be seriously disturbed.

The interaction of the tip to move the Fe is likely to involve the divergent electric field from the tip, which can induce an electric dipole moment,

$$\mathbf{p} = \alpha \mathbf{E}, \tag{3.7}$$

in the electron cloud of the atom. There is then a dipole interaction energy,

$$U = -\mathbf{p}.\mathbf{E}. \tag{3.8}$$

Because the E field is stronger close to the tip, the dipole is pulled closer to the tip:

$$F = -\mathrm{d}U/\mathrm{d}z = p_z \mathrm{d}E_z/\mathrm{d}z. \tag{3.9}$$

The strength of the force depends on the electric field approximately $E = V/d$, where V is the tip bias voltage and d is its spacing above the surface. Both V and d can be adjusted through the controls of the STM. With the upward force suitably adjusted, it is found that the Fe atom will follow a horizontal displacement of the tip, thus moving the atom. Other forces may contribute here, possibly related to the tunneling current.

This situation therefore does allow a variable force of attraction between the tip and an atom. However, if the atom actually jumps onto the surface of the tip, then further control of that atom is lost.

This situation is a prototype for the imagined molecular assembler tip. It should be realized that STM tips in practice have radii very much larger than an atomic radius. So there is no way that a carried atom could be placed into a recessed position, because access would be blocked by the large radius of the tip. (Tips of smaller radius are found to vibrate excessively.)

Incidentally, the circular ripples, peaking at the center of the circle of Fe atoms are evidence for the wave nature of the electrons in the (111) surface of the Cu sam-

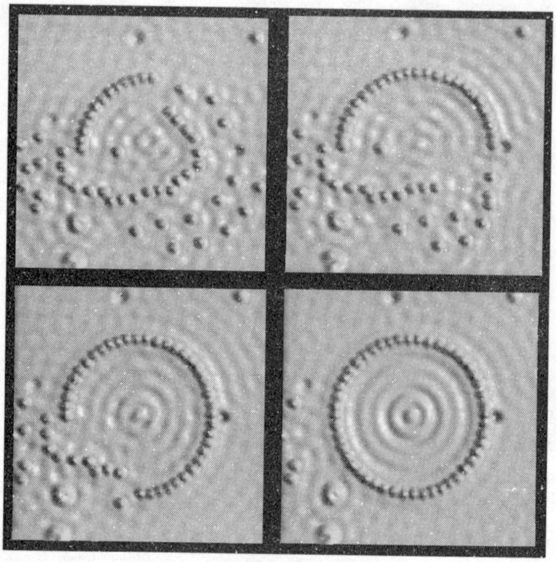

Figure 3.8 Assembling a ring of 48 Fe atoms on a (111) Cu surface with an STM [29]. The diameter of the ring is 14.3 nm.

ple. These standing waves result from reflection of the electron waves from the barrier represented by the row of Fe atoms. This situation resembles the reflection of water waves from a solid surface such as a pier or the edge of a canal. The researchers [29] found that the spacing of the crests of the waves was in agreement with the nanophysics of electron waves which will be taken up in Chapter 4. From the nanophysics point of view these ripples represent electrons, acting independently, caught in a two-dimensional circular potential well.

A second example of assembly of a molecular scale object using an STM tip is shown in Figure 3.8 ([30] after Hopkinson, Lutz and Eigler). Here is a grouping of 8 cesium and 8 iodine atoms which have formed a molecule on the (111) surface of copper. It appears that the strong ionic bond of Cs and I has dominated the arrangement, illustrating that an STM tip may not dictate the individual locations of atoms in a structure that is being assembled.

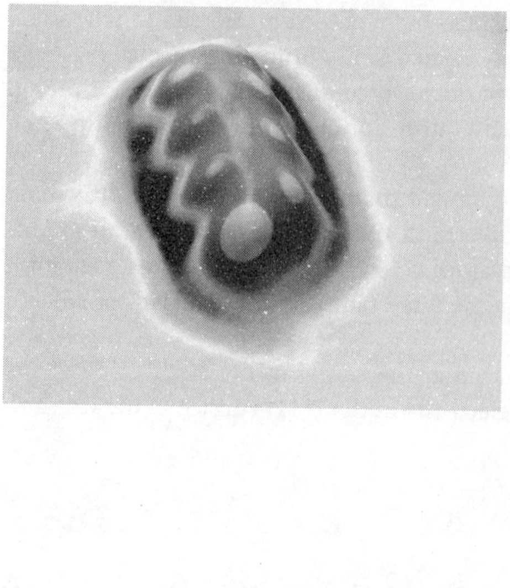

Figure 3.9 Cesium and iodine on Cu 111 [30] This pattern represents a molecule on the copper surface which contains eight cesium and eight iodine atoms. This image illustrates that the assembled atoms may choose their own structure, beyond control of the tip, in this case driven by the strong ionic bond.

In this situation one can speculate that the strongest interactions are between Cs and I ions, leading to eight molecules. These molecules are polar and probably arrange themselves in a structure dominated by the dipole-dipole forces between the CsI molecules.

In this case it not clear how strongly the molecules are attached to the copper surface.

3.3.3
How Can you Connect it to the Outside World?

A complex machine to be useful may need multiple connections to its outer environment. An example is the array of wires attached to the edges of a computer chip in a computer. If the premise is that the engineered nanomachine is to perform a function useful to the human-scale world, a large number of connections scaling up from the nanoscale to the centimeter scale may be needed. At the same time, the wiring must be such that unwanted signals or noise from the outer world do not propagate back into the nanoscale device to destroy it.

3.3.4
If you Can't See it or Connect to it, Can you Make it Self-assemble and Work on its Own?

The genius of biology is that complex structures assemble and operate autonomously, or nearly so. Self-assembly of a *complex* nano-structure is completely beyond present engineering approaches, but the example of DNA-directed assembly in biology is understood as an example of what is possible.

3.3.5
Approaches to Assembly of Small Three-dimensional Objects

All of the small objects mentioned so far have been essentially planar. Computer chips have several layers of wiring built up sequentially, but they are essentially planar. The same is true of the MEMS (microelectromechanical) devices based on the semiconductor technology. There are many uses for such devices, of course, but the impact of a fully three-dimensional approach would be very positive.

This situation has been lucidly stated by G. M. Whitesides [31] "The fabrication of microstructures is one of the most pervasive of modern technologies. Almost all microfabrication is now based on photolithography and its dependent technologies, and the dominance of this family of technologies is genuinely remarkable. Photolithography is intrinsically planar, although it can, with difficulty, be induced to produce certain types of nonplanar structures. The development of flexible, economical methods that would have the power of lithography, but would build 3D microstructures, would open the door to a host of applications in microfluidic systems, MEMS, optical devices, and structural systems...."

Variable thickness electroplating

A pioneering approach to making small 3D structures is summarized by Angus and Landau [31]. The approach [32] involves electroplating, with variable thickness images enabled by exposing a photo-sensitized gel through a gray-scale mask, which serves to cross-link the gelatin in proportion to the exposure [31]. The resistance to ionic transport through the gelatin increases with the cross-linking. "Therefore, on electroplating through the gelatin, the gray scale of the original optical mask is translated into thickness variations on the final surface; that is, darker areas on the optical mask lead to thicker electrodeposits. The method provides a convenient additive method for generating 3D surface relief".

Lithography onto curved surfaces

A recent three-dimensional method is summarized in Figure 3.10 [33].

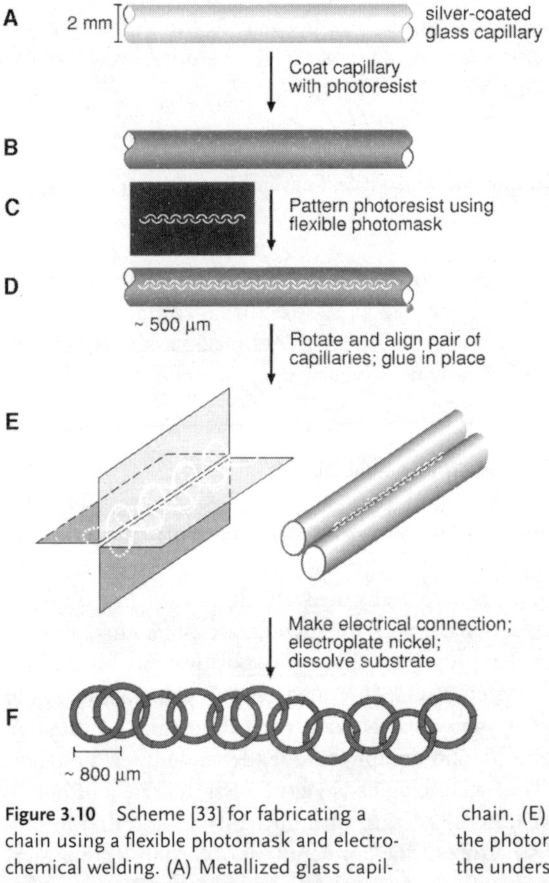

Figure 3.10 Scheme [33] for fabricating a chain using a flexible photomask and electrochemical welding. (A) Metallized glass capillaries coated with photoresist, by pulling them slowly from bulk solution. (B) Capillaries hardbaked at 105 °C for 3 min. Exposure (8s) of coated capillary to UV light through a flexible mask (design shown in (C)) wrapped around its surface. (D) Under optical microscope align two patterned capillaries to be in close proximity with their patterns matched to form a chain. (E) (Links correspond to openings in the photoresist. Dotted lines represent links on the undersides of the capillaries that are not visible from the top). Electroplating nickel for 30 min. at density 20 mA/cm^2 welded together the ends of the chain links, in those areas defined by the photoresist. Finally, release the nickel chain by dissolving photoresist in acetone, dissolving silver metallization in aqueous ferricyanide bath, and dissolving the titanium and glass in concentrated HF [33].

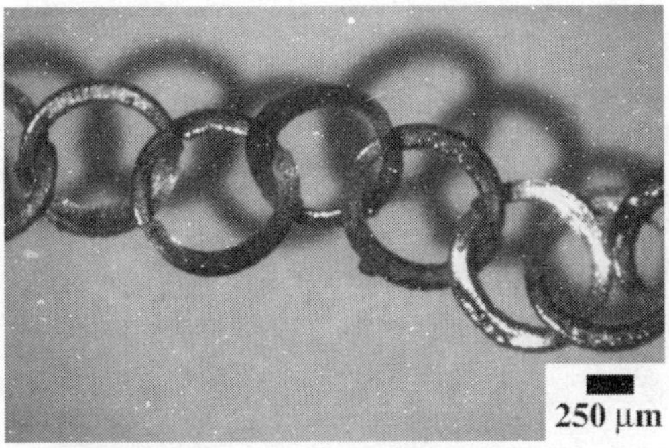

Figure 3.11 Optical micrograph of a free-jointed nickel chain formed by the process shown in Figure 3.10 [33]. The final thickness of the nickel was about 50 μm.

This method depends upon a flexible mask, that can be curved around the glass capillary tubes. The size of objects that can be made by this method is limited in part by the relative alignment of the two cylinders that is needed. Generalizations are discussed in [33].

The authors suggest that applications of such fully three-dimensional structures could include ultralight structures for micro air and space vehicles, components for microelectromechanical systems, 3D metallic membranes and electrodes, and, at smaller dimensions, dielectric structures for photonic band gap materials.

Optical tweezers

"Optical tweezers" is the name associated with the use of a focused laser beam to trap and pull a dielectric particle. The effect is somewhat similar to that described in equations (3.7–3.9), but the inhomogeneous electric field is at optical frequency and is most intense at the focus ("waist") of the laser beam. The optical tweezer is typically applied to dielectric spheres, such as polystyrene, of about 1 μm size, which are chemically attached to some biological system of interest. The laser power can be adjusted so that the force on the sphere is appreciable, while the force on single molecules, which are smaller, is negligible. The dielectric sphere will be stably positioned at the center of the trap (waist of the focused laser beam) and as it is displaced from the center, a restoring force proportional to the displacement will be exerted by the trap. An application of this biological technique is indicated in Figure 3.12 [34]. In the left panel (A) the bead, attached to the free DNA strand (template) by the RNA polymerase engine in its copying function, is shown at succeeding times t_1, t_1, t_3. In panel (B) the increasing length of the RNA copy, calibrated in nucleotides (nt), is plotted vs. time [34].

The enzyme RNA polymerase (RNAP) transcribes a DNA template into messenger RNA. In doing so it moves like an engine along the DNA template. (In the experiment shown, the RNAP engine is fixed to the glass slide, and the DNA is pulled to the left through it, in the process of making the copy (RNA, shown as coiling upward in (A) of Figure 3.12). In the experiment the trap position was fixed, and

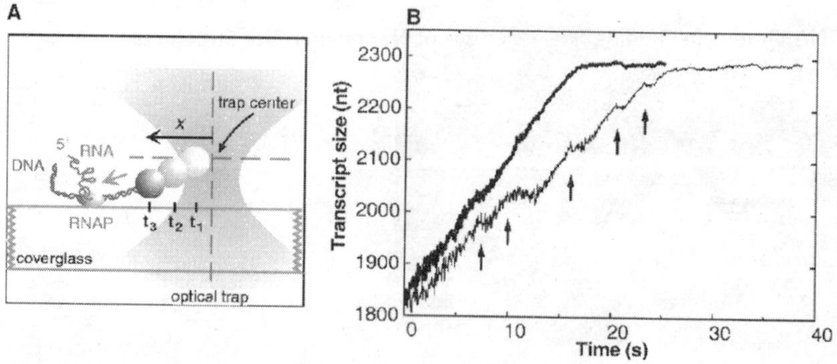

Figure 3.12 Use of optical tweezers to observe single RNAP molecule pulling DNA [34].

the motion of the bead was monitored by an optical interferometer. As the motion proceeded, the restoring force of the trap increased, and at positions marked by arrows in Figure 3.12 (B) the motor stalled, allowing a measurement of the maximum force available. The forces are in the range 21 – 27 pN. After stalling, the trap was repositioned and transcription continued. The upper trace in (B) was constructed by removing the stall phenomena artificially, to determine (from this graph) a replication rate of about 26 nucleotides per second.

Arrays of optical traps

A sophisticated scheme of interfering focused laser light beams has allowed creation and manipulation of 3D optically trapped structures [35]. Figure 3.13 schematically shows the nature of this optical trap and some of the configurations of dielectric spheres that have been stabilized. In Fig 3.13 (A) a detail schematic of the waist of the four-fold beam, which is shown stabilizing an array of eight dielectric spheres. (B) shows six other arrangements of identical dielectric spheres which were stably trapped in this device [35].

These traps are generally useful for micrometer sized particles, as far as individual positioning is concerned. Many particles can be loaded into such traps, encouraging them to condense in a self-organizing fashion. Arrays of more than four traps seem possible, based upon splitting a coherent light beam up into parallel beams.

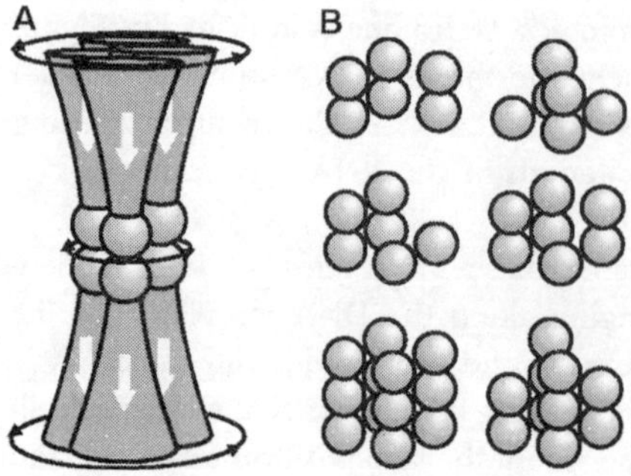

Figure 3.13 Four-fold rotating optical trap stabilizes 3D arrays [35].

3.3.6
Use of DNA Strands in Guiding Self-assembly of Nanometer Size Structures

One approach to nanofabrication, following the methods of biology, is to uniquely assemble pre-existing nanometer or micrometer scale structures (rather than building structures atom-by-atom). Biology has a large inventory of useful molecules (look at the ingredients of vitamin pills), which are needed at specific locations. The present state of nanotechnology, as distinct from biology and biotechnology, offers a limited but promising inventory of useful nanoscale structures. Some of the most promising "nano building blocks" are carbon nanotubes; nanowires of Si, CdS, and other semiconductors; and the family of symmetric carbon molecules such as C_{60} "Buckyballs".

These molecular objects are *atomically perfect*: the average diameter, for example, of a Buckyball, is absolutely certain. This makes the capacitance of an isolated Buckyball a reliable constant. In the ongoing miniaturization of semiconductor electronics, a central problem is that the fractional uncertainty in the size of a transistor gate, for example, tends to increase as the size of the gate is reduced. Building such a well-defined object as a C_{60} molecule into a nanoscale electronic circuit (for example, as a gate or a single transistor island) could be an avenue to successful miniaturization. In regard to the miniaturization of semiconductor devices, the possibility exists of incorporating such nanoscale, molecular structures into microelectronic chips, to do specific functions. But how can the molecule be accurately positioned in the semiconductor structure?

At present, one of the most promising ways to incorporate such an element onto specific sites in a microelectronic structure is to use complementary DNA strands. (See Figures 3.12 and 5.7 for some information on DNA.) A DNA single strand with a specific order of the bases C,G,A,T can be attached to particular sites in the structure. The desired imported structure, e.g., a carbon nanotube, can be tagged with the specific complement to the site tag. In a suitable aqueous environment, the tagged nanotubes will deposit at (van der Waals bond to) the complementarily tagged sites. The solvent can then be evaporated, leaving an accurately ordered nanostructure. It appears that this scheme would allow incorporation onto the substrate of extended nano building blocks, such as nanotubes, if DNA tags were attached to each end of the object (and to each end of the target location). It would seem possible in principle to appropriately assemble polar objects, perhaps bistable molecules that may act as logic elements, by attaching different DNA strands to the respective ends of the object.

This scheme for assembly of a carbon nanotube field-effect transistor has recently been reported [36]. The steps in the assembly are indicated in Figure 3.14. A series of microscope images verifying the stages of the assembly are shown in Figure 3.15 [36]. A sketch of the completed device is given in the lower right of Figure 3.14, showing a side view of the SWNT contacted at each end with an evaporated gold electrode, the whole assembly resting on a passivated oxidized (conductive) p-type silicon substrate that acts as the gate electrode.

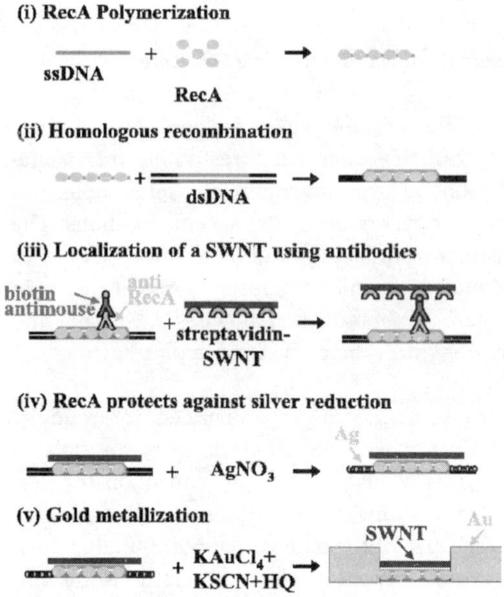

Figure 3.14 Assembly of a DNA-templated FET and Au contacts [36]. (i) Short single-strand DNA (ssDNA) is reacted with RecA protein derived from *E. coli* bacteria to form a nucleoprotein filament, which is 500 bases (250 nm) in length. (ii) Long double-stranded DNA (dsDNA) molecule serves as scaffold. The assembly process is guided by the information encoded in ssDNA and dsDNA molecules: the dsDNA was synthesized so that its "sequence" (see Figure 5.7) is identical to the dsDNA at the designated location of the FET. "Homologous recombination" locates the short dsDNA (RecA-coated) segment at the proper location on the longer dsDNA, ending step (ii). (The RecA coating later helps locate a (streptavidin-coated) SWNT, and protects the covered DNA segment against metallization.) (iii) A streptavidin-coated SWNT is bound to the scaffold using a primary antibody to RecA and a biotin-conjugated secondary antibody. (iv) Incubation in an $AgNO_3$ solution leads to Ag clusters on sections not protected by RecA. (v) "Electroless" gold deposition, using the Ag clusters as nucleation centers, results in formation of two DNA-templated Au wires contacting the SWNT bound at the gap. (Reprinted with permission from Science, copyright AAAS.)

In the above, the SWNT is longer than the gap dictated by the RecA, so that the deposited metals cover the ends of the nanotube and make contacts to it. The authors [36] go on to describe e-beam evaporation of wires to the Au contacts, and display in their paper the electrical characteristics of the DNA-templated carbon nanotube FET.

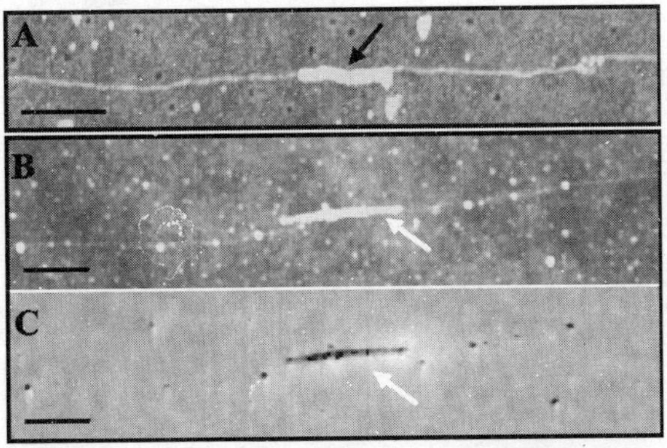

Figure 3.15 Localization of a SWNT at a specific address on the scaffold dsDNA using RecA [36]. (A) AFM topograph of 250-base (250 nm) RecA nucleoprotein filament (black arrow) located at matching segment on DNA scaffold molecule. Bar, 200 nm. (B) AFM image of streptavidin-coated SWNT (white arrow) bound to 500-base-long nucleoprotein filament localized on DNA scaffold molecule. Bar, 300 nm. (C) Scanning conductance image of the same region as in (B). The conductive SWNT (white arrow) yields a signal (black indicates conductivity) while the insulating DNA is hardly resolved. Bar, 300 nm. (Reprinted with permission from Science, copyright AAAS.)

References

[1] M. Planck, Annalen der Physik **4**, 553 (1901).

[2] A. Einstein, Annalen der Physik **17**, 132 (1905).

[3] R. A. Millikan, Physical Review **32**, 349 (1911).

[4] L. Pauling and E. B. Wilson, Jr., *Introduction to Quantum Mechanics with applications to Chemistry*, (Dover, Mineola, N.Y., 1985).

[5] F. L. Pilar, *Elementary Quantum Chemistry*, (Dover, Mineola, N.Y., 2001).

[6] Reprinted with permission from L. Mahadevan and P. Matsudaira, Science **288**, 95 (2000). Copyright 2000 AAAS.

[7] H. Stebbings and J. S. Hyams, *Cell Motility*, (Longman, London, 1979).

[8] Reprinted with permission from R. D. Vale and R. A. Milligan, Science **288**, 88 (2000). Copyright 2000 AAAS.

[9] V. C. Abraham, V. Krishnamurthi, D. L. Taylor, and F. Lanni, Biophys. J. **77**, 1721 (1999).

[10] M. Dogterom and B. Yurke, Science **278**, 856 (1997).

[11] J. E. Molloy, J. E. Burns, J. Kendrick-Jones, R. T. Treager, and D. C. White, Nature **378**, 209 (1995).

[12] S. M. Block, Cell **93**, 5 (1998).

[13] W. B. Amos, Nature **229**, 127 (1971).

[14] R. Niklas, Annu. Rev. Biophys. Biophys. Chem. **17**, 431 (1988).

[15] D. J. DeRosier, Cell **93**, 17 (1998).

[16] R. M. Berry and J. P. Armitage, Adv. Microb. Physiol. **41**, 291 (1999).

[17] J. Tilney and S. Inoue, J. Cell. Biol. **93**, 820 (1982).

[18] A. Yildiz, J. N. Forkey, S. A. McKinney, T. Ha, Y. E. Goldman, and P. R. Selvin, Science **300**, 2061 (2003).

[19] J.E. Molloy and C. Veigel, Science **300**, 2045 (2003).

[20] Reprinted with permission from Y. Sambongi, U. Iko, M. Tanabe, H. Omote, A. Iwamoto-Kihara, I. Ueda, T. Yanagida, Y. Wada, and M. Futai, Science **286**, 1722 (1999). Copyright 1999 AAAS.

[21] Reprinted with permission from R. K. Soong, G. D. Bachand, H. P. Neves, A. G. Olkhovets, H. G. Craighead, and C. D. Montemagno, Science **290**, 1555 (2000). Copyright 2000 AAAS.

[22] Reprinted with permission from Nature: M. Schumacher and J. P.Adelman, Nature **417**, 501 (2002). Copyright 2002, Macmillan Publishers Ltd.

[23] Y. Jiang, A. Lee, J. Chen, M. Cadene, B. T. Chait, and R. MacKinnon, Nature **417**, 515 (2002).

[24] Y. Jiang, A. Lee, J. Chen, M. Cadene, B. T. Chait, and R. MacKinnon, Nature **417**, 523 (2002).

[25] A. L. Hodgkin and A. F. Huxley, J. Physiol. (London) **117**, 500 (1952).

[26] Reprinted with permission from Nature: Y. Jiang, A. Lee, J. Chen, V. Ruta, M. Cadene, B. T. Chait, and R. MacKinnon, Nature **423**, 33 (2003). Copyright 2003, Macmillan Publishers Ltd.

[27] Y. Jiang, V. Ruta, J. Chen, A. Lee, and R. MacKinnon, Nature **423**, 42 (2003).

[28] Courtesy Martin A. Schmidt, Microsystems Technology Laboratories at Massachusetts Institute of Technology.

[29] Courtesy IBM Research, Almaden Research Center. Unauthorized use not permitted.

[30] Courtesy IBM Research, Almaden Research Center. Unauthorized use not permitted.

[31] J. C. Angus, U. Landau, and G. M. Whitesides, Science **281**, 1143 (1998).

[32] J. C. Angus, U. Landau, S. H. Liao, and M. C. Yang, J. Electrochem. Soc. **133**, 1152 (1986).

[33] Reprinted with permission from R. J. Jackman, S. T. Brittain, A. Adams, M. G. Prentiss and G. M. Whitesides, Science **280**, 2089–2091 (1998). Copyright 1998 AAAS.

[34] Reprinted with permission from M. D. Wang *et al.*, Science **282**, 902–907 (1998). Copyright 1998 AAAS.

[35] Reprinted with permission from M. P. MacDonald, L. Paterson, K. Volke-Sepulveda, J. Arit, W. Sibbett, and K. Dholakia, Science **296**, 1101–1103 (2002). Copyright 2002 AAAS.

[36] K. Keren, R. S. Berman, E. Buchstab, U. Sivan and E. Braun, Science **302**, 1380 (2003).

4
Quantum Nature of the Nanoworld

The particles of matter (electrons, protons, neutrons) provide limits to the smallness of anything composed of chemical matter, which is itself composed of atoms. The rules that these particles obey are different from the rules of macroscopic matter. An understanding of the rules is useful to understand the structure of atoms and chemical matter. An understanding of these rules, and of the relation between the wave and particle natures of light, is also key to understanding deviations from behavior of devices and machines that will appear as their dimensions are reduced toward the atomic size in any process of scaling.

One of the areas in which anticipated failures of classical scaling are of intense interest is in microelectronic devices, where it is obvious that Moore's Law (Figure 1.2), of increasing silicon chip performance, will eventually be modified. For one thing, the new rules allow a particle to penetrate, or tunnel through, a barrier. Such an effect is becoming more likely in the gate oxide of field effect transistors, as scaling efforts reduce its thickness.

The other aspect of the new nanophysical rules is that they permit new device concepts, such as the Esaki tunnel diode, mentioned in Chapter 1.

The new rules say that all matter is made of particles, but that the particles have an associated wave property. That wave property was noted in connection with Figure 3.8, which exhibited electron ripples.

One of the first quantum rules discovered, which had a large impact on understanding of atoms and their absorption and emission of light, was the quantization of angular momentum and energy levels of an electron in a hydrogen atom. This was first discovered in 1913 by Nils Bohr [1], whose work also stimulated further progress until by 1926 or so a more complete understanding of the wave properties of matter was achieved by Schrodinger [2].

4.1
Bohr's Model of the Nuclear Atom

The structure of the atom is completely nanophysical, requiring quantum mechanics for its description. Bohr's semi-classical model of the atom was a giant step toward this understanding, and still provides much useful information. By in-

Nanophysics and Nanotechnology: An Introduction to Modern Concepts in Nanoscience. Second Edition.
Edward L. Wolf

ISBN: 3-527-40651-4

troducing, in 1913, a completely arbitrary "quantum number", Bohr [1] was able to break the long-standing failure to understand how an atom could have sharply defined energy levels. These levels were suggested by the optical spectra, which were composed of sharp lines.

Bohr's model is planetary in nature, with the electron circling the nucleus. The model was based on information obtained earlier: that the nucleus of the atom was a tiny object, much smaller in size than the atom itself, containing positive charge Ze, with Z the atomic number, and e the electron charge, 1.6×10^{-19} C. The nucleus is much more massive than the electron, so that its motion will be neglected.

Bohr's model describes a single electron orbiting a massive nucleus of charge $+Ze$. The attractive Coulomb force $F = kZe^2/r^2$, where $k = (4\pi\varepsilon_o)^{-1} = 9 \times 10^9\ \mathrm{Nm^2/C^2}$, balances $m_e v^2/r$, which is the mass of the electron, $m_e = 9.1 \times 10^{-31}$ kg, times the required acceleration to the center, v^2/r. The total energy of the motion, $E = mv^2/2 - kZe^2/r$, adds up to $-kZe^2/2r$. This is true because the kinetic energy is always -0.5 times the (negative) potential energy in a circular orbit, as can be deduced from the mentioned force balance.

There is thus a crucial relation between the total energy of the electron in the orbit, E, and the radius of the orbit, r:

$$E = -kZe^2/2r. \tag{4.1}$$

This classical relation predicts collapse (of atoms, of all matter): for small r the energy is increasingly favorable (negative). So the classical electron will spiral in toward $r = 0$, giving off energy in the form of electromagnetic radiation. All chemical matter is unstable to collapse in this firm prediction of classical physics.

4.1.1
Quantization of Angular Momentum

Bohr, recognizing that such a collapse does not occur, was emboldened to impose an arbitrary quantum condition to stabilize his model of the atom. Bohr's postulate of 1913 was of the quantization of the angular momentum L of the electron of mass m circling the nucleus, in an orbit of radius r and speed v:

$$L = mvr = n\hbar = nh/2\pi. \tag{4.2}$$

Here n is the arbitrary integer quantum number $n = 1,2\ldots$. Note that the units of Planck's constant, $\mathrm{J \cdot s}$, are also the units of angular momentum. This additional constraint leads easily to the basic and confirmed properties of the "Bohr orbits" of electrons in hydrogen and similar one-electron atoms:

$$E_n = -kZe^2/2r_n, \quad r_n = n^2 a_o/Z, \quad \text{where } a_o = \hbar^2/mke^2 = 0.053\ \text{nm}. \tag{4.3}$$

Here, and elsewhere, k is used as a shorthand symbol for the Coulomb constant $k = (4\pi\varepsilon_o)^{-1}$.

The energy of the electron in the nth orbit can thus be given as $E_n = -E_o Z^2/n^2$, $n = 1,2,\ldots$, where

$$E_o = mk^2e^4/2\hbar^2 = 13.6\,\text{eV}. \tag{4.3a}$$

All of the spectroscopic observations of anomalous discrete light emissions and light absorptions of the one-electron atom were nicely predicted by the simple quantum condition

$$h\nu = hc/\lambda = E_o(1/n_1{}^2 - 1/n_2{}^2). \tag{4.4}$$

The energy of the light is exactly the difference of the energy of two electron states, n_1, n_2 in the atom. This was a breakthrough in the understanding of atoms, and stimulated work toward a more complete theory of nanophysics which was provided by Schrodinger in 1926 [2].

The Bohr model, which does not incorporate the basic wavelike nature of microscopic matter, fails to precisely predict some aspects of the motion and location of electrons. (It is found that the idea of an electron orbit, in the planetary sense, is wrong, in nanophysics.)

In spite of this, the electron energies $E_n = -E_o Z^2/n^2$, spectral line wavelengths, and the characteristic size of the electron motion, $a_o = \hbar^2/mke^2 = 0.053\,\text{nm}$, are all exactly preserved in the fully correct treatment based on nanophysics, to be described below.

4.1.2
Extensions of Bohr's Model

The Bohr model also remains useful in predicting the properties of "hydrogenic" electrons bound to donor impurity ions in semiconductors. This analysis explains how the carrier concentrations and electrical conductivities of industrial semiconductors are related to the intentionally introduced donor and acceptor impurity concentrations, N_D and N_A, respectively. The additional ideas needed are of the relative dielectric constant of the semiconductor and the "effective mass" that an electron exhibits as it moves in a semiconductor.

The Bohr model is also useful in analyzing the optical spectra of semiconductors exposed to radiation, having energy $E = hc/\lambda > E_g$ which produces electron–hole pairs. (E_g is the symbol for the "energy gap" of a semiconductor, which is typically about 1 eV.) It is found that such electrons and holes, attracted by the Coulomb force, momentarily orbit around each other, described by the mathematics of the Bohr model, and emit photons whose energies are predicted by the relevant Bohr model. Such states, called "excitons", are well documented in experiments measuring the spectra of fluorescent light from optically irradiated semiconductors.

A relevant topic in nanophysics is the alteration, from the exciton spectrum, of the fluorescent light emitted by a semiconductor particle as its size, L, is reduced. It is found that the correct light emission wavelengths for small sample sizes L, are obtained from the energies of electrons and holes contained in three-dimensional potentials, using the Schrodinger equation. This understanding is the basis for the behavior of "quantum dots", marketed as fluorescent markers in biological experiments, as will be described below.

4.2
Particle-wave Nature of Light and Matter, DeBroglie Formulas $\lambda = h/p$, $E = h\nu$

One of the most direct indications of the wave nature of light is the sinusoidal interference pattern of coherent light falling on a screen behind two linear slits of small spacing, d. The rule for appearance of maxima at angular position θ in the interference pattern,

$$n\lambda = d\sin\theta, \tag{4.5}$$

is that the difference in the path length of the light from the two slits shall be an integral number n of light wavelengths, $n\lambda$. Dark regions in the interference pattern occur at locations where the light waves from the two slits arrive 180 degrees out of phase, so that they exactly cancel.

The first prediction of a wave nature of matter was given by Louis DeBroglie [3]. This young physics student postulated that since light, historically considered to be wavelike, was established to have a particle nature, it might be that matter, considered to be made of particles, might have a wave nature. The appropriate wavelength for matter, DeBroglie suggested, is

$$\lambda = h/p, \tag{4.6}$$

where h is Planck's constant, and $p = mv$ is the momentum. For light $p = E/c$, so the relation $\lambda = h/p$ can be read as $\lambda = hc/E = c/\nu$. Filling out his vision of the symmetry between light and matter, DeBroglie also said that the frequency ν associated with matter is given by the same relation,

$$E = h\nu, \tag{4.7}$$

as had been established for light by Planck.

This postulated wave property of matter was confirmed by observation of electron diffraction by Davisson and Germer [4]. The details of the observed diffraction patterns could be fitted if the wavelength of the electrons was exactly given by h/p. For a classical particle, $\lambda = h/(2mE)^{1/2}$, since $p^2 = 2mE$. So there was no doubt that a wave nature for matter particles is correct, as suggested by DeBroglie [3]. The question then became one of finding an equation to determine the wave properties in a given situation.

4.3 Wavefunction Ψ for Electron, Probability Density $\Psi^*\Psi$, Traveling and Standing Waves

The behavior of atomic scale particles is guided by a wavefunction, $\Psi(r,t)$, which is usually a complex number. The probability of finding the particle is given by the square of the absolute value

$$\Psi^*(r,t)\Psi(r,t) = P(r,t) \tag{4.8}$$

of the wavefunction. This quantity P is a probability density, so that the chance of finding the particle in a particular small region $dxdydz$ is $Pdxdydz$.

Complex Numbers x+iy. The complex number is a notation for a point in the xy plane, where the symbol "i" acts like a unit vector in the y direction, formally obtained by rotating a unit vector along the x-axis in the ccw direction by $\pi/2$ radians. Thus $i^2 = -1$. The *complex conjugate* of the complex number, $x + iy$, is obtained by changing the sign of y, and is thus $x - iy$. The absolute value of the complex number is the distance r from the origin to the point x,y, namely

$$r = (x^2 + y^2)^{1/2} = [(x + iy)(x - iy)]^{1/2}. \tag{4.9}$$

A convenient representation of a complex number is

$$r\exp(i\theta) = r(\cos\theta + i\sin\theta), \text{ where } \theta = \tan^{-1}(y/x). \tag{4.10}$$

The wavefunction should be chosen so that P is normalized. That is,

$$\iiint P(x,y,z)dxdydz = 1 \tag{4.11}$$

if the integral covers the whole region where the electron or other particle may possibly exist.

There are other properties that a suitable wavefunction must have, as we will later discuss. Similarly, a many-particle $\Psi(r_1,r_2,\ldots,r_n,t)$ and probability $P(r_1,r_2,\ldots,r_n,t)$ can be defined.

The wavefunction for a beam of particles of identical energy $p^2/2m$ in one dimension is a traveling wave

$$\Psi(x,t) = L^{-1/2}\exp(ikx - i\omega t) = L^{-1/2}[\cos(kx - \omega t) + i\sin(kx - \omega t)], \tag{4.12}$$

where $k = 2\pi/\lambda$ and $\omega = 2\pi\nu$. According to the DeBroglie relation (4.6)

$$p = h/\lambda = \hbar k, \tag{4.13}$$

where $\hbar = h/2\pi$. Similarly, from (4.7),

$$\omega = 2\pi\nu = E/\hbar. \tag{4.14}$$

The normalization gives one particle in each length L, along an infinite x-axis. A point of fixed phase (such as a peak in the real part of the wave) moves as $x = (\omega/k)t$, so (ω/k) is called the phase velocity

$$v_{ph} = (\omega/k) \tag{4.15}$$

of the wave. Note that

$$\Psi(x,t) = L^{-1/2}e^{(ikx-i\omega t)} = L^{-1/2}\exp(ikx-i\omega t) \tag{4.16}$$

has a constant absolute value at any x, describing a particle equally likely to be at any position on the infinite x-axis. There is no localization in this wavefunction since the momentum is perfectly described, implying $\Delta x = \infty$.

Maxwell's Equations and Waveguide Modes

The radially diverging Coulomb electric field of a point charge Q, $E = kQ/r^2$, leads to Gauss's Law of electrostatics: outward electric flux

$$\Phi_E = \int \mathbf{E}\cdot d\mathbf{S} = Q/\varepsilon_o, \tag{4.17a}$$

where Q is the total electric charge inside the closed surface S (and the Coulomb constant $k_C = 1/(4\pi\varepsilon_o)$). This law can be restated as

$$\nabla\cdot\mathbf{E} = \rho/\varepsilon_o, \tag{4.17b}$$

where ρ is the charge density. $\nabla\cdot\mathbf{E} = \partial E_x/\partial x + \partial E_y/\partial y + \partial E_z/\partial z$ is suggestively called the divergence of E: div E. We see that the divergence of E is ρ/ε_o.

Turning to the magnetic field, **B**, the corresponding Gauss's Law type of statement is

$$\nabla\cdot\mathbf{B} = 0, \tag{4.18}$$

since nature includes no free magnetic charges. The basic sources of **B** are spin magnetic moments and electric currents I, most frequently in the form of an electron in an atomic orbit, which forms a current loop, and leads to a magnetic moment and a dipole magnetic field. The magnetic field **B** circling a long wire carrying current I is easily calculated using Ampere's Law:

$$\int \mathbf{B}\cdot d\mathbf{L} = \mu_o I, \tag{4.19a}$$

where the integral follows a closed path L enclosing the current I. The magnetic field vector **B** curls in a circular fashion around the wire:

$$\text{curl } \mathbf{B} = \nabla\times\mathbf{B} = \mu_o\mathbf{J}, \tag{4.19b}$$

where **J** is the current density vector, of units A/m^2. The curl of vector **B**, $\nabla\times\mathbf{B}$, is again a vector. It is usually presented (taking **i**, **j**, **k** as unit vectors along x, y, z) as a determinant whose rows are **i**, **j**, **k**; $\partial/\partial x$, $\partial/\partial y$, $\partial/\partial z$; and B_x, B_y, B_z. For example, if we assume $\mathbf{J} = \mathbf{k}J_0$, then $\mathbf{k}\mu_o J_0 = \mathbf{k}(\partial B_y/\partial x - \partial B_x/\partial y)$, leading to the expected circling **B** vector field in the x,y plane.

Further, Faraday's Law states that an electric field **E** appears along a path enclosing a changing magnetic flux:

$$\int \mathbf{E}\cdot d\mathbf{L} = -\partial(\Phi_M = \int \mathbf{B}\cdot d\mathbf{S})/\partial t. \tag{4.20a}$$

This circling electric field is an analog of the magnetic field circling the current mentioned above. By analogy, the corresponding differential statement of Faraday's Law is

$$\text{curl } \mathbf{E} = \nabla\times\mathbf{E} = -\partial\mathbf{B}/\partial t. \tag{4.20b}$$

J. C. Maxwell discovered (by restoring a missing current density term, $\varepsilon_o\partial\mathbf{E}/\partial t$, into the Ampere's Law expression) that propagating electromagnetic waves are implied by these experimental laws. The missing current density is called Maxwell's "displacement current". The expanded (corrected) Ampere's Law expression is

$$\text{curl } \mathbf{B} = \nabla\times\mathbf{B} = \mu_o\mathbf{J} + \varepsilon_o\mu_o\partial\mathbf{E}/\partial t. \tag{4.19c}$$

The new term provides a current density between the plates of a charging capacitor, making the charging current continuous around the loop, as is sensible.

Waves Propagating in Free Space (Vacuum)

Following Maxwell, in free space, the relevant equations (assuming no currents and no charge density) are (i) $\nabla\times\mathbf{E} = -\partial\mathbf{B}/\partial t$, and (ii) $\nabla\times\mathbf{B} = \varepsilon_o\mu_o\partial\mathbf{E}/\partial t$. Forming the curl of equation (i), $\nabla\times[\nabla\times\mathbf{E} = -\partial\mathbf{B}/\partial t]$, and then using equation (ii), we find $\nabla\times\nabla\times\mathbf{E} + \varepsilon_o\mu_o\partial^2\mathbf{E}/\partial t^2 = 0$. The mathematical identity $\nabla\times\nabla\times\mathbf{E} = \nabla(\nabla\cdot\mathbf{E}) - \nabla^2\mathbf{E}$, plus the fact that $\nabla\cdot\mathbf{E} = 0$ for vacuum, lead to the important Maxwell wave equation for free space:

$$\nabla^2\mathbf{E} - \varepsilon_o\mu_o\partial^2\mathbf{E}/\partial t^2 = 0. \tag{4.21}$$

This equation has traveling plane wave solutions with phase velocity in vacuum $c = (\varepsilon_o\mu_o)^{-1/2} = 2.998\times10^8$ m/s. An example of such a wave is $E_y = E_{y0} \cos[2\pi(z-ct)/\lambda] = E_{y0}\cos(kz-\omega t)$, where $\lambda = c/f = 2\pi/k$ is the wavelength and f the frequency. (The polarization direction and the propagation direction are at right angles in these waves.)

These results explain various electromagnetic radiation phenomena, from gamma rays to ultraviolet light to 60 kHz (5 km wavelength) radio waves used to synchronize "atomic watches" over thousand mile distances. Maxwell's equations also describe the flow of electromagnetic power (about 1 kW/m^2 on the Earth's surface) from the Sun.

Waveguide Modes for Microwaves and for Light

Maxwell's equations also accurately describe electromagnetic modes in confined geometries such as waveguides for microwaves. Figure 4.1 shows the electric **E** and magnetic **H** fields in a rectangular waveguide, for propagation of the lowest frequency TE10 mode, in the positive *z*-direction. One wavelength is sketched. This solution is obtained with the wave equation, div $E = 0$ and requiring that the tangential component of **E** and the normal component of **H** are zero at each metallic surface.

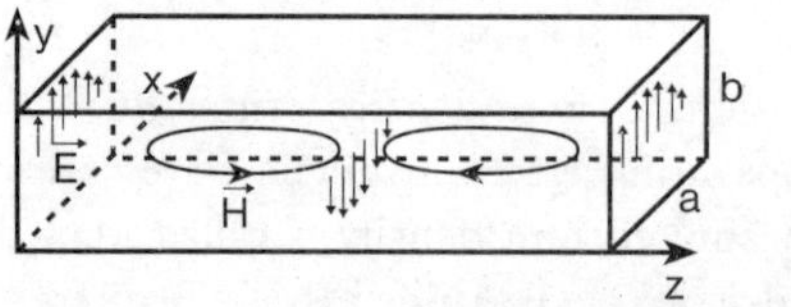

Figure 4.1 Sketch of transverse electric mode TE10 in a rectangular metallic waveguide.

The TE10 transverse electric mode is the lowest frequency mode for a rectangular waveguide with $a > b$. The electric and magnetic fields are of the form

$$\psi = \psi_{x,y} \exp[i(kz - \omega t)], \quad \text{where} \quad k = 2\pi/\lambda. \tag{4.22}$$

The radian frequency is related to the wavelength $2\pi/k$ as

$$\omega_{10} = 2\pi f_{10} = [\omega_c{}^2 + (ck)^2]^{1/2}, \tag{4.23}$$

where c is the speed of light, and cutoff $\omega_c = c\pi/a$.

The cutoff frequency ω_c is the lowest frequency (longest wavelength) that will propagate, and e.g. for $a = 3$ cm the cutoff frequency is 5 GHz. The phase velocity is $v_p = \omega_{10}/k$ and the group velocity $v_g = \partial\omega_{10}/\partial k$. The group velocity becomes zero at frequency ω_c and approaches the vacuum speed of light c at high frequency, $\omega >> \omega_c$. Detailed understanding of these electromagnetic modes was essential in the historical development of radar.

The variety of boundary conditions imposed on the Maxwell differential equations in specific cases leads to a wide variety of solutions. For example, a cylindrical optical fiber, with much different (non-metallic) boundary conditions, also supports modes which propagate in the *z*-direction. Figure 4.2 shows a set of calculated modes for a particular type of optical fiber.

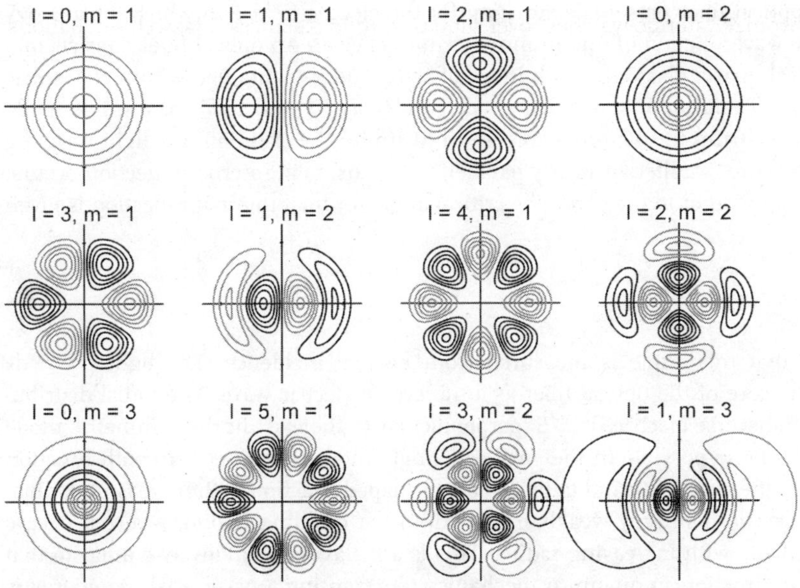

Figure 4.2 Calculated modes for cylindrical dielectric waveguide, assumed to extend out of the page. (By courtesy of RP Photonics Consulting GmbH, from: Encyclopedia of Laser Physics and Technology. http://www.rp-photonics.com/fibers.html.)

A similar approach is needed to solve Schrodinger's equation for matter waves, as we will see below. In one branch of modern silicon device development, "wave transistor" devices are fabricated in nearly perfect semiconductors, where electrons behave as waves predicted by Schrodinger's equation (in the absence of scattering) and have electron probability plots that qualitatively resemble those shown for electromagnetic waves in Figure 4.2.

4.4 Maxwell's Equations; *E* and *B* as Wavefunctions for Photons, Optical Fiber Modes

The laws of electricity and magnetism give the values of electric field E and magnetic field B, as functions of position, in various circumstances. In classical electricity and magnetism it is known that the energy density in the electromagnetic field is

$$(\varepsilon_o E^2 + \mu_o B^2)/2. \tag{4.24}$$

Since the electromagnetic energy resides in particles called photons, this classical energy density can also be thought of as a probability function for finding photons. The analogy is most direct if the E and B fields represent traveling waves, and thus transport energy.

The optical fiber presents a situation for the electric field E in which the analogy with the wavefunction of quantum mechanics is clear. An optical fiber consists of a cylindrical quartz glass core with index of refraction n_1 contained within concentric silica glass cladding of somewhat smaller index n_2. Typical radii for a single mode fiber are 5–10 μm for the inner core and 120 μm for the cladding. For light traveling along the core, parallel, or nearly parallel, to its axis, total internal reflection occurs, confining the light to the core. The critical angle for total internal reflection is given by

$$\theta_r = \sin^{-1}(n_2/n_1). \tag{4.25}$$

Note that this angle is measured from normal incidence. The light proceeds down the core of the optical fiber as a transverse electric wave. The radial distribution of transverse electric field E (perpendicular to the axis) in the symmetric mode (the only mode possible in the smallest, single mode, fibers) is a smooth function peaked at the axis, $r = 0$, and falling in a bell shaped function with increasing radius.

The electric field only weakly extends into the cladding region, where it decays exponentially with increasing radius. This is a behavior qualitatively similar to that of a wave function in quantum mechanics representing a particle whose energy is less than the potential energy.

4.5 The Heisenberg Uncertainty Principle

The uncertainty principle [5] is a consequence of a wave description of the location of a particle. It states that the position x and the momentum p of a particle can both be simultaneously known only to minimum levels of uncertainty, Δx and Δp, respectively, where

$$\Delta x\, \Delta p \geq \hbar/2. \tag{4.26}$$

The free particle of precise momentum $p = \hbar k$, described by the wavefunction $\Psi(x,t) = L^{-1/2}\exp(ikx - i\omega t)$ (4.16) represents a case of $\Delta x = \infty$ and $\Delta p \cong 0$.

A wavefunction that describes a localized particle can be constructed by adding waves having a distribution of k and ω values. An example of such a linear combiniation, using a trigonometric identity, is

$$\Psi = A[\cos(k_1x - \omega_1t) + \cos(k_2x - \omega_2t)] = 2A\cos[(1/2)\Delta kx - (1/2)\Delta\omega t]\cos(k_{av}x - \omega_{av}t), \tag{4.27}$$

where $\Delta k = k_2 - k_1$, $\Delta\omega = \omega_2 - \omega_1$, $k_{av} = (k_2 + k_1)/2$, and $\omega_{av} = (\omega_2 + \omega_1)/2$.

Assuming the two k's and ω's are closely spaced, so that $\Delta k < k_{av}$ and $\Delta\omega < \omega_{av}$, then the $\cos[(1/2)\Delta kx - (1/2)\Delta\omega t]$ factor acts like an *envelope function* of long wave-

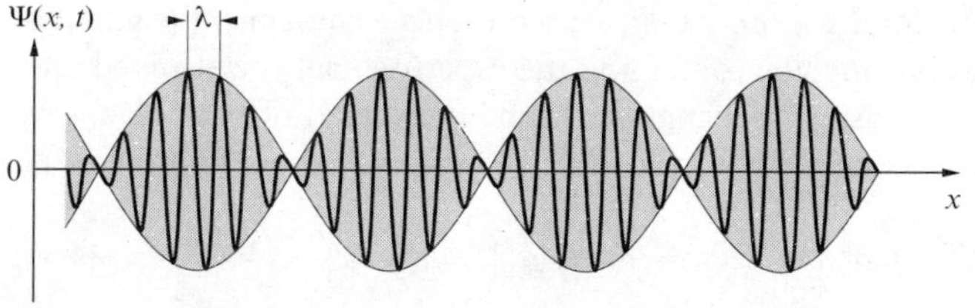

Figure 4.3 Adding waves creates regions of localization that move at the group velocity.

length and low frequency which modulates the original wave. At fixed t, the spacing Δx between adjacent nodes of $\cos(\Delta kx/2)$ is

$$\Delta x = 2\pi/\Delta k. \tag{4.28}$$

This width can be considered as a length of localization for the particle represented by the superposition of two waves. The equivalent relation $\Delta k \Delta x = 2\pi$ is a form of the uncertainty principle, which we see is simply a wave property. Since $\Delta p = \hbar\Delta k$, from the DeBroglie relation, this gives $\Delta p \Delta x = h$. A tighter localization, achieved by a wavepacket composed of a Gaussian distribution of k values, gives the smaller uncertainty product quoted above, $\Delta x\, \Delta p \geq \hbar/2$.

Finally, the motion of the envelope function in (4.20) is described by $x = (\Delta\omega/\Delta k)t$. The new velocity is called the group velocity

$$v_g = (\partial\omega/\partial k). \tag{4.29}$$

To apply this simple analysis to the motion of a localized particle of energy $E = \hbar\omega = mv^2/2m = \hbar^2k^2/2m$, let us calculate the expected classical velocity. From the foregoing, $\omega = \hbar k^2/2m$, so $\partial\omega/\partial k = \hbar k/m = v$. Thus, the group velocity $v = \partial\omega/\partial k$, and DeBroglie's relations correctly reproduce the classical velocity p/m of a particle.

4.6 Schrodinger Equation, Quantum States and Energies, Barrier Tunneling

The equation to provide the wavefunction $\Psi(x,t)$ describing the location of a particle in a given physical situation was provided by Schrodinger [2]. This equation may seem mysterious, but actually it is not so, at least in hindsight. Any successful wave equation has to reflect the DeBroglie matter-wave relations [3], $E = \hbar\omega$ and $p = \hbar k$ $(\lambda = h/p)$. The correct equation must provide a traveling wave solution $\Psi(x,t) = L^{-1/2}\exp(ikx - i\omega t)$ for a free particle, in order to match the electron diffraction observations of Davisson and Germer [4].

Further guidance in finding the correct matter-wave equation is afforded by Maxwell's wave equation for all electromagnetic waves, from elementary physics,

$$\partial^2\, \Psi(x,t)/\, \partial x^2 - \varepsilon_o\mu_o\, \partial^2\, \Psi(x,t)/\, \partial t^2 = 0. \tag{4.30}$$

This equation, where, for the moment, $\Psi(x,t)$ represents a component (e.g., E_y) of the **E** or **B** vectors, was obtained by combining the experimentally determined laws of electricity and of magnetism. Applying the second derivatives of Maxwell's wave equation (4.30) to the traveling wavefunction $\Psi(x,t) = \exp(ikx - i\omega t)$, we find

$$\partial^2 \Psi(x,t)/\partial x^2 = -k^2 \Psi(x,t),\ \partial^2 \Psi(x,t)/\partial t^2 = -\omega^2 \Psi(x,t). \tag{4.31}$$

Substitution of the relations (4.31) into the equation (4.30) produces the condition

$$[-k^2 + \omega^2 \varepsilon_o \mu_o]\ \Psi(x,t) = 0. \tag{4.32}$$

This equality requires $\omega/k = (\varepsilon_o \mu_o)^{-1/2} = 2.99793 \times 10^8$ m/s. This speed is the measured speed of light, and makes clear then the origin of all electromagnetic waves as similar to light.

The factor in brackets above (4.32) relates to the speed of the light wave. The question now is how to generate a matter wave equation so that an analogous bracket term will give some condition on matter.

4.6.1 Schrodinger Equations in one Dimension

A good guess for the corresponding bracketed factor in the matter wave equation is a statement of the energy of the particle, $K + U = E$, or, using the DeBroglie relations:

$$[\hbar^2 k^2/2m + U - \hbar\omega]\ \Psi(x,t) = 0. \tag{4.33}$$

Based on this correct statement of conservation of energy, and knowing the solution $\Psi(x,t) = \exp(ikx - i\omega t)$, the equation has to involve $\partial^2 \Psi(x,t)/\partial x^2$, as before. In addition, the first time-derivative $\partial \Psi(x,t)/\partial t = -i\omega\ \Psi(x,t)$ is needed, in order to produce the $\hbar\omega$ in the statement of conservation of energy, which in (4.33) appears in the bracket factor.

Time-dependent Equation

On this basis, the Schrodinger equation in the one-dimensional case, with time-dependent potential $U(x,t)$, is

$$-\hbar^2/2m\ \partial^2 \Psi(x,t)/\partial x^2 + U(x,t)\ \Psi(x,t) = i\hbar\ \partial \Psi(x,t)/\partial t = \mathscr{H}\Psi. \tag{4.34}$$

The left-hand side of the equation is sometimes written $\mathscr{H}\Psi$, with $\mathscr{H}$ the operator which represents the energy terms on the left side of the equation.

Time-independent Equation

In the frequent event that the potential U is time-independent, a product wavefunction

$$\Psi(x,t) = \psi(x)\varphi(t), \tag{4.35}$$

when substituted into the time-dependent equation above, yields

$$\varphi(t) = \exp(-iEt/\hbar), \tag{4.36}$$

and the time-independent Schrodinger equation,

$$-(\hbar^2/2m)d^2\psi(x)/dx^2 + U\psi(x) = E\psi(x), \tag{4.37}$$

to be solved for $\psi(x)$ and energy E. The solution $\psi(x)$ must satisfy the equation and also boundary conditions, as well as physical requirements.

The physical requirements are that $\psi(x)$ be continuous, and have a continuous derivative except in cases where the U is infinite. A second requirement is that the integral of $\psi^*(x)\psi(x)$ over the whole range of x must be finite, so that a normalization can be found. This may mean, for example, in cases where real exponential solutions $\exp(\kappa x)$, $\exp(-\kappa x)$ satisfy the equation, that the positive exponential solution can be rejected as non-physical.

The momentum of a particle has been associated in our treatment with $-i\hbar\partial/\partial x$, which when operated on $\exp(ikx)$ gives the momentum $\hbar k$ times $\exp(ikx)$. One can see from this that a real, as opposed to a complex, wavefunction, will not represent a particle with real momentum.

4.6.2 The Trapped Particle in one Dimension

The simplest problem is a trapped particle in one dimension. Suppose $U = 0$ for $0 < x < L$, and $U = \infty$ elsewhere, where $\psi(x) = 0$. For $0 < x < L$, the equation becomes

$$d^2\psi(x)/dx^2 + (2mE/\hbar^2)\,\psi(x) = 0. \tag{4.38}$$

This has the same form as the simple harmonic oscillator equation (2.1) discussed earlier, so the solutions (2.2) and (2.3), similarly, can be written as

$$\psi(x) = A\sin kx + B\cos kx, \text{ where} \tag{4.39}$$

$$k = (2mE/\hbar^2)^{1/2} = 2\pi/\lambda. \tag{4.40}$$

The infinite potential walls at $x = 0$ and $x = L$ require $\psi(0) = \psi(L) = 0$, which means that $B = 0$. Again, the boundary condition $\psi(L) = 0 = A\sin kL$ means that

$$kL = n\pi, \quad \text{with } n = 1,2\ldots \tag{4.41}$$

This, in turn, gives

$$E_n = \hbar^2(n\pi/L)^2/2m = n^2h^2/8mL^2, \qquad n = 1,2,3,\ldots \tag{4.42}$$

We see that the allowed energies increase as the square of the integer quantum number n, and that the energies increase quadratically as L is decreased.

Finally, the normalization constant A of the wavefunction $\psi_n(x) = A\sin(n\pi x/L)$ must be chosen to make the probability of finding the particle somewhere in the trap to be unity. Thus

$$A^2 \int_0^L \sin^2(n\pi x/L)\mathrm{d}x = 1 \tag{4.43}$$

implies $A = (2/L)^{1/2}$.

The condition for allowed values of $k = n\pi/L$ is equivalent to

$$L = n\lambda/2, \tag{4.44}$$

the same condition that applies to waves on a violin string.

The exact solution of this simple problem illustrates typical nanophysical behavior in which there are discrete allowed energies and corresponding wavefunctions. The wavefunctions do not precisely locate a particle, they only provide statements on the probability of finding a particle in a given range.

For example, the probability distribution for $n = 3$ is $P_3(x) = (2/L)\sin^2(3\pi x/L)$. This normalized function has three identical peaks on the interval $0, L$, and one can see by inspection that the probability of finding the particle in the range $L/3$ to $2L/3$ is exactly $1/3$. As the quantum number n becomes large, the function $P_n(x)$ approaches the classical probability distribution, $P = 1/L$, as the oscillations of $\sin^2(n\pi x/L)$ become too rapid to observe.

Linear Combinations of Solutions

An additional property of differential equations such as the Schrodinger equation is that linear combinations of solutions are also solutions. So a possible linear combination solution is

$$\Psi_{1+3}(x,t) = A[\sin(\pi x/L)\exp(-\mathrm{i}E_1t/\hbar) + \sin(3\pi x/L)\exp(-\mathrm{i}E_3t/\hbar)]. \tag{4.45}$$

This will lead to a time-dependent probability density.

Expectation Values

The $P(x)$ function can be used to obtain precise expectation values. For example, in the case of the trapped particle in one dimension,

$$\langle x_n^2 \rangle = (2/L)\int_0^L x^2 \sin^2(n\pi x/L)\mathrm{d}x = (L^2/3)\left(1 - \frac{3}{n^2\pi^2}\right), \quad n = 1,2\ldots \tag{4.46}$$

This result, exact for any quantum number n, approaches the classical value $(L^2/3)$ as the quantum number n becomes large. This is a general property in quantum mechanics, that the classical result is recovered in the limit of large quantum numbers.

The expectation value of the coordinate, x, between states m,n is

$$<x_{m,n}> = (2/L) \int_0^L x \sin(m\pi x/L) \exp(-iE_m t/\hbar) \sin(n\pi x/L) \exp(iE_n t/\hbar) dx$$

$$= x_{m,n} \exp(-i\omega_{m,n} t), \quad (4.47)$$

where $\omega_{m,n} = (E_m - E_n)/\hbar$.

For a charged particle, $p = e<x_{m,n}>$ represents an oscillating electric dipole moment which can lead to absorption or emission of a photon.

The expectation value of the energy in any particular quantum state is

$$<E> = (2/L) \int_0^L \psi^* \mathcal{H} \psi dx. \quad (4.48)$$

Two-particle Wavefunction

The wavefunction for two non-interacting particles simultaneously present in the 1D trap, in quantum states n,m, respectively, can be written as

$$\psi_{n,m}(x_1,x_2) = A^2 \sin(n\pi x_1/L) \sin(m\pi x_2/L) = \psi_n(x_1)\, \psi_m(x_2). \quad (4.49)$$

The corresponding probability density is

$$P_{n,m}(x_1,x_2) = \psi^*_{n,m}(x_1,x_2)\, \psi_{n,m}(x_1,x_2). \quad (4.50)$$

4.6.3 Reflection and Tunneling at a Potential Step

A particle moving toward a finite potential step U_o at $x = 0$ illustrates reflection and tunneling effects which are basic features of nanophysics. Suppose $U = 0$ for $x < 0$ and $U = U_o$ for $x \geq 0$.

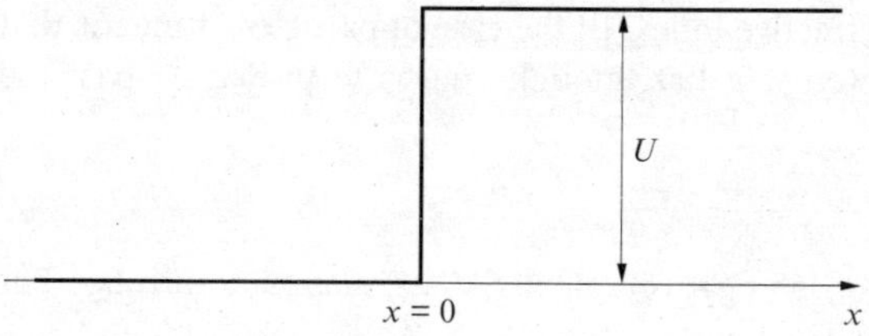

Figure 4.4 Finite potential step at $x = 0$.

Case 1: $E > U_o$

In case $E > U_o$, the particle incident from negative x is partially transmitted and partially reflected. For negative x,

$$\psi(x) = A\exp(ikx) + B\exp(-ikx), \text{ where } k = (2mE/\hbar^2)^{1/2}. \quad (4.51)$$

For positive x,

$$\psi(x) = C\exp(ik'x) + D\exp(-ik'x), \text{ where } k' = [(2m(E-U_o)/\hbar^2]^{1/2}. \quad (4.52)$$

Here A, B, C, and D are arbitrary complex numbers. The physical constraints on the allowable solutions are essential to solving this problem.

First, $D = 0$, since no particles are incident from the right.

Second, at $x = 0$ the required continuity of $\psi(x)$ implies $A + B = C$.

Third, at $x = 0$ the derivatives, $d\psi(x)/dx = Aik\exp(ikx) - Bik\exp(-ikx)$ on the left, and $d\psi(x)/dx = Cik'\exp(ik'x)$, on the right, must be equal. Thus

$$A + B = C \quad \text{and} \quad k'C = k(A-B). \quad (4.53)$$

The equations (4.53) are equivalent to

$$B = \frac{k-k'}{k+k'}A \quad \text{and} \quad C = \frac{2k}{k+k'}A. \quad (4.54)$$

The reflection and transmission probabilities, R and T, respectively, for the particle flux are then

$$R = \frac{|B|^2}{|A|^2} = \left(\frac{k-k'}{k+k'}\right)^2 \quad \text{and} \quad T = \frac{k'|C|^2}{k|A|^2} = \frac{4kk'}{(k+k)^2}. \quad (4.55)$$

It is easy to see that equation (4.55) implies

$$R + T = 1. \quad (4.56)$$

These equations are equivalent to the laws for reflection and transmission of light waves at a discontinuity in the refractive index. In the case of particles, it means that a particle is partially reflected even if it has enough energy to propagate past the jump in potential U_o.

Case 2: $E < U_o$

The only change is that now $E-U_o$ is negative, making k' an imaginary number. For this reason k' is now written as $k' = i\kappa$, where

$$\kappa = [2m(U_o-E)/\hbar^2]^{1/2} \quad (4.57)$$

is a real decay constant. Now the solution for positive x becomes

$$\psi(x) = C\exp(-\kappa x) + D\exp(\kappa x), \text{ where } \kappa = [2m(U_o - E)/\hbar^2]^{1/2}. \quad (4.58)$$

In this case, $D = 0$, to prevent the particle from unphysically collecting at large positive x. Equations (4.54) and the first of equations (4.55) remain valid setting $k' = i\kappa$. It is seen that $R = 1$, because the numerator and denominator in the first equation (4.55) are complex conjugates of each other, and thus have the same absolute value.

The solution for positive x is now an exponentially decaying function, and is not automatically zero in the region of negative energy. Find the probability at $x = 0$, $|C|^2$, by setting $k' = i\kappa$ in equation (4.54) and forming $|C|^2 = C^*C$, to be

$$|C|^2 = |A|^2 \frac{4k^2}{(k^2+\kappa^2)} = |A|^2 \frac{4E}{[E+(U_o-E)]} = |A|^2 \frac{4E}{U_o}, \quad (4.59)$$

where $E = (\hbar^2 k^2/2m) < U_o$. Note that $|C|^2 = 0$ for an infinite potential, as in our treatment of the trapped particle. Also, this expression agrees in the limit $E = U_o$ with equation (4.54).

Thus, the probability of finding the particle in the forbidden region of positive x is

$$P(x>0) = |2A|^2 (E/U_o) \int_0^{\infty} \exp(-2\kappa x)dx = 2|A|^2 E/(\kappa\, U_o), \quad (4.60)$$

where A is the amplitude of the incoming wave and $E < U_o$.

This is an example of the nanophysical tunneling effect, a particle can be sometimes found where its classical energy is negative, $E < U$.

The electron waves on the left of the barrier, $x < 0$, from (4.51) with $|A| = |B|$, add up to

$$\psi(x) = 2|A|\cos(kx + \delta), \quad (4.61)$$

which is a standing wave. Here the phase shift δ of the standing wave brings equation (4.61), when evaluated at $x = 0$, into agreement with equation (4.59). This standing wave will give "ripples" whenever an electron wave is reflected from a barrier, and the wavelength will be $\lambda = 2\pi/k$, where $k = (2mE/\hbar^2)^{1/2}$, given in (4.51). Actually, the observable quantity is

$$P(x) = \psi^*(x)\,\psi(x) = 4|A|^2\cos^2(kx + \delta), \quad (4.62)$$

so the observed wavelength will be halved. Such ripples are evident in Figure 3.8.

4.6.4
Penetration of a Barrier, Escape Time from a Well, Resonant Tunneling Diode

If the problem is turned into a tunneling barrier penetration problem by limiting the region of potential U_o to a width t: $0 < x < t$, the essential dependence of the tunneling transmission probability $|T|^2$ is proportional to $\exp(-2\kappa t)$. In this case there will be a traveling wave, say of amplitude F, for large positive x: $\psi(x) = F\exp(ikx)$, and the relevant transmission probability is defined as the ratio

$$|T|^2 = |F|^2/|A|^2, \tag{4.63}$$

where A is the amplitude of the wave coming from negative x values. To solve this problem the wavefunctions in the three separate regions: $x < 0$, $0 < x < t$, and $x > t$ are written and adjusted to match in value and in slope at $x = 0$ and at $x = t$. The result in case $\kappa t >> 1$ is

$$|T|^2 = \frac{16k^2\kappa^2\exp(-2\kappa t)}{(k^2+\kappa^2)^2}, \tag{4.64}$$

where the symbols have the same meanings as above, and the exponential function provides the dominant dependence.

For many purposes, the prefactor to the exponential term in (4.64) can be set to unity, so that

$$T^2 = \exp(-2\kappa t). \tag{4.64a}$$

Important cases in electronics and in radioactive decay involve a transiently trapped particle enclosed in a potential well of length L, between barriers of finite width t, for a mean residence time τ. As an instructive approximation, think of two barriers of height V, widths t, and spacing L. If V and t are large, though finite, one may reasonably approximate the lowest members of the set of bound states in the $V = 0$ $(0 \le x \le L)$ region between the two barriers by E_n from (4.42), and the wavefunctions ψ as

$$\psi_n = (2/L)^{1/2}\sin(n\pi x/L), \quad \text{for} \quad n = 1, 2, \ldots. \tag{4.39a}$$

(This *approximation* makes sense only for states of small n such that E_n is much smaller than V, the barrier height).

However, for such strongly bound states, one can approximate the escape rate τ^{-1} as the product of the rate of collisions of the particle with a barrier (v/L) multiplied by the chance per encounter, T^2, of transmission and escape. Such a method was long ago used by Gamow to successfully model the systematics of alpha-particle decay of radioactive nuclei.

4.6.5
Trapped Particles in Two and Three Dimensions: Quantum Dot

The generalization of the Schrodinger equation to three dimensions is

$$-(\hbar^2/2m)\nabla^2\psi(\mathbf{r}) + U(\mathbf{r})\,\psi(\mathbf{r}) = E\psi(\mathbf{r}), \tag{4.65}$$

where $\nabla^2 = \partial^2/\partial x^2 + \partial^2/\partial y^2 + \partial^2/\partial z^2$ and $\mathbf{r}$ is a vector with components x,y,z.

It is not difficult to see that the solutions for a particle in a three-dimensional infinite trap of volume L^3 with impenetrable walls are given as

$$\psi_n(x,y,z) = (2/L)^{3/2} \sin(n_x \pi x/L)\ \sin(n_y \pi y/L)\ \sin(n_z \pi z/L), \quad \text{where } n_x = 1,2\ldots, \text{etc.,} \tag{4.66}$$

and

$$E_n = [h^2/8mL^2](n_x{}^2 + n_y{}^2 + n_z{}^2). \tag{4.67}$$

These simple results can be easily adapted to two-dimensional boxes and also to boxes of unequal dimensions $L_x\ L_y\ L_z$.

Electrons Trapped in a Two-dimensional Box

A scanning tunneling microscope image, shown in Figure 4.5, reveals some aspects of the trapping of electrons in a 2D rectangular potential well. The well in this case is generated by the rectangular array of iron atoms (silver-colored dots in this image), which reflect electrons, much as in equation (4.62). The electrons are free electrons on the surface of (111) oriented single crystal copper. These electrons, by a quirk of the solid state physics of (111) copper, are essentially confined to the surface, and are further confined by the iron atoms to stay inside (or outside) the box. The traces of "ripples" which are seen here (also seen more clearly in Figure 3.8 in a circular geometry) are visible because the STM measurement, on a specifically 2D electron system, is sensitive to the density of states [7] as well as to the height of the surface.

Electron ripples inside a rectangular 2D box can be predicted from $P(x,y)$ from equation (4.67) (setting $n_z = 0$ to recover a two-dimensional case). The values of n_x, n_y

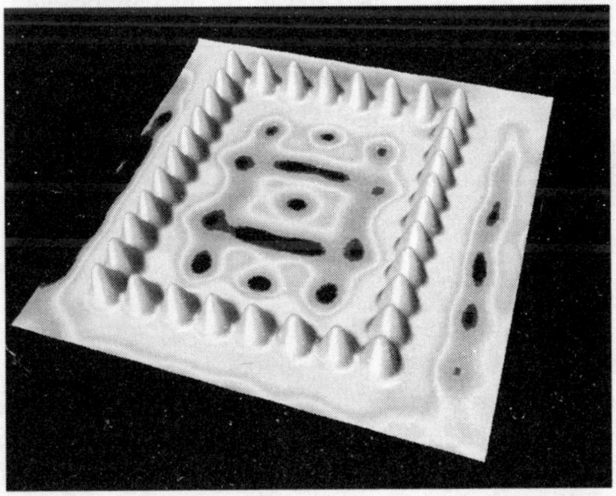

Figure 4.5 Electrons trapped in a small two-dimensional box on the (111) surface of copper. A rectangular array of iron atoms serves as a barrier, reflecting electron waves [6].

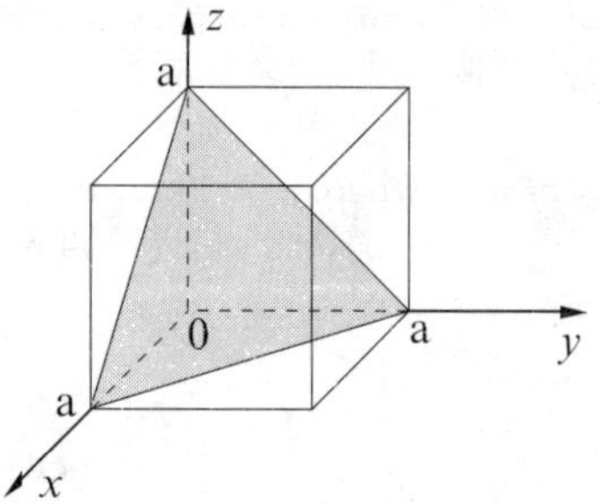

Figure 4.6 Geometry of a (111) plane, shown shaded. Copper is a face-centered cubic crystal, but only if the surface is cut to consist of the indicated (111) plane will the 2D electron effects be present.

appropriate to the ripples in Figure 4.5 are not known, but will be influenced by two facts:

1) First, the energy of the electrons being seen by the STM tip is the Fermi energy of the copper, plus or minus the voltage applied to the tip.
2) Second, the energy scale, suggested by equation (4.67): $E_n = [h^2/8mL^2](n_x^{\ 2} + n_y^{\ 2})$ (which needs a small modification for unequal sides L_x, L_y) of the trapped electron levels, should be offset below the Fermi energy of copper by 0.44 eV. This is known to be the location of the bottom of the band of the 2D surface electrons on the (111) face of copper.

Electrons in a 3D "Quantum Dot"

Equations (4.66) and (4.67) are applicable to the electron and hole states in semiconductor "quantum dots", which are used in biological research as color-coded fluorescent markers. Typical semiconductors for this application are CdSe and CdTe.

A "hole" (missing electron) in a full energy band behaves very much like an electron, except that it has a positive charge, and tends to float to the top of the band. That is, the energy of the hole increases oppositely to the energy of an electron.

The rules of nanophysics that have been developed so far are also applicable to holes in semiconductors. To create an electron-hole pair in a semiconductor requires an energy at least equal to the energy bandgap, E_g, of the semiconductor.

This application to semiconductor quantum dots requires L in the range of 3–5 nm, the mass m must be interpreted as an effective mass m^*, which may be as small as 0.1 m_e. The electron and hole particles are generated by light of energy

$$hc/\lambda = E_{n,\text{electron}} + E_{n,\text{hole}} + E_g. \tag{4.68}$$

Here the first two terms depend strongly on particle size L, as L^{-2}, which allows the color of the light to be adjusted by adjusting the particle size. The bandgap energy, E_g, is the minimum energy to create an electron and a hole in a pure semiconductor. The electron and hole generated by light in a bulk semiconductor may form a bound state along the lines of the Bohr model, described above, called an exciton. However, as the size of the sample is reduced, the Bohr orbit becomes inappropriate and the states of the particle in the 3D trap, as described here, provide a correct description of the behavior of quantum dots.

4.6.6
2D Bands and Quantum Wires

2D Band

A second physical situation that often arises in modern semiconductor devices is a carrier confined in one dimension, say z, to a thickness d and free in two dimensions, say x and y. This is sometimes called a quantum well.

In this case

$$\psi_n(x,y,z) = (2/d)^{1/2} \sin(n_z \pi z/d) \exp(ik_x x) \exp(ik_y y), \tag{4.69}$$

and the energy of the carrier in the nth band is

$$E_n = (h^2/8md^2)\, n_z^2 + \hbar^2 k_x^2/2m + \hbar^2 k_y^2/2m. \tag{4.70}$$

In this situation, the quantum number n_z is called the sub-band index and for $n = 1$ the carrier is in the first sub-band. We discuss later how a basic change in the electron's motion in a semiconductor band is conveniently described with the introduction of an effective mass m^*.

Quantum Wire

The term Quantum Wire describes a carrier confined in two dimensions, say z and y, to a small dimension d (wire cross section d^2) and free to move along the length of the wire, x. (Qualitatively this situation resembles the situation of a carrier moving along a carbon nanotube, or silicon nanowire, although the details of the bound state wavefunctions are different.)

In the case of a quantum wire of square cross section,

$$\psi_{nn}(x,y,z) = (2/d) \sin(n_y \pi y/d) \sin(n_z \pi z/d) \exp(ik_x x), \tag{4.71}$$

and the energy is

$$E_n = (h^2/8md^2)\,(n_y^2 + n_z^2) + \hbar^2 k_x^2/2m. \tag{4.72}$$

It is possible to grow nanowires of a variety of semiconductors by a laser assisted catalytic process, and an example of nanowires of indium phosphide is shown in Figure 4.7 [8].

The InP wires in this experiment are single crystals whose lengths are hundreds to thousands of times their diameters. The diameters are in the nm range, 10–50 nm. The extremely anisotropic shape is shown in Figure 4.4 to lead to extremely polarization-dependent optical absorption.

Since these wires have one long dimension, they do not behave like quantum dots, and the light energy is not shifted from the bandgap energy. It is found that the nanowires can be doped to produce electrical conductivity of N- and P-types, and they can be used to make electron devices.

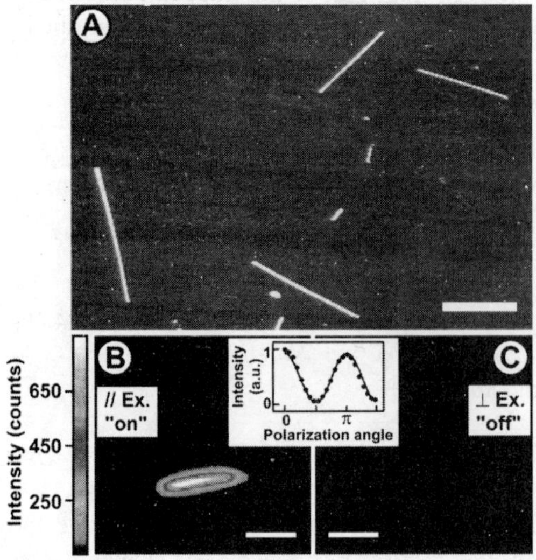

Figure 4.7 Indium phosphide nanowires[8]. InP nanowires grown by laser-assisted catalytic growth, in 10, 20, 30 and 50 nm diameters, were studied by Atomic Force Microscope image (A) and also by observation of photoluminescence (B) and (C) under illumination by light with energy $hc/\lambda > E_g$. The bandgap is about 1.4 eV. In (A), the white scale bar is 5 µm, so the wires are up to 10 µm in length. (B) and (C) gray scale represantation of light emitted from 20 nm diameter InP nanowire excitation in panel (B) with bandgap light linearly polarized along the axis of the wire produces a large photoluminescence, but (C) no light is emitted when excited by bandgap light linearly polarized *perpendicular* to wire axis. Inset shows dependence of emission intensity on polarization angle between light and wire axis.

4.6.7 The Simple Harmonic Oscillator

The simple harmonic oscillator (SHO) represents a mass on a spring, and also the relative motions of the masses of a diatomic molecule. (It turns out that it is also relevant to many other cases, including the oscillations of the electromagnetic field between fixed mirror surfaces.) In the first two cases the mass is m, the spring constant can be taken as k, and the resonant frequency in radians per second is

$$\omega = (k/m)^{1/2}. \qquad (4.73)$$

Treated as a nanophysics problem, one needs to solve the Schrodinger equation (4.37) for the corresponding potential energy (to stretch a spring a distance x):

$$U(x) = kx^2/2. \qquad (4.74)$$

This is a more typical nanophysics problem than that of the trapped particle, in that the solutions are difficult, but, because of their relevance, have been studied and tabulated by mathematicians.

The solutions to the SHO in nanophysics are:

$$\psi_n = A_n \exp(-m\omega x^2/2\hbar) H_n(x) \tag{4.75}$$

where the $H_n(x)$ are well-studied polynomial functions of x, and

$$E_n = (n + 1/2)\,\hbar\omega, \quad n = 0,1,2,3,\ldots \tag{4.76}$$

These wavefunctions give an oscillatory probability distribution $P_n(x) = \psi_n^{*}\psi_n$ that has $n + 1$ peaks, the largest occurring near the classical turning points, $x = \pm\sqrt{2E/k}$. For large n, $P_n(x)$ approaches the classical $P(x)$:

$$P(x)\mathrm{d}x = \mathrm{d}x/v = \frac{\mathrm{d}x}{\sqrt{(2/m)(E - m\omega^2 x^2/2)}}, \tag{4.77}$$

where v is the classical velocity. Since v goes to zero at the turning points, $P(x)$ becomes very large at those points. This feature is approached for large n in the solutions (4.75).

For small n, these functions differ substantially from the classical expectation, especially in giving a range of positions for the mass even in the lowest energy configuration, $n = 0$.

An important point here is that there is an energy $\hbar\omega/2$ (*zero point energy*) for this oscillator even in its lowest energy state, $n = 0$. Also, from equation (4.75), this oscillating mass does not reside solely at $x = 0$ even in its ground state. It has an unavoid-

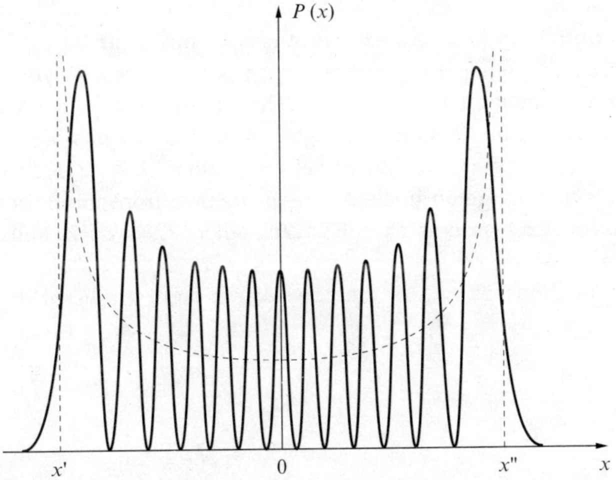

Figure 4.8 Probability $P_n(x)$ density for $n = 13$ state of simple harmonic oscillator. Dashed line is classical $P(x)$, with classical turning points x' and x'' indicated.

able fluctuation in position, given by the Gaussian function $\exp(-m\omega x^2/2\hbar)$, which is close to that predicted by the Uncertainty Principle, equation (4.29).

4.6.8
Schrodinger Equation in Spherical Polar Coordinates

A more substantive change in the appearance of the Schrodinger equation occurs in the case of spherical polar coordinates, which are appropriate to motion of an electron in an atom. In this case, the energy U depends only on the radius r, making the problem spherically symmetric. In standard notation, where $x = r\sin\theta\cos\varphi$, $y = r\sin\theta\sin\varphi$, $z = r\cos\theta$, with θ and φ, respectively, the polar and azimuthal angles:

$$\frac{-\hbar^2}{2m}\frac{1}{r^2}\frac{\partial}{\partial r}\left(\frac{r^2\partial\psi}{\partial r}\right)-\frac{\hbar^2}{2mr^2}\left[\frac{1}{\sin\theta}\frac{\partial}{\partial\theta}\left(\sin\theta\frac{\partial\psi}{\partial\theta}\right)+\frac{1}{\sin^2\theta}\frac{\partial^2\psi}{\partial\varphi^2}\right]+U(r)\psi=E\psi \quad (4.78)$$

4.7
The Hydrogen Atom, One-electron Atoms, Excitons

The Schrodinger equation is applied to the hydrogen atom, and any one-electron atom with nuclear charge Z, by choosing $U = -kZe^2/r$, where k is the Coulomb constant. It is found in such cases of spherical symmetry that the equation separates into three equations in the single variables r, θ, and φ, by setting

$$\psi = R(r)f(\theta)g(\varphi). \quad (4.79)$$

The solutions are conventionally described as the quantum states $\Psi_{n,l,m,m}$, specified by quantum numbers n, l, m, m_s.

The principal quantum number n is associated with the solutions $R_{n,l}(r) = (r/a_o)^l \exp(-r/na_o)\mathscr{L}_{n,l}(r/a_o)$ of the radial equation. Here $\mathscr{L}_{n,l}(r/a_o)$ is a Laguerre polynomial in $\rho = r/a_o$, and the radial function has $n-l-1$ nodes. The parameter a_o is identical to its value in the Bohr model, but it no longer signifies the exact radius of an orbit. The energies of the electron states of the one-electron atom, $E_n = -Z^2E_o/n^2$ (where $E_o = 13.6$ eV, and Z is the charge on nucleus) are unchanged from the Bohr model. The energy can still be expressed as $E_n = -kZe^2/2r_n$, where $r_n = n^2a_o/Z$, and $a_o = 0.0529$ nm is the Bohr radius.

The lowest energy wavefunctions $\Psi_{n,i,m,m}$ of the one-electron atom are listed in Table 4.1 [9].

Table 4.1 One-electron wavefunctions in real form [9]

Wavefunction designation	Wavefunction name, real form	Equation for real form of wavefunction*, where $\rho = Zr/a_o$ and $C_1 = Z^{3/2}/\sqrt{\pi}$
Ψ_{100}	1s	$C_1 e^{-\rho}$
Ψ_{200}	2s	$C_2\,(2-\rho)\,e^{-\rho/2}$
$\Psi_{21,\cos\varphi}$	$2p_x$	$C_2\,\rho\,\sin\theta\,\cos\varphi\,e^{-\rho/2}$
$\Psi_{21,\sin\varphi}$	$2p_y$	$C_2\,\rho\,\sin\theta\,\sin\varphi\,e^{-\rho/2}$
Ψ_{210}	$2p_z$	$C_2\,\rho\,\cos\theta\,e^{-\rho/2}$
Ψ_{300}	3s	$C_3\,(27-18\rho+2\rho^2)\,e^{-\rho/3}$
$\Psi_{31,\cos\varphi}$	$3p_x$	$C_3\,(6\rho-\rho^2)\,\sin\theta\,\cos\varphi\,e^{-\rho/3}$
$\Psi_{31,\sin\varphi}$	$3p_y$	$C_3(6\rho-\rho^2)\,\sin\theta\,\sin\varphi\,e^{-\rho/3}$
Ψ_{310}	$3p_z$	$C_3(6\rho-\rho^2)\,\cos\theta\,e^{-\rho/3}$
Ψ_{320}	$3d_z{}^2$	$C_4\,\rho^2\,(3\cos^2\theta-1)\,e^{-\rho/3}$
$\Psi_{32,\cos\varphi}$	$3d_{xz}$	$C_5\,\rho^2\,\sin\theta\,\cos\theta\,\cos\varphi\,e^{-\rho/3}$
$\Psi_{32,\sin\varphi}$	$3d_{yz}$	$C_5\,\rho^2\,\sin\theta\,\cos\theta\,\sin\varphi\,e^{-\rho/3}$
$\Psi_{32,\cos2\varphi}$	$3d_{x^2-y^2}$	$C_6\,\rho^2\,\sin^2\theta\,\cos2\varphi\,e^{-\rho/3}$
$\Psi_{32,\sin2\varphi}$	$3d_{xy}$	$C_6\,\rho^2\,\sin^2\theta\,\sin2\varphi\,e^{-\rho/3}$

* $C_2 = C_1/4\sqrt{2}$, $C_3 = 2C_1/81\sqrt{3}$, $C_4 = C_3/2$, $C_5 = \sqrt{6}C_4$, $C_6 = C_5/2$.

$$\Psi_{100} = (Z^{3/2}/\sqrt{\pi})\exp(-Zr/a_o) \tag{4.80}$$

represents the ground state.

The probability of finding the electron at a radius r is given by $P(r) = 4\pi r^2\Psi^2{}_{100}$, which is a smooth function easily seen to have a maximum at $r = a_o/Z$. This is not an orbit of radius a_o, but a spherical probability cloud in which the electron's most

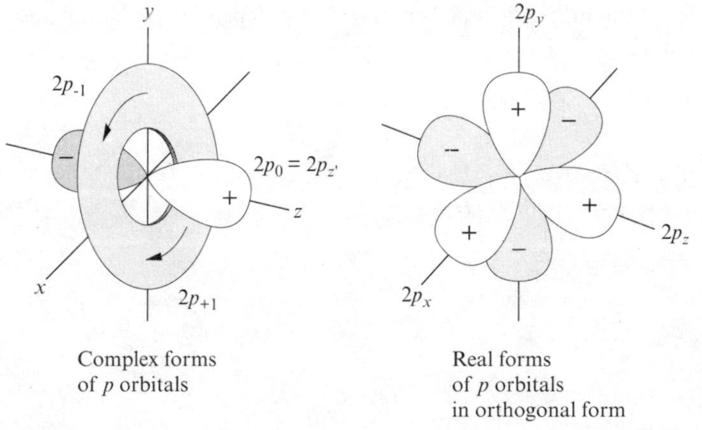

Figure 4.9 2p wavefunctions in schematic form. Left panel, complex forms carry angular momentum. Right panel, linear combinations have the same energy, now assume aspect of bonds.

probable radius from the origin is a_o. There is no angular momentum in this wavefunction!

This represents a great correction in concept, and in numerics, to the Bohr model. Note that Ψ_{100} is real, as opposed to complex, and therefore the electron in this state has no orbital angular momentum. Both of these features correct errors of the Bohr model.

The $n = 2$ wavefunctions start with Ψ_{200}, which exhibits a node in r, but is spherically symmetric like Ψ_{100}. The first anisotropic wavefunctions are:

$$\Psi_{21,\pm1} = R(r)f(\theta)g(\varphi) = C_2\,\rho\,\sin\theta\,e^{-\rho/2}\exp(\pm i\varphi), \tag{4.81}$$

where $\rho = Zr/a_o$.

These are the first two wavefunctions to exhibit orbital angular momentum, here $\pm\hbar$ along the z-axis. Generally

$$g(\varphi) = \exp(\pm im\varphi), \tag{4.82}$$

where m, known as the magnetic quantum number, represents the projection of the orbital angular momentum vector of the electron along the z-direction, in units of $\hbar$. The orbital angular momentum **L** of the electron motion is described by the quantum numbers l and m.

The orbital angular momentum quantum number l has a restricted range of allowed integer values:

$$l = 0,1,2\ldots,n-1. \tag{4.83}$$

This rule confirms that the ground state, $n = 1$, has zero angular momentum. In the literature the letters s,p,d,f,g, respectively, are often used to indicate $l = 0,1,2,3,$ and 4. So a 2s wavefunction has $n = 2$ and $l = 0$.

The allowed values of the magnetic quantum number m depend upon both n and l according to the scheme

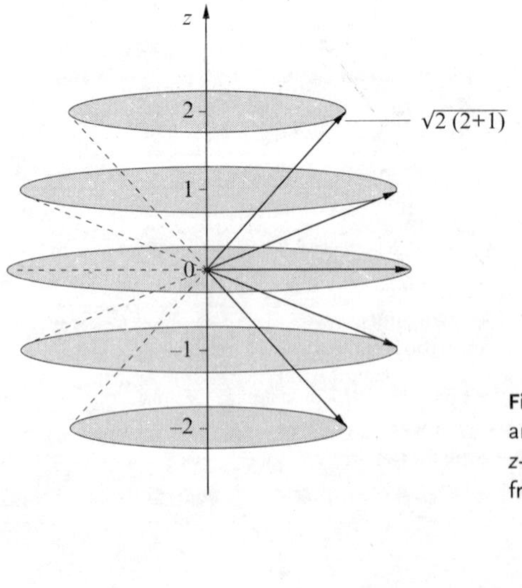

Figure 4.10 Five allowed orientations of angular momentum $l = 2$, length of vector and z-projections in units of $\hbar$. Azimuthal angle is free to take any value.

$$m = -l, -l + 1, \ldots, (l-1), l. \qquad (4.84)$$

There are $2\,l + 1$ possibilities. For $l = 1$, for example, there three values of m: –1,0, and 1, and this is referred to as a "triplet state". In this situation the angular momentum vector has three distinct orientations with respect to the z-axis: $\theta = 45°$, 90° and 135°. In this common notation, the $n = 2$ state (containing 4 distinct sets of quantum numbers) separates into a "singlet" (2s) and a "triplet" (2p).

For each electron there is also a spin quantum number S with projection

$$m_s = \pm 1/2. \qquad (4.85)$$

These strange rules, which are known to accurately describe the behavior of electrons in atoms, enumerate the possible distinct quantum states for a given energy state, n.

Following these rules one can see that the number of distinct quantum states for a given n is $2n^2$. Since the Pauli exclusion principle for electrons (and other Fermi particles) allows only one electron in each distinct quantum state, $2n^2$ is also the number of electrons that can be accommodated in the nth electron shell of an atom. For $n = 3$ this gives 18, which is seen to be twice the number of entries in Table 4.1 for $n = 3$.

A further peculiarity of angular momentum in nanophysics is that the vector **L** has length $L = \sqrt{(l(l + 1))}\ \hbar$ and projection $L_z = m\hbar$. A similar situation occurs for the spin vector **S**, with magnitude $S = \sqrt{(s(s + 1))}\ \hbar$ and projection $m_s\hbar$. For a single electron $m_s = \pm 1/2$. In cases where an electron has both orbital and spin angular momenta (for example, the electron in the $n = 1$ state of the one-electron atom has only S, but no L), these two forms of angular momentum combine as $J = L + S$, which again has a strange rule for its magnitude: $J = \sqrt{(j(j + 1))}\ \hbar$.

The wavefunctions $\Psi_{21,\pm 1} = C_2\ \rho\ \sin\theta\ e^{-\rho/2} \exp(\pm i\varphi)$ are the first two states having angular momentum. A polar plot of $\Psi_{21,\pm 1}$ has a node along z, and looks a bit like a doughnut flat in the xy plane.

The sum and difference of these states are also solutions to Schrodinger's equation, for example

$$\Psi_{211} + \Psi_{21-1} = C_2\ \rho\ \sin\theta\ e^{-\rho/2} [\exp(i\varphi) + \exp(-i\varphi)] = C_2\ \rho\ \sin\theta\ e^{-\rho/2}\, 2\cos\varphi. \qquad (4.86)$$

This is just twice the $2p_x$ wavefunction in Table 4.1. This linear combination is exemplary of all the real wavefunctions in Table 4.1, where linear combinations have canceled the angular momenta to provide a preferred direction for the wavefunction.

A polar plot of the $2p_x$ wavefunction (4.86), shows a node in the z-direction from the $\sin\theta$ and a maximum along the x-direction from the $\cos\varphi$, so it is a bit like a dumb-bell at the origin oriented along the x-axis. Similarly the $2p_y$ resembles a dumb-bell at the origin oriented along the y-axis.

These real wavefunctions, in which the $\exp(im\varphi)$ factors have been combined to form $\sin\varphi$ and $\cos\varphi$, are more suitable for constructing bonds between atoms in

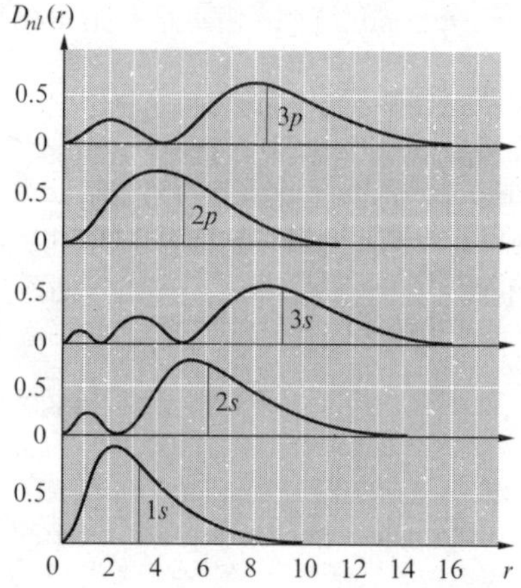

Figure 4.11 Radial wavefunctions of one-electron atoms exhibit n–ℓ–1 nodes, as illustrated in sketch for (bottom to top) 1s, 2s, 3s, 2p and 3p wavefunctions.

molecules or in solids, than are the equally valid (complex) angular momentum wavefunctions. The complex wavefunctions which carry the exp($im\varphi$) factors are essential for describing orbital magnetic moments such as occur in iron and similar atoms. The electrons that carry orbital magnetic moments usually lie in inner shells of their atoms.

4.7.1 Magnetic Moments

The magnetic properties of atoms are important in nanotechnology, and lie behind the function of the hard disk magnetic memories, for example. A magnetic moment

$$\mu = i\mathbf{A} \tag{4.87}$$

is generated by a current loop bounding an oriented area **A**. The magnetization $\boldsymbol{M}$ is the magnetic moment per unit volume. A magnetic moment produces a dipole magnetic field similar to that of an ordinary bar magnet. The magnetization of a bar of magnetite or of iron is caused by internal alignment of huge numbers of the atomic magnetic moments. These, as we will see, are generated by the orbital motion of electrons in each atom of these materials.

How does an orbiting electron produce a magnetic moment? The basic formula $\mu = i\mathbf{A}$ for a magnetic moment can be rewritten as $(dq/dt)\pi r^2 = (ev/2\pi r)\pi r^2 = (e/2m)mrv$. Finally, since mrv is the angular momentum L, we get

$$\mu = (e/2m)\mathbf{L}. \tag{4.88}$$

This is the basic gyro-magnetic relation accounting for atomic orbital magnetism. Recalling that the z-component of **L** is $L_z = m\hbar$, we see that the electron orbital motion leads to $\mu_z = (e\hbar/2m_e)\ m$. The Bohr Magneton μ_B is defined as

$\mu_B = e\hbar/2m_e = 5.79 \times 10^{-5}$ eV/T. The magnetic moment arising from the electron spin angular momentum is described by a similar formula,

$$\mu_z = g(e\hbar/2m_e)\ m_s. \tag{4.89}$$

Here g, called the gyromagnetic ratio, is $g = 2.002$ for the electron spin and $m_s = \pm 1/2$. The energy of orientation of a magnetic moment μ in a magnetic field B is

$$U = -\mu \cdot \mathbf{B}, \tag{4.90}$$

where the dot product of the vectors introduces the cosine of the angle between them. Thus, the difference in energy between states $m_s = 1/2$ and $m_s = -1/2$ for an electron spin moment in a magnetic field B (taken in the z-direction) is $\Delta E = g\mu_B B$.

The reasons for the alignment of atomic magnetic moments in iron and other ferromagnetic substances, which underlies the operation of the magnetic disk memory, will be taken up later in connection with basic symmetry of particles under exchange of positions.

4.7.2
Magnetization and Magnetic Susceptibility

A system of N non-interacting magnetic moments per unit volume in an applied magnetic field $\mathbf{B}$ will develop a magnetic polarization, because the magnetic moment has a lower energy when pointing along the $\mathbf{B}$ field direction. If the splitting is $2\mu_B B$ (setting $g = 2$ for the electron), and assuming a Boltzmann thermal occupation $\exp(\mu_B B/kT)$ for spin up and $\exp(-\mu_B B/kT)$ for spin down, then the polarization P, the fractional unbalance of spin up vs. spin down, is given by

$$P = (n\uparrow - n\downarrow)/(n\uparrow + n\downarrow) = \tanh(\mu_B B/kT). \tag{4.91}$$

The consequence will be a volume magnetization $\mathbf{M}$ (defined as magnetic moment per unit volume)

$$\mathbf{M} = N\mu_B \mathbf{P} = N\boldsymbol{\mu}_B \tanh(\mu_B B/kT). \tag{4.92}$$

In the frequent (low field) case of $\mu_B B/kT << 1$, one gets Curie's Law:

$$\mathbf{M} = N(\mu_B{}^2/kT)B. \tag{4.93}$$

These equations are based on spin $S = ½$, and can be generalized to systems of larger spin S, which we omit.

If a region of uniform magnetization $\mathbf{M}$ exists, as a consequence of a uniform applied magnetic field $\mathbf{B}$, then an added magnetic field will result, whose magnitude is $\mu_o\mathbf{M}$, where $\mu_o/4\pi = 10^{-7}$ in SI units. (This result is closely related to the formula for the magnetic field inside a long solenoid of n turns per unit length, which is $B = \mu_o nI$, where I is the current through the coil of n turns per unit length).

The magnetic susceptibility is defined as $\chi_m = M/B$. The local magnetic field in a magnetically susceptible material is thus

$$B = B_{app}(1 + \chi_m). \tag{4.94}$$

There are cases (involving ferromagnetism, see below) in which a very large local magnetic field can result from a small applied field; in these cases $(1 + \chi_m) = \kappa_m$ is called the "magnetic permeability". Large values of κ_m exceeding 10^4 are obtained in alloys of Ni and Fe (e.g. Permalloy and Supermalloy) and are fundamental to the operation of magnetic sensors used in computer hard disks and in iPod memory devices.

4.7.3
Positronium and Excitons

In the treatment of the hydrogen and one-electron atoms, it has been assumed that the proton or nucleus of charge Z is fixed, or infinitely massive. This simplification is not hard to correct, and a strong example is that of positronium, a bound state of a positron (positive electron) and an electron. These two particles have opposite charges but equal masses, and then jointly orbit around their center of mass, half the distance between the plus and minus charges. It is an exercise to see that the Bohr radius for positronium is $2a_o$ and its binding energy is $E_o/2$. All of the results for the hydrogenic atom can be transcribed to this case by interpreting r as the interparticle distance and interpreting m to the central mass:

$$1/m_{cm} = 1/m_1 + 1/m_2. \tag{4.95}$$

The classic case is positronium, but a more relevant case is the exciton.

The exciton is the bound state of a photoexcited electron and photoexcited hole in a semiconductor, produced momentarily by light illumination of energy greater than the band gap. The electron and hole particles thus produced orbit around each other, having exactly equal and opposite charges. (In a short time the charges unite, giving off a flash of light.) In the medium of the semiconductor, however, the electron and hole assume effective masses, respectively, m^*_e and m^*_h, which are typically significantly different from the free electron mass.

If, as is typical, one mass, say the electron mass, is small, say 0.1 of the free electron mass, then, from equation (4.95), m_{cm} becomes small. Because the interparticle Bohr radius is inversely dependent on mass m_{cm}, this makes the Bohr radius large, roughly $10\,a_o$. In turn, because the bound state energy is $-kZe^2/2r$, this makes the energy small, roughly $0.1\,E_o$.

4.8 Fermions, Bosons and Occupation Rules

Schrodinger's equation tells us that, in physical situations described by potential energy functions U, there are quantum energy states available for particles to occupy. It turns out that particles in nature divide into two classes, called fermions and bosons, which differ in the way they can occupy quantum states. Fermions follow a rule that only one particle can occupy a fully described quantum state, such as a hydrogen electron state described by n, l, m_l, and m_s. This rule was first recognized by Pauli, and is called the Pauli exclusion principle. In a system with many states and many fermion particles to fill these states, the particles first fill the lowest energy states, increasing in energy until all particles are placed. The highest filled energy is called the Fermi energy E_F.

For bosons, there is no such rule, and any number of particles can fall into exactly the same quantum state. Such a condensation of many photons into a single quantum state is what happens in the operation of a laser.

When a nanophysical system is in equilibrium with a thermal environment at temperature T, then average occupation probabilities for electron states are found to exist. In the case of fermions, the occupation function (the Fermi–Dirac distribution function, f_{FD}), is

$$f_{FD} = \left[\exp\left(\frac{E-E_F}{kT}\right) + 1\right]^{-1}. \tag{4.96}$$

For photons, the corresponding Bose–Einstein distribution function, f_{BE}, is

$$f_{BE} = \left[\exp\left(\frac{hc}{\lambda kT}\right) - 1\right]^{-1}. \tag{4.97}$$

References

[1] N. Bohr, Phil. Mag. **26**, 1 (1913).
[2] E. Schrodinger, Annalen der Physik **79**, 361 (1926).
[3] L. de Broglie, Ann. de phys. **3**, 22 (1925).
[4] C. Davisson and L. Germer, Proc. Natl. Acad. Sci. **14**, 619 (1928).
[5] W. Heisenberg, Zeit. fur. Phys. **43**, 172 (1927).
[6] Courtesy IBM Research, Almaden Research Center. Unauthorized use not permitted.
[7] E. L. Wolf, *Principles of Electron Tunneling Spectroscopy*, (Oxford, New York, 1989), page 319.
[8] Reprinted with permission from Wang, J.S. Gudiksen, X. Duan, Y. Cui, and C.M. Lieber, Science **293**, 1455 (2001). Copyright 2001 AAAS.
[9] F. J. Pilar, Elementary Quantum Chemistry, (Dover, New York., 2001), p. 125.

5
Quantum Consequences for the Macroworld

Chemical matter is made of atoms, which are constructed according to the strange rules of nanophysics. The stability of atoms against radiative collapse was first predicted in the Bohr hydrogen atom model, by the arbitrary imposition of a quantum condition on angular momentum. The Schrodinger equation explains the stability of chemical matter in more detail. An essential additional principle is the Pauli statement that only one electron can occupy a fully described quantum state. This is the "building-up principle" that describes the shell structure of atoms, and the chemical table of the elements.

5.1
Chemical Table of the Elements

The rules governing the one-electron atom wavefunction $\Psi_{n,l,m,m}$ and the Pauli exclusion principle, which states that only one electron can be accommodated in a completely described quantum state, are the basis for the Chemical Table of the Elements. As we have seen, the strange rules of nanophysics allow $2n^2$ distinct states for each value of the principal quantum number, n. There are several notations to describe this situation. The "K shell" of an atom comprises the two electrons of $n = 1$ ($1s^2$), followed by the "L shell" with $n = 2$ ($2s^2\,2p^6$); and the "M shell" with $n = 3$ ($3s^2\,3p^6\,3d^{10}$). These closed shells contain, respectively, 2, 8, and 18 electrons.

In the Chemical Table of Mendeleyev, one notable feature is the stability of filled "electron cores", such as those which occur at $Z = 2$ (He, with a filled K shell), and $Z = 10$ (Ne, with filled K and L shells).

The situation of a single electron beyond a full shell configuration, such as sodium, potassium, rubidium, and cesium, can be roughly modeled as an ns electron moving around the rare gas core described with an effective charge Z', less than Z. That Z' is reduced results from the shielding of the full nuclear charge Z by the inner closed shell electrons. It is remarkable that interactions between electrons in large atoms can be in many cases ignored.

These rules of nanophysics are believed to account for the schematics of the chemical table of the elements, and, as well, to the properties of chemical compounds. It is a logical progression to expect that the larger aggregations of molecules

Nanophysics and Nanotechnology: An Introduction to Modern Concepts in Nanoscience. Second Edition.
Edward L. Wolf

ISBN: 3-527-40651-4

that are characteristic of biology are also understandable in the framework of Schrodinger quantum mechanics.

5.2
Nano-symmetry, Di-atoms, and Ferromagnets

A further and profound nanophysical behavior, with large consequences for the macroworld, is based on the identical nature of the elementary particles, such as electrons.

It is easy to understand that ionic solids like KCl are held together by electrostatic forces, as the outer electron of K fills the outer electron shell of Cl, leading to an electrostatic bond between K^+ and Cl^-.

But what about the diatomic gases of the atmosphere, H_2, O_2, and N_2? The bonding of these symmetric structures, called covalent, is entirely nanophysical in its origin. It is strange but true that the same symmetry-driven electrostatic force that binds these di-atoms is also involved in the spontaneous magnetism of iron, cobalt and other ferromagnetic metals.

5.2.1
Indistinguishable Particles, and their Exchange

The origin of the symmetric covalent bond is the indistinguishable nature of electrons, for which labels are impossible. If two electrons are present in a system, the probability distributions $P(x_1,x_2)$ and $P(x_2,x_1)$ must be identical.

No observable change can occur from exchanging the two electrons. That is,

$$P(x_1,x_2) = P(x_2,x_1) = |\psi_{n,m}(x_1,x_2)|^2, \tag{5.1}$$

from which it follows that either

$$\psi_{n,m}(x_2,x_1) = \psi_{n,m}(x_1,x_2) \qquad \text{(symmetric case), or} \tag{5.2}$$

$$\psi_{n,m}(x_2,x_1) = -\psi_{n,m}(x_1,x_2) \qquad \text{(antisymmetric case).} \tag{5.3}$$

If we apply this idea to the two non-interacting electrons in, for example, a 1D trap, with the wavefunction (4.49)

$$\psi_{n,m}(x_1,x_2) = A^2 \sin(n\pi x_1/L)\,\sin(m\pi x_2/L) = \psi_n(x_1)\,\psi_m(x_2),$$

we find that this particular two-particle wavefunction is neither symmetric nor antisymmetric. However, the combinations of symmetric and antisymmetric two-particle wavefunctions

$$\psi_S(1,2) = [\psi_n(x_1)\,\psi_m(x_2) + \psi_n(x_2)\,\psi_m(x_1)]/\sqrt{2}, \text{ and} \tag{5.4}$$

$$\psi_A(1,2) = [\psi_n(x_1)\ \psi_m(x_2) - \psi_n(x_2)\ \psi_m(x_1)]\ /\sqrt{2}, \tag{5.5}$$

respectively, are correctly symmetric and anti-symmetric.

Fermions

The antisymmetric combination ψ_A, equation (5.5), is found to apply to electrons, and to other particles, including protons and neutrons, which are called *fermions.* By looking at ψ_A in the case $m = n$, one finds $\psi_A = 0$.

The wavefunction for two fermions in exactly the same state, is zero! This is a statement of the Pauli exclusion principle: only one Fermi particle can occupy a completely specified quantum state.

Bosons

For other particles, notably photons of electromagnetic radiation, the symmetric combination $\psi_S(1,2)$, equation (5.4), is found to occur in nature. Macroscopically large numbers of photons can have exactly the same quantum state, and this is important in the functioning of lasers. Photons, alpha particles, and helium atoms are examples of Bose particles, or *bosons.*

The Bose condensation of rare gas atoms has been an important research area in the past several years. The superfluid state of liquid helium, and the superconducting ground state of electron pairs, can also be regarded as representing Bose condensates, in which macroscopically large numbers of particles have exactly the same quantum numbers.

Electron pairs and the Josephson effect are available only at low temperatures, but they enable a whole new technology of computation, in which heat generation (energy consumption) is almost zero. So the question, is a refrigerator worth its expense, is becoming more of a realistic economic choice, as energy density in the silicon technology continues to rise with Moore's Law.

Orbital and Spin Components of Wavefunctions

To return to covalent bonds of di-atoms, it is necessary to complete the description of the electron state, by adding its spin projection $m_s = \pm 1/2$. It is useful to separate the space part $\phi(x)$ and the spin part χ of the wavefunction, as

$$\psi = \phi(x)\chi. \tag{5.6}$$

For a single electron $\chi = \uparrow$ (for $m_s = 1/2$) or $\chi = \downarrow$ (for $m_s = -1/2$). For two electrons there are two categories, $S = 1$ (parallel spins) or $S = 0$ (antiparallel spins). While the $S = 0$ case allows only $m_s = 0$, the $S = 1$ case has three possibilities, $m_s = 1,0,-1$, which are therefore referred to as constituting a "spin triplet".

A good notation for the spin state is $\chi_{S,m}$, so that the spin triplet states are

$$\chi_{1,1} = \uparrow_1\uparrow_2,\quad \chi_{1,-1} = \downarrow_1\downarrow_2,\quad \text{and}\quad \chi_{1,0} = 1/\sqrt{2}\ (\uparrow_1\downarrow_2 + \downarrow_1\uparrow_2)\quad \text{(spin triplet)} \tag{5.7}$$

For the singlet spin state, $S = 0$, one has

$$\chi_{0,0} = 1/\sqrt{2}\,(\uparrow_1\downarrow_2 - \downarrow_1\uparrow_2) \quad \text{(spin singlet)} \tag{5.8}$$

Inspection of these makes clear that the spin triplet ($S = 1$) is symmetric on exchange, and the spin singlet ($S = 0$) is antisymmetric on exchange of the two electrons.

Since the *complete* wavefunction (for a fermion like an electron) must be antisymmetric for exchange of the two electrons, this can be achieved in two separate ways:

$$\psi_A(1,2) = \phi_{sym}\,(1,2)\,\chi_{anti}\,(1,2) = \phi_{sym}\,(1,2)\;S = 0\;\text{(spin singlet)} \tag{5.9}$$

$$\psi_A(1,2) = \phi_{anti}\,(1,2)\,\chi_{sym}\,(1,2) = \phi_{anti}\,(1,2)\;S = 1\;\text{(spin triplet).} \tag{5.10}$$

The structure of the orbital or space wavefunctions $\phi_{sym,anti}$ (1,2) here is identical to those shown above for $\psi_{S,A}(1,2)$, equations (5.4) and (5.5).

5.2.2
The Hydrogen Molecule, Di-hydrogen: the Covalent Bond

Consider two protons, (labeled a, and b, assume they are massive and fixed) a distance R apart, with two electrons. If R is large and we can neglect interaction between the two atoms, then, following the discussion of Tanner [1],

$$[-(\hbar^2/2m)\,(\nabla_1{}^2 + \nabla_2{}^2) + U(r_1) + U(r_2)]\psi = (\mathscr{H}_1 + \mathscr{H}_2)\,\psi = E\psi \tag{5.11}$$

where the rs represent the space coordinates of electrons 1 and 2. Solutions to this problem, with no interactions, can be $\psi = \psi_a(x_1)\psi_b(x_2)$ or $\psi = \psi_a(x_2)\psi_b(x_1)$ (with $\psi_{a,b}$ the wavefunction centered at proton a,b) and the energy in either case is $E = E_a + E_b$.

The interaction between the two atoms is the main focus, of course. The interactions are basically of two types. First, the repulsive interaction $ke^2/r_{1,2}$, with $r_{1,2}$ the spacing between the two electrons. Secondly, the attractive interactions of each electron with the 'second' proton, proportional to $(1/r_{a,2} + 1/r_{b,1})$. The latter attractive interactions, primarily occurring when the electron is in the region between the two protons, and can derive binding from both nuclear sites at once, stabilize the hydrogen molecule (but destabilize a ferromagnet).

Altogether we can write the interatom interaction as

$$\mathscr{H}_{int} = ke^2[1/R + 1/r_{1,2} - 1/r_{a,2} - 1/r_{b,1}]. \tag{5.12}$$

To get the expectation value of the interaction energy the integration, below, following equation (4.48) must extend over all six relevant position variables q. (Spin variables are not acted on by the interaction.)

$$<E_{int}> = \int \psi^* \mathscr{H}_{int}\, \psi dq. \tag{5.13}$$

The appropriate wavefunctions have to be overall antisymmetric. Thus, following equations (5.9) and (5.10), the symmetric $\phi_{sym}(1,2)$ orbital for the antisymmetric $S = 0$ (singlet) spin state, and the anti-symmetric $\phi_{anti}(1,2)$ orbital for the symmetric $S = 1$ (triplet) spin state.

The interaction energies are

$$<E_{int}> = A^2\,(K_{1,2} + J_{1,2}) \quad \text{for} \quad S = 0 \text{ (spin singlet)} \tag{5.14}$$

$$<E_{int}> = B^2\,(K_{1,2} - J_{1,2}) \quad \text{for} \quad S = 1 \text{ (spin triplet), where} \tag{5.15}$$

$$K_{1,2}(R) = \iint \phi_a^*(x_1)\phi_b^*(x_2)\mathscr{H}_{int}\, \phi_b(x_2)\phi_a(x_1)d^3x_1d^3x_2 \tag{5.16}$$

$$J_{1,2}(R) = \iint \phi_a^*(x_1)\phi_b^*(x_2)\mathscr{H}_{int}\, \phi_a(x_2)\phi_b(x_1)d^3x_1d^3x_2 \tag{5.17}$$

The physical system will choose, for each spacing R, the state providing the most negative of the two interaction energies. For the hydrogen molecule, the exchange integral $J_{1,2}$ is negative, so that the covalent bonding occurs when the spins are anti-parallel, in the spin singlet case.

Covalent bonding and covalent anti-bonding, purely nanophysical effects
The qualitative understanding of the result is that in the spin singlet case the orbital wavefunction is symmetric, allowing more electron charge to locate halfway between the protons, where their electrostatic energy is most favorable. A big change in electrostatic energy (about $2J_{1,2}$) is linked (through the exchange symmetry requirement) to the relative orientation of the magnetic moments of the two electrons. This effect can be summarized in what is called the exchange interaction

$$\mathscr{H}_e = -2J_e\, \mathbf{S}_1 \bullet \mathbf{S}_2 \tag{5.18}$$

In the case of the hydrogen molecule, J is negative, giving a negative *bonding* interaction for *antiparallel* spins. The parallel spin configuration is repulsive or *anti-bonding*. The difference in energy between the bonding and antbonding states is about 9 eV for the hydrogen molecule at its equilibrium spacing, $R = 0.074$ nm. The bonding energy is about 4.5 eV.

This is a huge effect, ostensibly a magnetic effect, but actually a combination of fundamental symmetry and electrostatics. The covalent bond is what makes much of matter stick together.

The covalent bond as described here is a short-range effect, because it is controlled by the overlap of the exponentially decaying wavefunctions from each nucleus. This is true even though the underlying Coulomb force is a long-range effect, proportional to $1/r^2$.

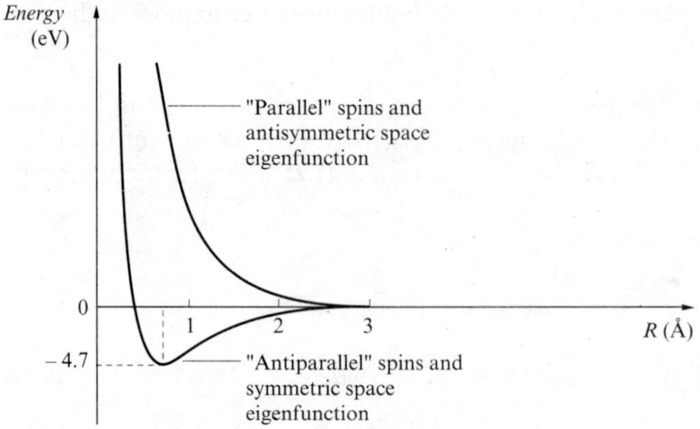

Figure 5.1 Energy curves for bonding and anti-bonding states of the hydrogen molecule. The bonding state requires anti-parallel spins. The equilibrium separation is 0.074 nm.

Covalent bonds exist not only between two s-states ($\ell = 0$), as in hydrogen, but also between two p-states ($\ell = 1$) as well as between an s and a p state. Organic molecules are, on the whole, bonded together in a covalent fashion.

Ferromagnetism, a purely nanophysical effect

In other cases J is positive, making the energy a favorable arrangement one of *parallel spins.* This leads to *ferromagnetism,* a cooperative state of matter in which huge numbers of magnetic moments are all locked parallel, leading to a macroscopic magnetization, M. It is important to recognize that the driving force for the spin alignment is electrostatic, the exchange interaction, represented by equation (5.18).

5.3
More Purely Nanophysical Forces: van der Waals, Casimir, and Hydrogen Bonding

We have just seen that the nanophysical exchange interaction leads both to the strong electrostatic covalent bond and to ferromagnetism. These are completely non-classical effects, depending on the wave-particle nature of matter and the symmetry of the wavefunction against exchange of particles. An ionic bond, such as that which stabilizes NaCl, table salt, is a more transparent, classical Coulomb force effect. The ionic bond is characteristic of many solids that dissolve in water. The covalent bond dominates the internal bonding of most of the molecules of organic chemistry, which includes the molecules of living things. To be sure, bonds are often partially ionic and partially covalent. For example, the basically covalent H_2O molecule has a partial negative charge on O, and thus has a permanent electric dipole moment. Covalent and ionic bonds are strong, giving several electron volts of binding energy. For example, the 6 eV bandgap energy of diamond is the energy to remove one electron from (create a hole in) an $n = 2$ tetrahedral sp^3 bond, a process in which the electron does not leave the solid but becomes free to move in the conduction band of the solid.

We now consider further bonding forces that occur in the macro-world that stem from its nanophysical basis. These are the polar and van der Waals (dispersion) forces, the Casimir force, and hydrogen bonding. These are all weaker forces than the covalent and ionic bonding forces. The Casimir force in particular is vanishingly small except for extremely closely spaced surfaces; yet it is that regime that nanotechnology may exploit in the future. However, for two metal surfaces spaced by 10 nm, there is an attractive pressure of about 1 atmosphere, 100 kPa, from the Casimir effect.

The polar, van der Waals, and hydrogen bonding forces are important in allowing molecules to bond together into larger, if less strongly bonded, assemblies such as proteins, and they play large roles in biology.

5.3.1
The Polar and van der Waals Fluctuation Forces

Many molecules, for example water, H_2O, are partially polar and have a net electric dipole moment, $\mathbf{p}=q\mathbf{a}$. Here q is the electric charge and $\mathbf{a}$ is its displacement vector.

The vector electric field $\mathbf{E}_d$ produced by an electric dipole $\mathbf{p}$ can be written

$$\mathbf{E}_d = k\left[\frac{\mathbf{p}}{r^3} - \frac{3(\mathbf{p}\bullet\mathbf{r})}{r^5}r\right]. \tag{5.19}$$

This dipolar field $\mathbf{E}_d$ resembles the magnetic $\mathbf{B}$ field of a bar magnet. Here k is the Coulomb constant $k=(4\pi\varepsilon_o)^{-1}$, and the overall strength of the dipole field is kp/r^3.

A second dipole $\mathbf{p}_2$ will interact with this field as $U=-\mathbf{p}_2\bullet\mathbf{E}_d$. So the interaction energy U is about

$$U \sim -kp_1p_2/r^3 \sim -ke^2a_o{}^2/r^3, \tag{5.20}$$

taking a value $p=ea_o$.

The force associated with this interaction is $-\mathrm{d}U/\mathrm{d}r$; this dipole-dipole force is attractive and the strength is

$$F \sim -3\,kp_1p_2/r^4. \tag{5.21}$$

For this reason polar molecules, for example, water molecules, will attract each other and may condense to a liquid or solid, giving an attractive interaction of about 0.42 eV per molecule (the heat of vaporization of liquid water). (Since this cohesive energy involves one molecule interacting with its group of near neighbors, the pairwise interaction at liquid water density is around 0.1 eV.) Another effect of importance is that polar molecules will cluster around an ion (an ion–dipole interaction), creating a sheath which decreases the strength of the Coulomb field of the ion at large distances. This is a screening effect. Hydration sheaths around ions in water are well known, and for example, will affect the rate at which an ion will diffuse in a liquid.

Electric polarizability of neutral atoms and molecules
An atom in an electric field develops an induced dipole moment

$$\mathbf{p} = \alpha \mathbf{E}, \tag{5.22}$$

where α is the electric polarizability. The simplest case is the hydrogen atom, for which

$$\alpha = (4.5)\, a_o{}^3/k, \tag{5.23}$$

where k, the Coulomb constant, appears because of the S.I. Units. The polarizability α increases with the number of electrons, in a complicated fashion. An estimate of the effect by London [2, 3] gives

$$\alpha \sim 2ne^2 z^2{}_{av}/I_1. \tag{5.24}$$

Here n is the number of electrons, $z^2{}_{av}$ is the average of the square of the position of the electron along the applied electric field, and I_1 is the first (lowest) ionization energy.

The important point is that the polarizability α becomes very large as the number of electrons on the atom increases. (This is probably why the IBM group chose xenon ($Z = 54$) as the atom to move along the surface of (111) Cu in their early STM nanofabrication work.)

This effect then gives an interaction energy between a dipolar atom and a non-polar atom, because the field of the dipolar atom will induce a moment in the non-polar atom. This interaction energy varies as $U \sim -\alpha k^2 p_1{}^2/r^6$.

The effects described so far come from static dipoles inherent in the bonding of molecules such as water.

Dipolar fluctuations of neutral and symmetric atoms
The purely quantum-mechanical van der Waals effect, in contrast, is a dynamic interaction, resulting from the fact that the electron in an atom is a charged particle in chaotic motion (not in a regular orbit). As such it represents a fluctuating electric dipole of strength $p = ea_o$. All of the electrons in an atom or molecule take part in this effect, not just those in the bonding, and the result is a fluctuating dipole electric field in the region around the atom or molecule, described as above.

The van der Waals effect comes from the *induced* dipoles in *neighboring* matter, arising from the fluctuating dipolar field and the polarizability of all its electrons. In this way fluctuations on one atom become correlated with fluctuations on neighboring atoms, and an attractive interaction U results. Calculations of the interaction energy U for two hydrogen atoms $U_{vdw} = -6.5\, ke^2 a_o{}^5/r^6$. (This result can be approached from the interaction between two hydrogen atoms considered above in connection with covalent bonding, by expanding $1/R$ in powers of x, y, and z.) (There is a correction to this energy estimate relating to the transit time for light between the two interacting electrons, treated by Casimir and Polder, which can be neglected.)

For two many-electron atoms the result is

$$U_{\mathrm{vdw}} = -1.5\, k^2 \frac{\alpha_A \alpha_B}{r^6} \frac{I_A I_B}{I_A + I_B}, \tag{5.25}$$

where I_A and I_B are the first ionization energies and the α_A and α_B are the polarizabilities. It is noteworthy that these polarizabilities are proportional to the numbers n of electrons in each atom, so this energy can be hundreds of times larger than the van der Waals interaction between two hydrogen atoms. This interaction is important in the liquefaction of symmetric atoms, such as rare gases, and in molecular solids, as shown in Figure 5.2 [after 4].

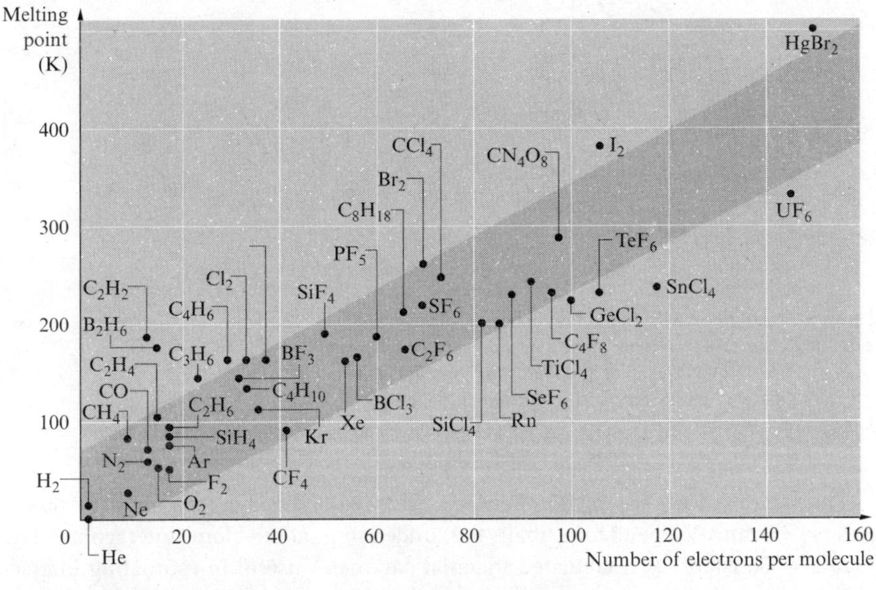

Figure 5.2 Melting points of molecular solids versus number of electrons per molecule. The melting point is a measure of the cohesive energy, presumably here largely originating in van der Waals interactions [after 4].

The van der Waals effect is a short-range interaction, varying as r^{-6}, while the Coulomb energy varies as r^{-1}. However, if a single molecule or atom is interacting with a large body such as a plane or large solid sphere, the dependence of the interaction on the spacing Z from the point to plane or sphere, evident after integrating the interaction energy over the locations in the extended body, will become $1/Z^3$. A nice summary of dipolar and van der Waals forces in chemical and biological contexts is given by Rietman [5].

A number of extended geometries have been analyzed based on the r^{-6} interaction. Figure 5.3 shows a few of these schematically. For more information see Rietman [5].

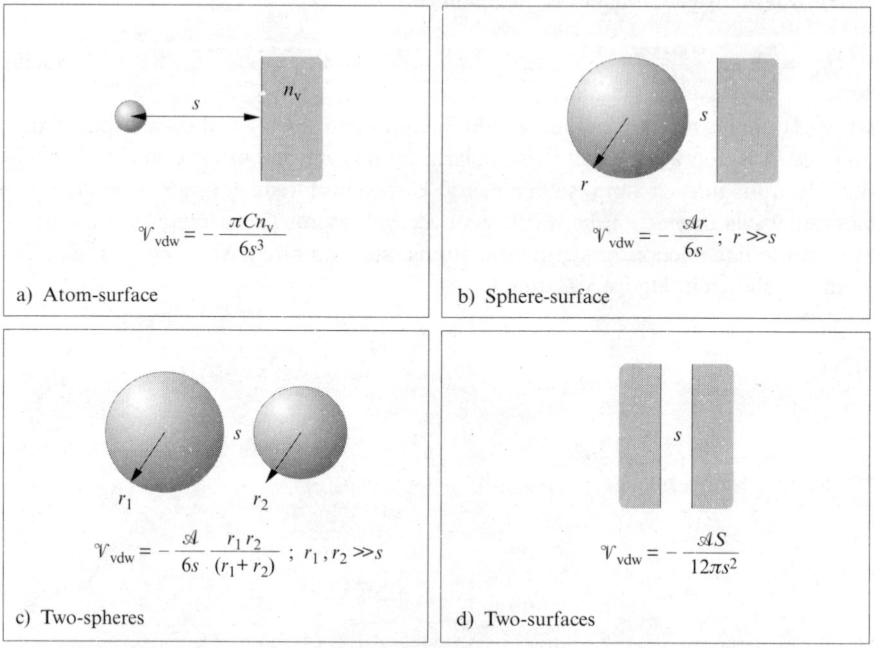

Figure 5.3 van der Waals interaction energies for several extended geometries. In these panels n_v represents the number of atoms per unit volume, $\mathcal{A}$ is the Hamaker constant, a tabulated material parameter, and S represents the area of two facing surfaces [after 6].

The van der Waals interaction between extended bodies typically shows a slower inverse distance dependence than the underlying atom–atom interaction. The Hamaker constant $\mathcal{A}$, a tabulated material parameter useful in estimating interactions, is typically in the range of 0.2 eV – 3.12 eV, and increases with the number of atoms per unit volume in the material in question. For more details the reader is referred to Drexler [6].

5.3.2
The Casimir Force

The Casimir force [7] operates between metallic surfaces, forcing them together. It is an electromagnetic effect which has to do with the modes of oscillation of the electromagnetic field in an enclosed region.

Two parallel mirrors spaced a distance Z apart will allow standing electromagnetic waves to build up, propagating in the z-direction perpendicular to the surfaces, when $Z = n\lambda/2$. If we think of the transverse electric field $E(z,t)$, then E must be zero at each mirror surface, $z = 0$ and $z = Z$. Modes like this are what produce the external light beam in a laser pointer. (One of the mirrors is partially metallized and lets some light propagate out.) This resonance is similar to a standing wave on a violin string, which we can think of in terms of simple harmonic motion. For the violin string in its fundamental mode, the transverse motion of the string mass is a simple harmonic motion, in which the energy is fully kinetic when the string is at zero displacement, and fully potential when the string is fully displaced. For both the electromagnetic modes and the resonant frequencies on the violin string, the shorter the length (the mirror spacing Z or the length L of the violin string) the fewer modes that can be supported, because the longest wavelength available in either case is twice the cavity or string length.

Equations (4.75) and (4.76) indicate that the harmonic oscillator in nanophysics acquires two strange characteristics: a zero point motion and a zero point energy, features which are consistent with the uncertainty principle. *The lowest energy state of the oscillator is not zero, but $\hbar\omega/2$.*

It turns out that the electromagnetic modes in the cavity between the two mirrors act like nanophysical oscillators, and have for this reason a zero point energy and also fluctuations in electric and magnetic fields even at temperature zero. From this point of view, even an empty cavity between mirrors has an energy U, namely: $\hbar\omega/2$, summed over all the modes which are allowed in the cavity. If the cavity width is Z, the only available frequencies are $n(c/2L)$, where $n = 1,2,3,\ldots$ so that all frequencies less than $c/2L$ are unavailable. Equivalently, wavelengths greater than $2L$ are excluded, and as L is reduced, more wavelengths, and more zero point energies, are excluded.

The energy cost U for the electromagnetic fluctuations in the cavity is reduced as L is reduced, and this leads to a force $F = -\mathrm{d}U/\mathrm{d}Z$ to collapse the cavity. This scenario has been experimentally verified by recent accurate observations of the Casimir force.

A careful calculation of the attractive Casimir force between parallel mirrors [7,8] gives

$$F_C = -\frac{\pi^2 \hbar c}{240}\frac{1}{Z^4} \tag{5.26}$$

Here c is the speed of light and Z is the spacing of the mirrors.

This represents an attractive force per unit area (negative pressure) which is very small at large spacings Z, but rises to a value of about 1 atmosphere, 100 kPa, for $Z = 10$ nm. 10 nm is a spacing that might well arise in the nanomachines of the future, so the Casimir force may well be of engineering relevance.

An easier geometry to arrange reliably is a sphere of radius R spaced a distance z from a plane surface. In this case, the Casimir force on a sphere is [8]

$$F_{Cs} = -\frac{\pi^3 \hbar c}{360} R \frac{1}{z^3}. \tag{5.27}$$

This equation was confirmed in a silicon nanotechnology experiment sketched (not to scale) in Figure 5.4. A stage on which two polysilicon capacitor plates are delicately suspended to rotate an angle θ by two silicon torsion fibers, is moved upward (using piezoelectric stage, below) toward 100 μm radius sphere R, reducing the spacing z to as small as 75 nm. Plate spacing of the capacitors is 2 μm, established by etching away a grown 2 μm SiO_2 layer on which the two polysilicon upper plates had been deposited.

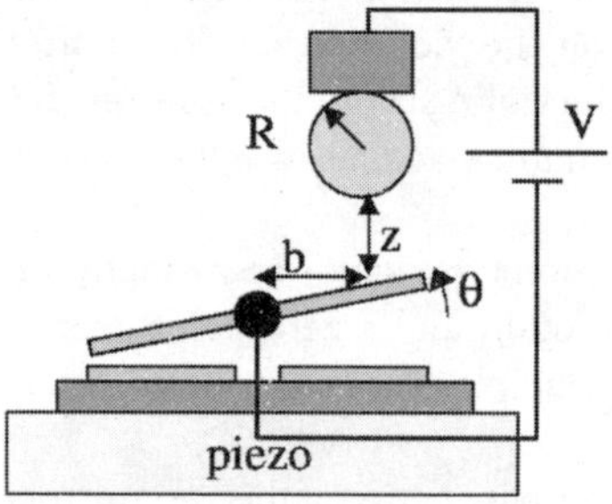

Figure 5.4 Schematic of measurement of the Casimir force between sphere and planar surface, measured with $V = 0$ [8]. Planar capacitor plates on paddle rotate against torsion fiber (center black dot) whose restoring force is calibrated using V and known Coulomb force between sphere and plane. Rotation angle θ is measured by capacitance difference between paddle plates and lower fixed plates, whose spacing t is 2 μm. Piezo stage lifts paddle assembly toward sphere, adjusting z down to 75 nm.

The micromachined balanced capacitor assembly, grown on a single crystal silicon chip, is shown in Figure 5.5 [8]. The upper capacitor plate is a 3.5 μm thick polysilicon rectangle deposited onto 2 μm thick silicon dioxide grown on a crystal immediately below. After deposition of the upper plate, the silicon dioxide is etched out, leaving the upper plate freely suspended.

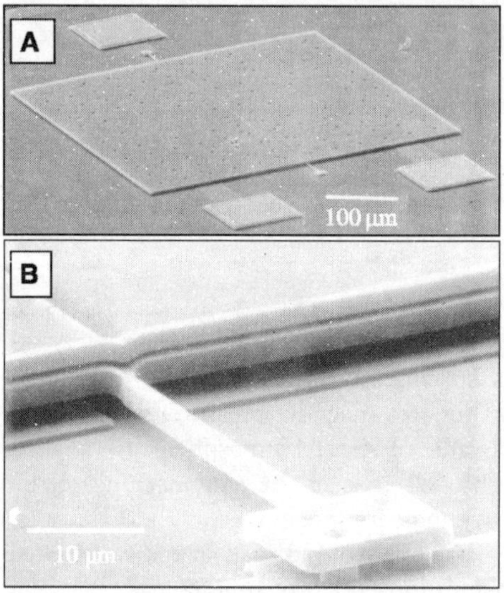

Figure 5.5 Micromachined planar capacitor plates freely suspended to rotate above silicon crystal on twin silicon torsion fibers [8]. (A) Rectangular paddle 500 μ on a side is freely suspended 2 μm above Si chip by twin Si fibers, front and back. (B) Detail of paddle (upper rotating capacitor plate) showing its separation from Si chip and showing the front Si suspending torsion fiber. 200 μ radius metallized sphere (not shown) is placed a variable distance z above center of right portion of the paddle (see Figure 5.4).

The force between the sphere and the right portion of the paddle is determined by the twist angle θ of the fiber, which in turn is determined from the difference in capacitances to the separate lower capacitor plates. (The capacitance of a parallel plate capacitor is $\varepsilon_o A/t$, where A is the area, t the plate spacing and ε_o the permittivity of vacuum, 8.854×10^{-12} Farads/meter.) The torsion constant of the fiber was calibrated by the twist angles θ observed with voltage biases V applied to the sphere, and with z large enough that the Casimir force is negligible. The Coulomb force between a sphere of radius R and a plane at spacing z is a known function of R, V and z, to which the measurements were carefully fitted, thus calibrating the force vs. twist angle θ.

Measured forces vs. sphere-plane spacing z are shown in Figure 5.6 [after 8].

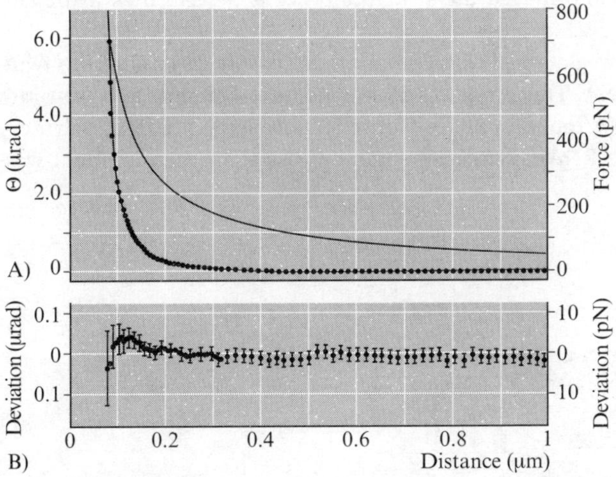

Figure 5.6 Measured Casimir force between sphere and plane vs. spacing z, shown by points fitted to theoretical curve [after 8]. (A) Data points and Casimir theory curve (lower trace) vs. spacing z. Upper curve is force vs. z for $V = 136$ mV, where Coulomb force matches Casimir force at closest approach (76 nm). (B) Deviation between measured points and fitted Casimir theory curve, on expanded scale.

Two important corrections to the theory equation (5.27) have been made in achieving the fit shown in Figure 5.6. The first is that the gold metal evaporated onto the surfaces is not in fact a perfect mirror, and allows electromagnetic waves of very high frequency to pass through. This effect was corrected for, by making use of the tabulated optical constants for gold. The second effect is that the gold mirror surfaces are not perfectly flat but have a roughness amplitude of 30 nm as measured with an atomic force microscope. These two corrections are carefully made and explained in [8], leaving no doubt that the Casimir force has been accurately measured and is an important effect for metal surfaces whose spacing is on the 100 nm scale. This scale is relevant to nanotechnology.

5.3.3
The Hydrogen Bond

The hydrogen bond occurs in situations where the slightly positive H in the polar water molecule locates midway between negative charges, which may occur on the negative end of other polar molecules. The sharing of this partially charged H atom with two negative entities is a bit like the sharing of a negative electron in a covalent bond between two protons. Hydrogen bonding occurs frequently when polar water molecules are important, and indeed the structure of ice is described as hydrogen bonded.

Hydrogen bonding is weak, which is appropriate to its role in joining the twin strands of double helix DNA. This weakness allows the helical strands to be separated easily in models of DNA replication (see Figure 5.7) involving the DNA polymerase engine, similar to the RNA polymerase engine whose force measurement was depicted in Figure 3.12.

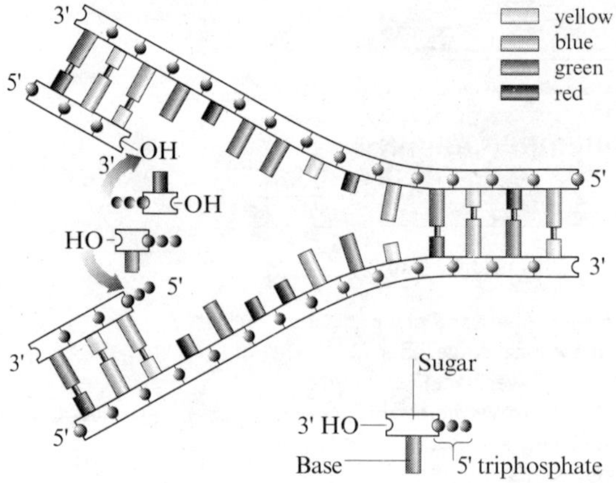

Figure 5.7 Models of DNA replication fork, showing breaking of hydrogen bonds between upper and lower bases (center of figure) as DNAP engine (left) pulls double helical DNA from right to left [after 9]. Hydrogen bonds are depicted here as smaller diameter cylinders connecting the larger cylinders representing the four different bases A, C, G, and T, which, however, bond only as complementary pairs AT and CG. The AT pairs form double hydrogen bonds, while the CG pairs form triple hydrogen bonds.

In Figure 5.7 [after 9] the "rails" depict the strong repeating nucleotide units that form the single DNA strands. The repeating unit consists of a deoxyribose cyclic sugar molecule, an ionic phosphate $(PO_4)^{2-}$, and the "base". The bases are Adenine, Cytosine, Guanine and Thymine, A, C, G, and T, of which only complementary combinations AT and CG hydrogen bond from one strand to the other. In Figure 5.7 the groups of three small spheres represent triphosphates. Adenosine triphosphate

(ATP) is well known as a source of energy, which is stored in the negatively charged phosphate bonds, and is released when these bonds are broken.

Following [9], nucleoside triphosphates serve as a substrate for DNA polymerase, according to the mechanism shown on the top strand. Each nucleoside triphosphate is made up of three phosphates (represented here by small spheres), a deoxyribose sugar (rectangle) and one of four bases (larger cylinders of differing lengths). The three phosphates are joined to each other by high-energy bonds, and the cleavage of these bonds during the polymerization reaction releases the free energy needed to drive the incorporation of each nucleotide into the growing DNA chain.

Hydrogen bonds are important in many other areas of large-molecule chemistry and biology. A good source of information on these topics is Ball [10].

5.4 Metals as Boxes of Free Electrons: Fermi Level, DOS, Dimensionality

To a first approximation, a metal or a semiconductor can be regarded as a 3D "box" containing the free electrons released from the outermost orbits of the atoms. Atoms such as Na or Ca give up one or two of their valence electrons when they form a solid. These freed electrons form a delocalized electron gas in the "box". To a remarkable degree the electrons do not interfere with each other, and the dominant effects come from the Pauli principle which means that each of the electrons go into a separate energy state, up to the Fermi energy.

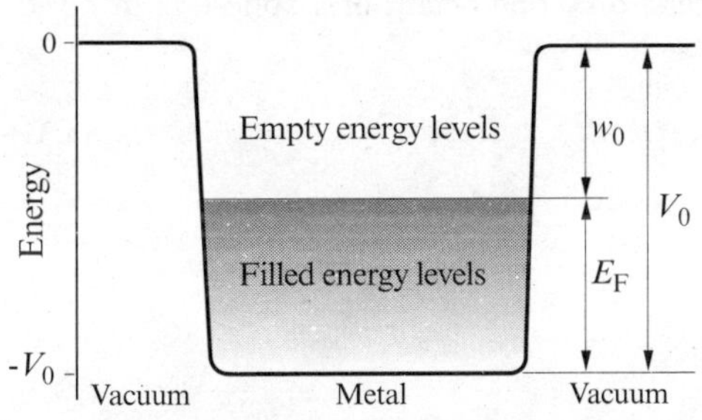

Figure 5.8 Metal as a 3D box filled with non-interacting electrons up to the Fermi energy E_F, following the Pauli exclusion principle. The total depth of the potential well is V_o, the sum of E_F and the work function, here labeled w_o.

Such a box is sketched in Figure 5.8, surrounded by the potential barrier described by the work function, described as *w* or ϕ, which contains the electrons inside the solid. The empty box is an unexpectedly good representation for the electron motion, it turns out, because of the *periodic spatial variation* of the realistic potential, $U(x,y,z)$, coming from the array of atoms in the real solid. The *periodic* arrangement of the charged ions very much reduces their disruption of electron motions, as we will discuss later.

The work function is usually several electron volts, and certainly contains the electrons in the metal. The states of the 3D trap considered earlier, are a good starting point.

$$\psi = (2/L)^{3/2} \sin(n_1\pi x/L) \sin(n_2\pi y/L) \sin(n_3\pi z/L). \tag{5.28}$$

Since $\sin kx = (e^{ikx} - e^{-ikx})/i2$, we can consider these states to be superpositions of oppositely directed traveling waves $\psi_+ = \exp(ikx)$ and $\psi_- = \exp(-ikx)$. This is of course the same as in the case of standing waves on a violin string. Here the moving waves $\psi_{+,-}$ are more fundamental for a description of conduction processes.

The singly most important aspect of nanophysics for this situation is the Pauli exclusion principle, which states that only one electron can occupy a fully described state. This means that if we add a large number of electrons to the box, the quantum numbers will be come very large, and also the energy of the successively filled states will be large. The energy is given by equation 4.67:

$$E_n = [h^2/8mL^2](n_x^{\ 2} + n_y^{\ 2} + n_z^{\ 2}). \tag{5.29}$$

It is convenient to rewrite this equation as

$$E_n = E_o\, r^2, \tag{5.30}$$

as an aid to counting the number of states filled out to an energy E, in connection with Figure 5.9. In coordinates labeled by integers n_x, n_y, and n_z, constant energy surfaces are spherical and exactly two electron states occupy a unit volume. Since the states are labeled by positive integers, only one octant of a sphere is involved. The number N of states out to radius r is

$$N = (2)\,(1/8)\,(4\pi r^3/3) = \pi r^3/3 = (\pi/3)\,(E/E_o)^{3/2}, \tag{5.31}$$

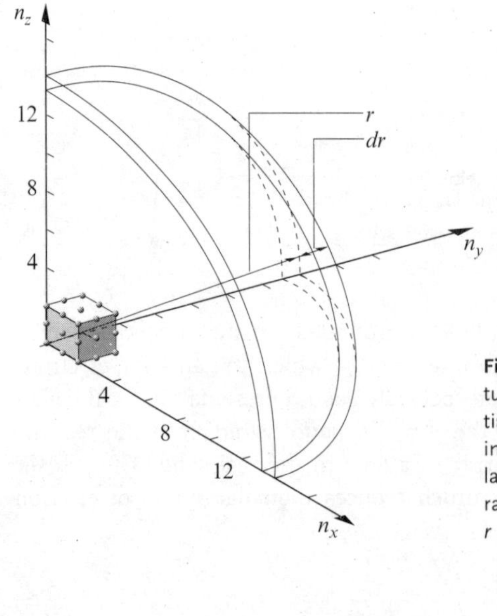

Figure 5.9 Constant energy surfaces in quantum number space for particle in a 3D potential well, a starting representation for electrons in a metal. There are two electron states per lattice point. The number of states N out to radius r is therefore $(2/8)(4\pi/3)r^3$, where $r = E/E_o$, and $E_o = h^2/8mL^2$ (see eq. 5.29).

where $E_o = h^2/8mL^2$, for a box of side L. This is equivalent to

$$E_F = (h^2/8m)\ (3N/\pi\ L^3)^{2/3} = (h^2/8m)\ (3N/\pi V)^{2/3}. \qquad (5.32)$$

Setting $N/V = n$, the number of free electrons/m^3, it is straightforward to see that

$$dN/dE = g(E) = (3n/2) E^{1/2} E_F^{-3/2} \qquad (5.33)$$

is the density of electron states per unit volume at energy E.

The Fermi velocity u_F is defined by $mu_F^2/2 = E_F$, and the Fermi temperature is defined by $kT_F = E_F$. The Fermi velocity and the Fermi temperature are much larger than their thermal counterparts, both important consequences of the Pauli principle in filling the electron states in a box representing the metal.

At zero temperature, states below E_F are filled and states above E_F are empty. At nonzero temperatures the occupation is given by the Fermi function $f = [\exp(\{E - E_F\}/kT) + 1]^{-1}$. The width in energy of transition of f from 1 to 0 is about kT.

Some of these features are sketched in Figure 5.10.

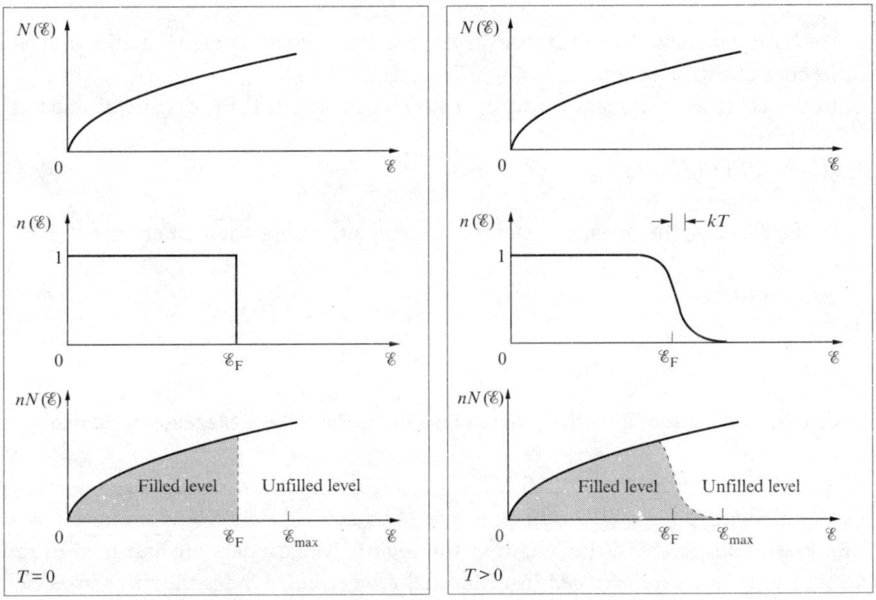

Figure 5.10 Density of states $g(E)$ and occupation $f(E)$ at $T = 0$ (left) and T nonzero (right) in the 3D case.

A collection of data on Fermi energies and Fermi temperatures is given in Table 5.1.

Table 5.1 Fermi energy E_F, Fermi temperature T_F, and free electron density $n = N/V$ for metals

Element	N/V ($\times 10^{28}$ m^{-3})	Fermi energy (eV)	Fermi temperature ($\times 10^4$ K)
Al	18.1	11.7	13.6
Ag	5.86	5.53	6.41
Au	5.90	5.55	6.43
Cu	8.47	7.06	8.19
Fe	17.0	11.2	13.0
K	1.40	2.13	2.47
Li	4.70	4.77	5.53
Mg	8.61	7.14	8.28
Mn	16.5	11.0	12.8
Na	2.65	3.26	3.78
Sn	14.8	10.3	11.9
Zn	13.2	9.50	11.0

We have considered the 3D case in the material above. Dimensionality makes a difference in the behavior.

In the 1D case, as a quantum wire, N atoms in a length L, the density of states is

$$g(E) = (L/\pi)\,(m/2\hbar^2 E)^{1/2}. \tag{5.34}$$

In the 2D case, the density of states is a constant, independent of energy.

$$g(E) = \text{constant} \tag{5.35}$$

5.4.1
Electronic Conduction, Resistivity, Mean Free Path, Hall Effect, Magnetoresistance

Electrical conduction occurs with N charge carriers per unit volume, each of charge Q, experiencing a force QE in the presence of an applied field $E = V/L$, where V is an impressed voltage and L the length of the region. The carriers are assumed in random chaotic motion described by a thermal energy k_BT. Under the impressed field $\mathbf{E}$, the carriers acquire a small drift velocity $\mathbf{v}_d = \mu\mathbf{E}$, where μ is the mobility, usually quoted in units m^2/V s. The mobility μ is easily seen to be $e\tau/m$, where τ is the time between collisions which randomize the velocity, and m is the mass of the particle, basically the electron mass. On this basis the conductivity σ is

$$\sigma = Ne^2\tau/m = 1/\rho, \tag{5.36}$$

where the resistivity ρ has units of ohm m. The formula for the resistance R of a linear specimen of cross sectional area A is $R = \rho L/A$, in ohms. The mean free time may be expressed as λ/v, where λ is the mean free path between elastic collisions and v is the relevant average velocity. In the simplest case v is the thermal velocity $(3kT/m)^{1/2}$. In the more usual case of a metal, v is the Fermi velocity $(2E_F/m)^{1/2}$, typically much larger than the thermal velocity.

The mean free path in a crystalline solid can be very long, much larger than the inter-atom spacing, limited only by the deviation of atoms from their lattice positions due to thermal motion. The electron mobility in high purity crystalline semiconductors can be many millions of V cm^2/s, reinstating the (Newtonian) idea of ballistic electrons, which are free and rarely collide.

The "Hall effect" is the basis in semiconductors of a method to measure the density of charge carriers, N. In a simple rectangular geometry, the Hall field is

$$E_H = R_H BJ, \tag{5.37a}$$

where B, J (with B and J the magnetic field and current density, respectively), and E are mutually orthogonal, and

$$R_H = -1/Ne \text{ (for electrons)} \tag{5.37b}$$

thus provides a measure of the carrier density N.

Magnetoresistance describes the dependence of electrical resistance $R = V/I$ on magnetic field, described by $(\Delta R/R)/B$. In modern technology, the sensor ("read head") of the magnetic field from individual ferromagnetic domains written on a magnetic disk (see Figure 1.5) is based on effects called GMR ("giant magnetoresistance") or TMR ("tunnel magnetoresistance"). Nanotechnology is needed to observe these effects, which require conducting layers smaller than the mean free path λ_{sf} of an electron to change its spin orientation (spin-flip). The GMR effect comes from the change of electron mobility observed between spin-up and spin-down electrons in traversing a ferromagnet: if the spin of the electron is parallel to the spins in the ferromagnet, the scattering (and thus the resistivity) is less than if the spin traverses a ferromagnetic region of antiparallel spins. This effect occurs for regions whose dimensions are so small that the electron spin direction is preserved during traversal. Thus, the inelastic mean free path for spin reversal must exceed the device dimension. In practice this means that the device has essential dimensions on the nanometer scale, true for both the GMR and the TMR magnetic sensors. The TMR effect is based on a tunnel junction between two ferromagnetic electrodes, as described below.

5.5
Periodic Structures (e.g. Si, GaAs, InSb, Cu): Kronig–Penney Model for Electron Bands and Gaps

Electrical conduction in metals and semiconductors is a topic of considerable importance in modern electronic technology, which is based on semiconductors such as silicon and gallium arsenide. An advanced understanding of the behavior of conduction electrons is available in these important cases. The "band theory" of solids is based on an assumption that electron behavior can be treated as a one-electron problem of motion in an average potential U which is primarily determined by the array of partially ionized atoms aligned on the crystalline lattice of the solid. Detailed predictions of physical properties are available for crystalline solid materials of interest, including many semiconductors and metals. It has even proven possible to predict the transition temperatures, T_c of superconducting metals.

This understanding of the behavior of electrons in such solids is based on concepts of nanophysics extended to include the periodic nature of the potential U that acts upon an electron in a crystalline solid. In large part, the basic formulas for current density $J = nev$, drift velocity $v = (e\tau/m)E$, mobility $\mu = e\tau/m$, resistivity $(1/\rho = ne^2\tau/m)$ etc. are retained, but the band theory allows improvement in their accuracy by providing an "effective mass" m^* which replaces the electron mass m_e, as well as a relative dielectric constant which multiplies ε_o. The band theory also provides a new basis for calculating the mean free path and scattering time τ in pure crystalline solids.

A basic description of a free electron of specified kinetic energy and momentum in one dimension is $\Psi_x = (1/L)^{1/2}\exp[i(kx-\omega t)]$, where $k = p/\hbar = 2\pi/\lambda$ and $E = \hbar\omega$. Here Ψ_x is normalized to provide unit probability to find a free electron in a length L. Ψ_x is a solution of the 1D Schrodinger equation with $U = 0$. How can such a solution be modified to describe an electron in a solid? Surely the correct wavefunction will have some periodic property to match the periodic potential.

It is easy to show that wavefunctions for such a periodic potential are of the form

$$\psi = u_k(x)\exp(ikx), \tag{5.38}$$

where $u_k(x)$ exhibits the periodicity of the atomic lattice, a, and where the allowed values of k are $n\pi/L$, where $n = 1,\ldots,N$. This wavefunction can still represent a particle of speed $\hbar k/m$.

The answer is that if $U(x)$ is a periodic function based on the atomic spacing of the crystalline solid, the free electron description still applies to certain ranges of energy E, while in other ranges of energy, called "energy gaps", no electron states are allowed.

Assume a row of $N = L/a$ atoms on the x-axis. Then $U(x + na) = U(x)$, with $n = 1,\ldots,L/a$.

With the periodic potential $U(x)$, new phenomena of "allowed energy bands" and "forbidden energy gaps" appear in the dependence $E(k)$ of electron energy on wavevector k. These effects are mathematically similar to the existence of *pass bands* in

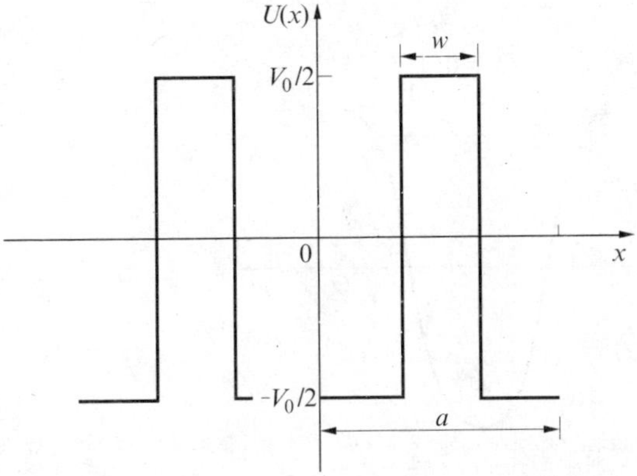

Figure 5.11 A simplified model potential, assumed extended periodically.

the impedance characteristics of periodic transmission lines. These new effects are best viewed in comparison to $E = \hbar^2k^2/2m = n^2h^2/8mL^2$ for the case $U = 0$, as in the 1D infinite square well.

An introduction to this new basic behavior is afforded by a simple soluble model, due to Kronig and Penney[11]. This model assumes a linear array of N atoms spaced by a along the x-axis, $0 < x < Na = L$. The 1D potential $U(x)$ is of a square-wave form, see Figure 5.11, with period a. We assume the origin of the x-axis is at the center of one atom;

$$U(x) = -V_o/2, \quad 0 < x < (a-w)/2$$
$$V_o/2, \quad 0 < x < (a+w)/2. \tag{5.39}$$

represented in this $U(x)$ by symmetric potential wells of depth $-V_o/2$ and width $a-w$, separated by barriers of height $V_o/2$ and width w. Note that this extended periodic $U(x)$ is symmetric about $x = 0$, i.e. $U(x) = U(-x)$, and that the average value of U is 0 as in the free particle case.

The solutions (5.38) are compatible with the 1D Schrodinger equation

$$(-\hbar^2/2m)\mathrm{d}^2\Psi/\mathrm{d}x^2 + [U(x) - E]\,\Psi = 0$$

containing the periodic $U(x)$ of (5.39), *only* if the following condition is satisfied:

$$\cos ka = \beta(\sin qa/qa) + \cos qa = R(E), \tag{5.40}$$

where $\beta = V_o wma/\hbar^2$ and $q = (2mE)^{1/2}/\hbar$. (The simplified form of (5.40) is actually obtained in a limiting process where the potential barriers are simultaneously made higher and narrower, preserving the value of β. This can be described as N δ-functions of strength β.) The parameter β is a dimensionless measure of the strength of the periodic variation.

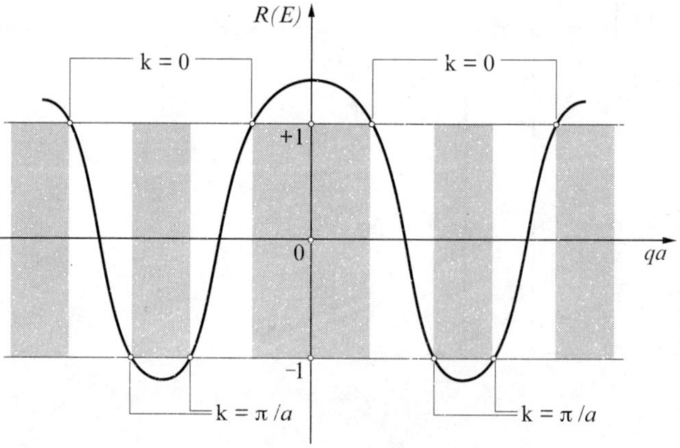

Figure 5.12 Kronig–Penney model, a schematic plot of ordinate $R(E)$ vs. abscissa qa. Allowed solutions (unshaded) occur only when ordinate $R(E)$ has magnitude unity or less.

One easily sees that in (5.40) in the limit $\beta = 0$, the condition $\cos ka = \cos \alpha a$ leads to $k = (2mE)^{1/2}/\hbar$, which recovers the free electron result, $E = \hbar^2 k^2/2m$. Next, if we let β become extremely large, the only way the term $\beta(\sin qa/qa)$ can remain finite, as the equation requires, is for $\sin qa$ to become zero. This requires $qa = n\pi$, or $a(2mE)^{1/2}/\hbar = n\pi$, which leads to $E = n^2h^2/8ma^2$. These are the levels for a 1D square well of width a (recall that in the limiting process the barrier width w goes to zero, so that each atom will occupy a potential well of width a).

The new interesting effects of band formation occur for finite values of β. Figure 5.12 gives a sketch of the right-hand side, $R(E)$, of (5.40), vs. qa. Solutions of this equation are only possible when $R(E)$ is between −1 and +1, the range of the $\cos ka$ term on the left. Solutions for E, limited to these regions, correspond to allowed energy bands (shown unshaded in sketch of Figure 5.12). Note that allowed solutions are possible for $-1 < \cos ka < 1$, which corresponds to $-\pi < ka < \pi$. More generally, boundaries of the allowed bands are at $k = (+/-)n\pi/a$, $n = 1,2,3\ldots.$ A sketch of the allowed $E(k)$ is shown dotted in Figure 5.13.

Bragg scattering of electron waves The special role of $k = \pi/a$ in these equations represents a fundamental physical process, Bragg scattering of the electron wave. When a traveling wave $\exp(ikx)$ is scattered by the atoms located at $x = na$, a coherent reflected wave, $\exp(-ikx)$, can be generated, leading to a standing wave.

The condition for coherent reflection from succeeding atom positions $x = na$ is the 1D analog of the Davisson–Germer experiment mentioned in Chapter 2. The path difference between a wave back-scattered at $x = 0$ and one back-scattered at $x = a$, is $2a$. Thus, for coherent reflection $2a$ must be an integer number of electron wavelengths. This Bragg condition, $n\lambda = 2a$, since $\lambda = 2\pi/k$, is equivalent to

$$k = n\pi/a \tag{5.41}$$

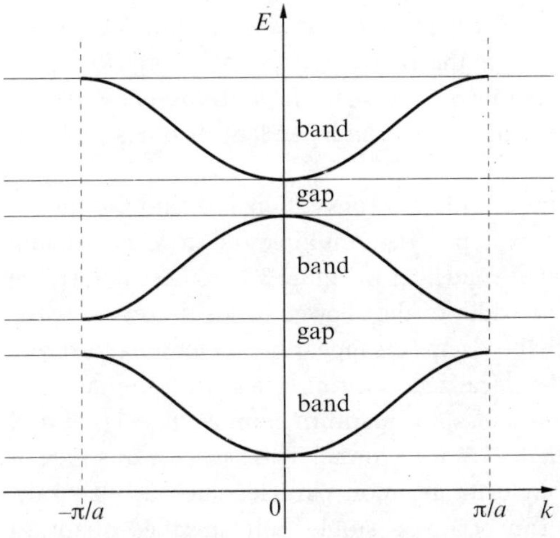

Figure 5.13 Schematic of bands E vs. k in a periodic potential, based on Kronig–Penney model. The bands are restricted in k to values less than π/a. Energy gaps occurring at $k = \pm(\pi/a)$ are also physically understood on the basis of Bragg reflections at $k = \pm(\pi/a)$. Physical arguments easily show that each band accommodates exactly $N/2$ electrons, so that one electron per atom gives a half-filled band and a metal, while two electrons per atom gives a filled band, and an insulator.

Any wave $\exp(in\pi x/a)$ thus generates a reflected wave $\exp(-in\pi x/a)$, and this means that the only waves present at $k = n\pi/a$ are the linear combinations $\exp(in\pi x/a) \pm \exp(-in\pi x/a)$:

$$\psi(k = n\pi/a) = \cos(n\pi x/a) \text{ or } \sin(n\pi x/a). \tag{5.42}$$

These are *standing waves*, corresponding, respectively to probability densities $P(x)$ peaking at atom positions ($x = na$), or midway between. The $\cos^2(n\pi x/a)$ solution will thus have a lower energy than the $\sin^2(n\pi x/a)$ solution (at the same k). The difference in these energies exactly provides the energy gap between the lower band and the next band.

A second result of Bragg reflection is that, at $k = n\pi/a$, the standing waves give no net particle transport (equal flows to right and left), so the group velocity, $\hbar^{-1}\partial E/\partial k = 0$. The energy bands thus have zero slope at the "zone boundaries", $k = \pm(\pi/a)$.

Returning to discussion of the Kronig–Penney model, it is easy to show that the slope of $E(k)$ is zero at the boundaries, i.e., the group velocity $\partial\omega/\partial k$ goes to zero at the edges of the bands. To show this, expand $\cos ka$ of (5.40) about the points $k = n\pi/a$; and then form the differential $\mathrm{d}E/\mathrm{d}k$. Let $ka = n\pi + x$, where x is small; so

$$\cos(ka) = \cos(n\pi + x) = \cos(n\pi)\cos(x) - \sin(n\pi)\sin(x) \approx \cos(n\pi)(1 - x^2/2) = R(E) \tag{5.43}$$

Taking the differentials of last two terms, $-\cos(n\pi)x\mathrm{d}x = (\mathrm{d}R/\mathrm{d}E)\mathrm{d}E$. Since $\mathrm{d}x = a\mathrm{d}k$, we have, near the points $k = n\pi/a$, the value $\mathrm{d}E/\mathrm{d}k \approx -a\cos(-\pi)[\mathrm{d}R/\mathrm{d}E]^{-1}x$. Thus, in the limit $x = 0$, (i.e., at the points $k = n\pi/a$) we have $\mathrm{d}E/\mathrm{d}k = \hbar\partial\omega/\partial k = 0$. This is actually always the case near a band edge, independent of our simplified model of the periodic $U(x)$.

As shown in the solid curves in Figure 5.13, it is conventional to shift the higher band segments back toward the origin by appropriate multiples of $2\pi/a$, assembling them in the "first zone" $-\pi/a < k < \pi/a$. The solid lines in Figure 5.13 represent the three lowest energy bands. It is clear that the width of the allowed bands decreases as the strength parameter β increases, and that the gaps disappear with β tending to zero.

For N sites the theory predicts $2N$ delocalized electron states, with the factor of 2 coming from the fact that N electrons of spin quantum number $m_s = 1/2$ and N electrons of spin quantum number $m_s = -1/2$ are allowed. The basis for this idea of "filling" the bands is the Pauli principle for fermion particles such as electrons, which states that only one fermion can occupy a single fully specified quantum state. For example, a hydrogen atom state specified by n,l,m,m_s can have only one electron. The electrons of course occupy lowest energy states preferentially. For this reason, N univalent atoms (such as Na) which donate only one electron, will lead to a half-filled band. Such a band gives metallic conduction, because empty $E(k)$ states are present immediately above filled states, allowing for acceleration of the electrons. Conventional metals Na, Cu, Ag, Au etc. have half-filled conduction bands. In these cases, the number of free electrons can be represented as νN, where ν is the valence. On the other hand, the theory predicts that a filled band will lead to no conduction, i.e., an insulator, because no empty states are available to allow motion of the electrons in response to an electric field. Such a situation is present in the pure semiconductors Si, GaAs, and Ge.

In these materials, four outer electrons (two each from 3s and 3p orbitals in Si and GaAs, and two each from 4s and 4p in Ge), are stabilized into four tetrahedral covalent bond orbitals which point from each atom to its four nearest neighbor atoms, located at apices of a tetrahedron.

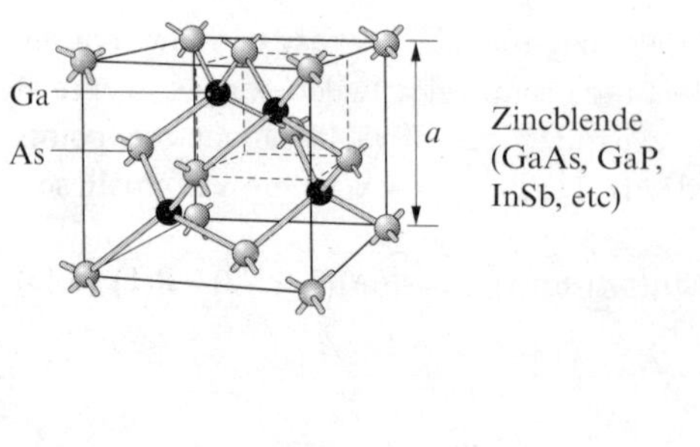

Figure 5.14 Diamond and zincblende crystal structures. Each atom is covalently bonded to four nearest neighbors in tetrahedral directions. The directed bonds are linear combinations of s and p orbitals (see Table 4.1), and analogous to directed orbitals sketched in Figure 4.9. Specifically, 2s and $2p^3$ for diamond (as in CH_4) and 3s and $3p^3$ for Si. There are four valence electrons per atom, leaving a band structure with filled bands.

5.6 Electron Bands and Conduction in Semiconductors and Insulators; Localization vs. Delocalization

The four electrons per atom (eight electrons per cell) which fill the covalent bonds of the diamond-like structure, completely fill the lowest two "valence" bands (because of the spin degeneracy, mentioned above), leaving the third band empty. In concept this would correspond to the first two bands in Figure 5.13 as being completely filled, and thus supporting no electrical conduction. Referring to Figure 5.13, one sees that at $k = 0$, just above the second band, there is a forbidden gap, E_g. There are no states allowing conduction until the bottom of band 3, which is about 1 eV higher in these materials. For this reason, at least at low temperatures, pure samples of these materials do not conduct electricity.

Electrical conductivity at low temperature and room temperature in these materials is accomplished by "doping"; substitution for the four-valent Si or Ge atoms either acceptor atoms of valence 3 or donor atoms of valence 5. In the case of five-valent donor atoms like P, As, or Sb, four electrons are incorporated into tetrahedral bonding and the extra electron becomes a free electron at the bottom of the next empty band. In useful cases, the number of free electrons, n, is just the number of donor atoms, N_D. This is termed an N-type semiconductor. In the case of boron, aluminum, and other three-valent dopant atoms, one of the tetrahedral bonding states is unfilled, creating a "hole". A hole acts like a positive charge carrier, it easily moves as an electron from an adjacent filled bond jumps into the vacant position. Electrical conductivity by holes is dominant in a "P-type semiconductor".

The band structures for Si and GaAs are sketched in Figure 5.15. These results are calculated from approximations to Schrodinger's equation using more realistic 3D forms for U. The curves shown have been verified over a period of years by various experiments.

In the case of these semiconductors, the charge carriers of importance are either electrons at a minimum in a nearly empty conduction band, or holes at the top of a nearly filled valence band. In either case, the mobility $\mu = e\tau/m^*$, such that $\mathbf{v} = \mu\mathbf{E}$, is an important performance parameter. A high mobility is desirable as increasing the frequency response of a device such as a transistor. A useful quantity, which can be accurately predicted from the band theory, is the effective mass, m^*. This parameter is related to the inverse of the curvature of the energy band. The curvature, $\mathrm{d}^2E/\mathrm{d}k^2$ can be calculated, and the formula, simply related to $E = \hbar^2k^2/2m^*$, is

$$m^* = \hbar^2/\mathrm{d}^2E/\mathrm{d}k^2. \tag{5.44}$$

In looking at the energy bands for Si and GaAs in Figure 5.15, one can see that generally the curvature is higher in the conduction band than in the valence band, meaning that the effective mass is smaller and the mobility therefore higher for electrons than for holes. Secondly, comparing Si and GaAs, the curvature in the conduction band minimum is higher in the latter case, leading to a higher mobility for GaAs electrons than in Si. A further aspect is that the conduction band minimum

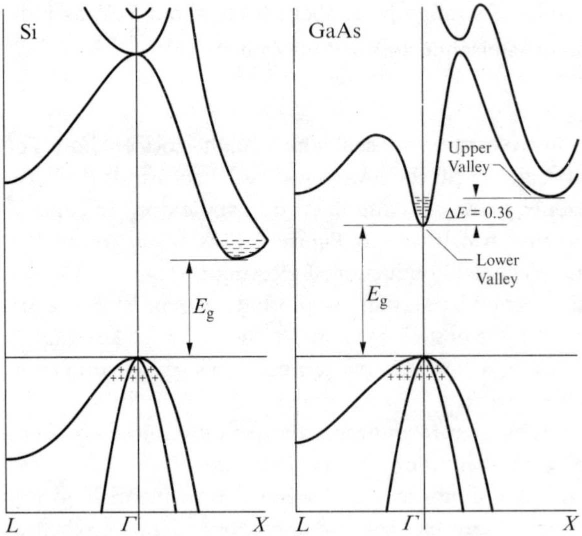

Figure 5.15 Energy band structures for Silicon (left) and GaAs (right). Energy is shown vertically, and k horizontally. The horizontal line marks the top of the filled "valence" bands; in pure samples the upper bands are empty except for thermal excitations (indicated by ++ and −-symbols.) The zero of momentum is indicated as "Γ", and separate sketches are given for E vs k in (111) left and (100) right directions.

in Si is shifted from $k = 0$, which has important effects especially on the absorption of photons by Si, and similar semiconductors having an "indirect" bandgap. Parameters describing several important semiconductors are collected in Table 5.2.

Table 5.2 Energy gaps and other electronic parameters of important semiconductors

Semiconductor	Band Gap (eV)		Mobility at 300 °K (cm^2/volt s)		Effective Mass m^*/m_o		Dielectric Constant ε
	300 °K	0 °K	Electrons	Holes	Electrons	Holes	
C	5.47	5.51	1800	1600	0.2	0.25	5.5
Ge	0.66	0.75	3900	1900	$m_l^* = 1.6$	$m_{lh}^* = 0.04$	16
					$m_l^* = 0.082$	$m_{hh}^* = 0.3$	
Si	1.12	1.16	1500	600	$m_l^* = 0.97$	$m_{lh}^* = 0.16$	11.8
					$m_l^* = 0.19$	$m_{hh}^* = 0.5$	
Grey Tin		~0.08					
AlSb	1.63	1.75	200	420	0.3	0.4	11
GaN	3.5						
GaSb	0.67	0.80	4000	1400	0.047	0.5	15
GaAs	1.43	1.52	8500	400	0.068	0.5	10.9
GaP	2.24	2.40	110	75	0.5	0.5	10
InSb	0.16	0.26	78 000	750	0.013	0.6	17
InAs	0.33	0.46	33 000	460	0.02	0.41	14.5
InP	1.29	1.34	4600	150	0.07	0.4	14

The energy gaps of semiconductors range from about 0.1 to about 5 eV, as indicated in Table 5.2. For devices that operate at room temperature a gap of at least 1 eV is needed to keep the number of thermally excited carriers sufficiently low.

The understanding of semiconductors represented in the sections above is sufficient to reconsider the case of the tunneling diode mentioned in Chapter 1. In more detail, the current voltage characteristic of a tunnel diode in forward bias is shown in Figure 5.16. The forward bias curve for a conventional diode is sketched in a dashed curve in Figure 5.16. At the peak of the anomalous current, marked b in the solid curve of Figure 5.16, electrons in the conduction band on the N-type side are able to tunnel directly into the filled hole states (empty electron states) at the top of the valence band on the P-side. This is possible only if the depletion layer width W is sufficiently small, on the order of 5 nm, and in turn this is possible only if the doping concentrations of donors on the n-side and acceptors on the p-side are sufficiently high. The planar junction between an N-type region and a P-type region in a semiconductor such as Ge contains a "depletion region" separating conductive regions filled with free electrons on the N-side and free holes on the P-side. The width W of the depletion region is

$$W = [2\varepsilon\varepsilon_o V_B(N_D + N_A)/e(N_D N_A)]^{1/2}. \tag{5.45}$$

Here $\varepsilon\varepsilon_o$ is the dielectric constant, e the electron charge, V_B is the energy shift in the bands across the junction, and N_D and N_A, respectively, are the concentrations of donor and acceptor impurities.

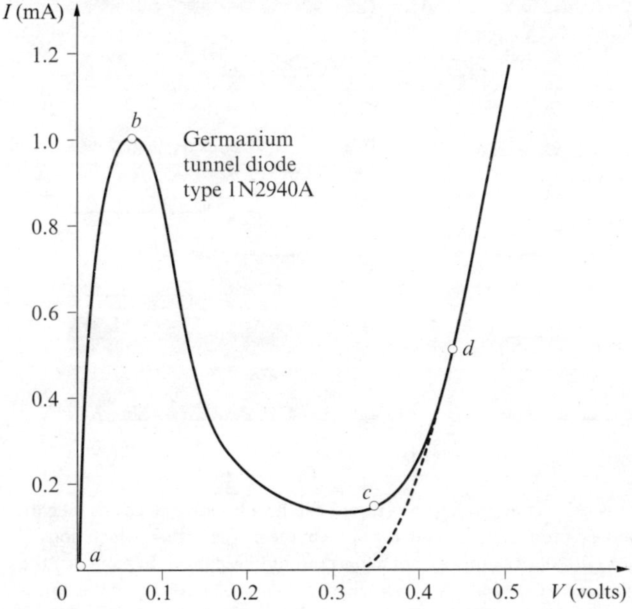

Figure 5.16 Current voltage measurement of germanium tunnel diode (Esaki diode), in forward bias, emphasizing anomalous current peak and negative resistance region (b to c). For an interpretation, see Figure 5.17.

Referring to Figure 5.17, the band configurations are illustrated corresponding to bias voltages labeled as a, b, c, and d in Figures 5.16 and 5.17. Positive or forward bias corresponds to raising the energy bands on the n-side relative to those on the p-side, so that electron current flows from right to left.

A final topic relates to the concept of resistivity, in the quantum picture. The electron states that we have been dealing with are perfectly conducting, in the sense that the electron wave of such a state maintains a velocity $v = \hbar k/m$. A perfectly periodic potential gives a perfect conductor. Indeed, very pure samples of GaAs, especially epitaxial films, measured at low temperatures, give mobility values of millions, in units of $cm^2/volt \cdot s$. The meaning of the lifetime, τ, in the expression for the mobility, is the lifetime of an electron in a particular k state. The cause of limited-scattering lifetime τ in pure metals and semiconductors at room temperature is loss of perfect periodicity as a consequence of thermal vibrations of the atoms on their lattice

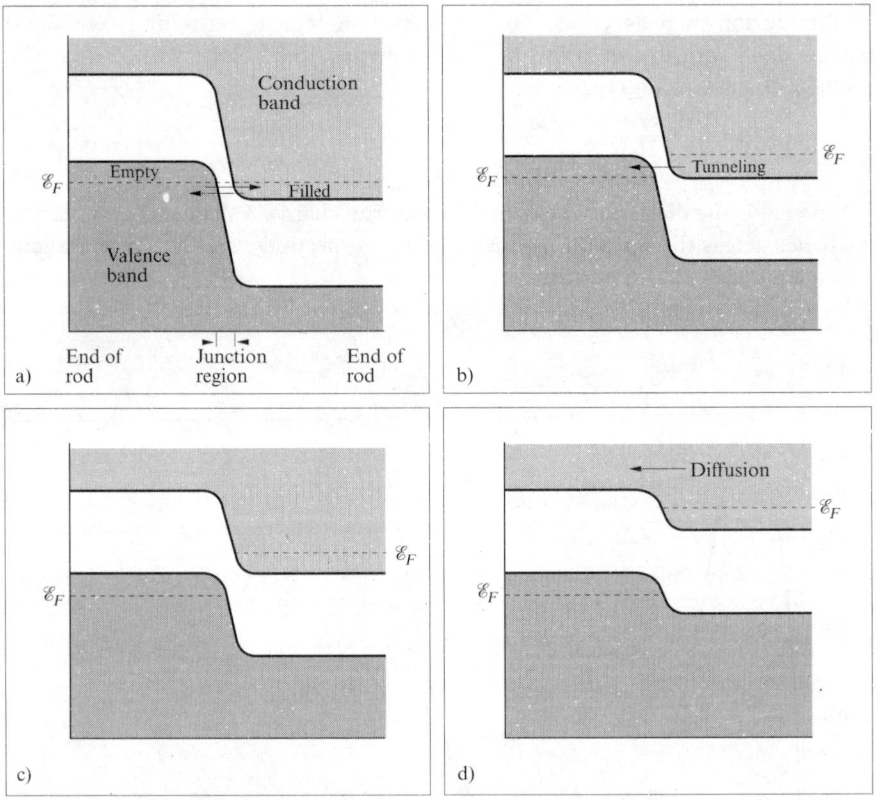

Figure 5.17 Band shifts in forward bias of heavily doped pn junction, showing tunneling as the origin of the anomalous current peak in the Esaki tunnel diode (see Figure 5.16). Panel (b) corresponds to forward bias in which electrons can tunnel directly into filled hole states on the left. At larger forward bias in panel (c) no current can flow because electrons face the energy gap of the p-type region, which does not allow current flow. Finally, in panel (d) the thermal activation of electrons over the barrier leads to the usual exponential growth of current with forward positive bias V.

positions. Calculations of this effect in metals, for example, lead to the observed linear dependence of the resistivity ρ on the absolute temperature T.

5.7 Hydrogenic Donors and Acceptors

Pure semiconductors have filled valence bands and empty conduction bands and thus have small electrical conductivity, depending on the size of the energy gap. As mentioned above, larger electrical conductivity is accomplished by "doping"; substitution for the four-valent Si or Ge atoms either acceptor atoms of valence 3 or donor atoms of valence 5.

In the case of five-valent *donor* atoms like P, As, or Sb, four electrons are incorporated into tetrahedral bonding and the fifth, extra, electron, which cannot be accommodated in the already-filled valence band, must occupy a state at the bottom of the next empty band, which can be close to the donor ion, in terms of its position. This free electron is attracted to the donor impurity site by the Coulomb force. The same physics that was described for the Bohr model should apply! However, in the semiconductor medium, the Coulomb force is reduced by the relative dielectric constant, ε. Referring to Table 5.1, values of ε are large, 11.8 and 16, respectively, for Si and Ge.

A second important consideration for the motion of the electron around the donor ion is the *effective mass* that it acquires because of the band curvature in the semiconductor conduction band. These are again large corrections, m^*/m is about 0.2 for Si and about 0.1 for Ge. So the Bohr model predictions must be *scaled* by the change in dielectric constant and also by the change of electron mass to m^*.

The energy and Bohr radius are, from equations (4.3), $E_n = -kZe^2/2r_n$, and $r_n = n^2 a_o/Z$, where $a_o = \hbar^2/mke^2 = 0.053$ nm. Consider the radius first, and notice that its equation contains both k, the Coulomb constant (which will be proportional to $1/\varepsilon$) and the mass. So the corrected Bohr radius will be

$$a_o^* = a_o(\varepsilon m/m^*). \tag{5.46}$$

Similarly considering the energy, $E_n = -kZe^2/2r_n = -E_o Z^2/n^2$, $n = 1,2,\ldots$, where $E_o = mk^2e^4/2\hbar^2$, it is evident that E scales as $m^*/(m\,\varepsilon^2)$:

$$E_n^* = E_n[m^*/(m\,\varepsilon^2)]. \tag{5.47}$$

For the case of donors in Si, we find $a^* = 59a_o = 3.13$ nm, and $E_o^* = 0.0014\,E_o = 0.0195$ eV. The large scaled Bohr radius is an indication that the continuum approximation is reasonable, and the small binding energy means that most of the electrons coming from donors in Si at room temperature escape the impurity site and are free electrons in the conduction band. An entirely analogous situation occurs with the holes circling acceptor sites. So for the case of the heavily-doped pn junction (Esaki diode), shown in Figure 5.16, the doping values N_D and N_A are essentially equal to numbers of free electrons and holes, as the analysis assumed.

5.7.1 Carrier Concentrations in Semiconductors, Metallic Doping

The objectives of this section are, first, to explain a standard method for finding N_e and N_h, the densities of conduction electrons and holes, respectively, in a semiconductor at temperature T. Secondly, an important special case occurs when the number of donors or acceptors becomes very large, and the semiconductor becomes "metallic".

At high doping, these systems behave like metals, with the Fermi energy actually lying above the conduction band edge in the N^+ case, or below the valence band edge in the P^+ case. These separate rules apply to the so-called N^+ regions used in making contacts, and also to the "two-dimensional electron gas" (2DEG) cases which are important in making injection lasers and in forming certain "charge-qubit" devices. In these "metallic" or "degenerate" cases the number of mobile carriers remains large even at extremely low temperature, as in a metal. (In the usual semiconductor case, the carrier density goes to zero at low temperature, making the material effectively insulating.)

The essential data needed to find N_e and N_h are:

a) the value of the Fermi energy E_F, the energy at which the available states have 0.5 occupation probability, i.e., $f_{FD}(E) = 0.5$, equation (4.96);
b) the bandgap energy E_g;
c) the temperature T; and
d) the numbers of shallow donor and shallow acceptor impurities, N_D and N_A, respectively (see Section 5.6).

A smaller influence on the carrier densities comes from the values of the effective masses (5.44) for electrons and holes.

In addition, it is important to recognize that the semiconductor as a whole must be *electrically neutral*, because each constituent atom, including the added donor and acceptor atoms, is an electrically neutral object.

To find the expected number of electrons in the conduction band requires integrating the density of states (DOS), multiplied by the occupation probability, over the energy range in the band. In a useful approximation, valid when the Fermi energy is several multiples of kT below the conduction band edge, a standard result is derived, which simplifies the task of finding the carrier densities N_e and N_h.

In the formulas for carrier concentrations, the energy E is always measured from the top of the valence band, as indicated in Figure 5.15. The standard formulas assume that the energies of electrons in the conduction band (and of holes in the valence band) vary with wavevector k as $E = \hbar^2k^2/2m^*$, where $\hbar = h/2\pi$, with h Planck's constant, and m^* the effective mass, as given by equation (5.44). In the formula for the energy E, m^* must be given in units of kg, while the common shorthand notation, as in Table 5.2, is to quote m^* as a number, in which case the reader must multiply that number by the electron mass m_e (in kg units), before entering it into a calculation. (The same applies to the mass of a hole $m_h{}^*$, recalling that the motion of a hole is really the motion of an adjacent electron falling into the vacant bond position.) Referring again to Figure 5.15, the bandgap E_g is measured from the top of the valence band $E_V = 0$, so that $E_C = E_g$.

The density of electron states (per unit energy per unit volume) in a semiconductor is given by

$$g(E) = C_e(E - E_C)^{1/2}, \qquad C_e = 4\pi(2m_e{}^*)^{3/2}/h^3. \tag{5.48a}$$

This formula (and the similar one for the density of hole states) is really just (5.33), which was derived for a three-dimensional metal. Similarly, the density of hole states in a semiconductor is given by

$$g(E) = C_h(-E)^{1/2}, \qquad C_h = 4\pi(2m_h{}^*)^{3/2}/h^3. \tag{5.48b}$$

The probability of occupation of an electron state is $f_{FD}(E)$ (equation (4.96)) and the probability that the state is occupied by a hole is $1 - f_{FD}(E)$. The shallow donor electron (see equation (5.47)) has a binding energy E_D (given as $E_n{}^*$ in (5.47)). Thus an electron at the bottom of the conduction band has energy E_g and an electron occupying a shallow donor site has energy $E = E_g - E_D$. The density of shallow donor states is designated N_D.

If the Fermi energy E_F is located toward the middle of the energy gap, or at least a few k_BT below the conduction band edge, the exponential term in the denominator of (4.96) exceeds unity, and the occupation probability is adequately given by $f_{FD}(E) \approx \exp[-(E-E_F)/k_BT]$. Thus, for the total number of electrons we can calculate

$$\begin{aligned} N_e &= \int_{E_C} C_e(E-E_C)^{1/2}\exp[-(E-E_F)/k_BT]\,dE \\ &= C_e \exp[-(E_g-E_F)/k_BT]\int_0^\infty x^{1/2}e^{-x}dx. \end{aligned} \tag{5.49a}$$

To get the second form of the integral, change variables to $E' = E-E_C$, and then $x = E'/kT$, recalling that $E_C = E_g$. Move the upper limit on x to infinity, which is safe since the exponential factor falls to zero rapidly at large x, and then the value of the integral is $\pi^{1/2}/2$. Making use of the definition $C_e = 4\pi(2m_e{}^*)^{3/2}/h^3$, we find

$$N_C = 2(2\pi m_e{}^* k_BT/h^2)^{3/2}. \tag{5.49b}$$

The number of electrons per unit volume in equilibrium at temperature T is then given as

$$N_e = N_C \exp[-(E_G-E_F)/k_BT]. \tag{5.49c}$$

(It is still necessary to know E_F in order to get an answer: always in such problems the central need is to find the value of E_F. In the case of a pure "intrinsic" material, where the number of electrons equals the number of holes, the Fermi level to a good approximation lies at the center of the gap, halfway between the filled and the empty states. As a first step in a pure sample one often will assume $E_F = E_g/2$.)

In the same way, the effective density of states for holes is

$$N_V = 2(2\pi m_h{}^* k_B T/h^2)^{3/2}. \quad (5.50a)$$

The number of holes per unit volume in equilibrium at temperature T is given as

$$N_h = N_V \exp[(-E_F)/k_B T]. \quad (5.50b)$$

In a pure semiconductor the number of holes must equal the number of electrons. By forcing the equality

$$N_e = N_C \exp[-(E_G - E_F)/k_B T] = N_h = N_V \exp[(-E_F)/k_B T],$$

one can easily solve for E_F:

$$E_F = E_g/2 + {}^3/_4\, kT \ln(m_h{}^*/m_e{}^*) \text{ for the pure sample.} \quad (5.51)$$

Finally, look at the algebraic form of the product:

$$N_e N_h = N_C N_V \exp(-E_G/k_B T). \quad (5.52)$$

This product, which is sometimes called N_i^2, is *independent* of E_F, meaning that it is unchanged by doping (impurity levels)!

As an implication, if we know that there are a large number of donors N_D of small binding energy $E_D << kT$, so that it is reasonable to assume that all electrons have left the donor sites to become free electrons (i.e., $N_e = N_D$), then we find that $N_h = N_i^2/N_D$, which can be very small! In this case, since there are electrons in large numbers near E_g (at the donor level or the conduction band edge, measured from the top of the valence band), the Fermi energy (where the occupation probability is ½) must be near the conduction band edge.

This implies, in the heavily electron-doped case, that the probability of occupation of a hole is extremely small, proportional to $\exp(-E_G/k_B T)$. These holes (in the N-doped case) are called "minority carriers", and we see that they are small in number and have strong temperature dependence, $\exp(-E_G/k_B T)$. This is the reason for the strong temperature dependence in the reverse current of the PN junction, below.

The Degenerate Semiconductor

Now consider an extreme case, which will lead to an understanding of a heavily doped, metallic semiconductor, as is typical of the important "2DEG" electron gas.

Consider pure InAs, for which data in Table 5.2 indicate a bandgap of 0.33 eV, electron mass 0.02, hole mass 0.41 and dielectric constant 14.5. First we estimate N_e at 300 K, assuming for simplicity that the Fermi energy is exactly at the center of the energy gap. We find

$$N_e = 2(2\pi m_e{}^* k_B T/h^2)^{3/2} \exp[-(E_G - E_F)/k_B T]$$
$$= 7.165 \times 10^{22} \times 1.7 \times 10^{-3} = 1.22 \times 10^{20}\,\text{m}^{-3},$$

and the same number applies to holes for the pure sample if we neglect the shift of the Fermi energy with effective mass difference.

Next, consider *heavily N-doped* InAs, with 10^{18} cm^{-3} = 1.0×10^{24} m^{-3} shallow donor impurities, at 300 K. This is a large doping, greatly exceeding the thermal intrinsic number of electrons, 1.22×10^{20} m^{-3}. What fraction of these electrons will remain on the donor sites at 300 K?

Calculate the value of the donor binding energy, E_D, from (5.47), by use of the effective mass $m^* = 0.02$ and the dielectric constant 14.5, and scale the energy from 13.6 eV for the hydrogen atom. The value is 0.00129 eV. This is small compared to kT at 300 K, which is 0.0259 eV. Thus, it is reasonable to assume that all of these electrons are in the conduction band!

The Degenerate, Metallic Semiconductor

To estimate what range of energies in the conduction band will be filled by these electrons, we use the formula for the Fermi energy for a metal, equation (5.32), again using the effective mass 0.02, taking N/V as 1.0×10^{24} m^{-3}. The result is $E_F = 0.181$ eV (measured from the conduction band edge), or 0.511 eV from the valence band edge! So, heavily doped InAs is really a metal, and we would expect that the number of free electrons will not change appreciably if this sample is cooled to low temperature. (On the other hand, the description as a metal is valid only for temperatures small enough that $kT << E_F$ as measured from the bottom of the band; otherwise the states in the band will not all be filled, as was assumed for a metal.)

This metallic behavior is indicated in Figure 5.17 for the Esaki diode (see also Section 1.3) by the dotted lines marked "E_F".

Finally, to complete the discussion of heavily doped InAs, making use of the "mass action law" $N_e N_h = N_i^2$ at 300 K, we find that

$$N_h = (1.22 \times 10^{20})^2/10^{24} = 1.49 \times 10^{16}\,\text{m}^{-3} = 1.49 \times 10^{10}\,\text{cm}^{-3}.$$

This is much smaller than the electron concentration, 10^{18} cm^{-3}. This value will change rapidly with varying temperature, according to $\exp(-0.33\ \text{eV}/kT)$.

5.7.2
PN Junction, Electrical Diode $I(V)$ Characteristic, Injection Laser

The current–voltage characteristic of the PN diode junction is shown in Figure 5.16 in the high impurity level case, which gives the anomalous initial peak in current, due to Esaki's tunneling effect (Section 1.3). It is useful to consider the properties of the common case of lower impurity levels, where the $I(V)$ (dashed line in Figure 5.16) rises exponentially with forward bias V (band configurations shown in panels c and d of Figure 5.17), and is small and constant for negative V (not shown in Figures 5.16 and 5.17), giving the familiar rectifying $I(V)$ curve. The origin of this rectifying behavior is important to understand, as it appears in many different situations of interest in nanotechnology.

Refer to Figure 5.17(d), but imagine reduced impurity concentrations N_D and N_A so that the Fermi levels E_F are in the energy gap, rather than (as depicted) in the conduction and valence bands. Reducing N_D and N_A increases the depletion region width, W, as seen in (5.45). Electron tunneling is no longer possible, the Esaki effect disappears, and current flow occurs by thermal excitation over the energy barrier. In conventional forward bias, $V \geq 0$, the energy barrier to diffusion of majority carriers is reduced from V_B to $V_B - V$, increasing the thermionic current, especially if kT is small. This leads to the characteristic $\exp(eV/kT)$ increase of the thermionic current density in forward bias V for the PN junction:

$$J = J_0[\exp(eV/kT) - 1]. \tag{5.53}$$

At $V = 0$ it is clear that zero current is obtained by a counter-flow $-J_0$ which balances the thermionic current just described. The counter-flow current density, $-J_0$, is the result of *diffusion of minority carriers* (holes of concentration p_n in the N-region and electrons of concentration n_p in the P-region). When minority carriers (in the field-free regions adjoining the junction) arrive by random thermal diffusion (diffusion length $L = (D\tau_r)^{1/2}$, see equation (2.12)) at the depletion region (junction), they are immediately swept across by the large electric field (on order $(V_B - V)/W$), providing the counter-flow current density, $-J_0$. In this expression, D is the diffusion constant and τ_r is the minority carrier lifetime (against recombination). Following this line of reasoning one obtains the formula

$$J_0 = e[D_n n_p/L_n + D_h p_n/L_h]. \tag{5.54}$$

J_0 is strongly dependent on temperature T. This is intuitively understood from the T-dependence of the minority carrier concentrations n_p and p_n, as described in Section 5.7.1. Approximately, e.g., $n_p \approx n_i^2/N_A$ (n_i the intrinsic carrier density), if we estimate the majority hole concentration as N_A. Similarly, $p_n \approx n_i^2/N_D$. Since $n_i^2 \propto \exp(-E_g/kT)$ (see Section 5.7.1), we conclude that the reverse current density $J_0 \propto \exp(-E_g/kT)$. This is the basis for the thermistor, a temperature sensor.

The PN junction acts as a voltage-dependent capacitance; normally this function is employed in reverse bias $V \leq 0$. In reverse bias the depletion width W is increased.

An intuitive understanding of the capacitance $C(V)$ of the junction is obtained from the usual parallel plate capacitance formula $C = \varepsilon\varepsilon_o A/W$, where A is the area, ε the relative dielectric constant, and W the voltage-dependent width of the depletion region, based on (5.45). A voltage-dependent capacitor $C(V)$ is called a "varactor".

Under forward bias, electrons flow into the junction from the N-region and holes flow into the junction from the P-region, and we have assumed that the holes and electrons do not recombine, but rather traverse the junction, leading to the external electrical diode current. Recombination in the junction, on the other hand, is possible, and is the basis for the injection laser.

Radiative Recombination and Emission of Light

In semiconductors such as GaAs which (unlike Si) has a *direct bandgap* (meaning that the minimum in the conduction band and the maximum in the valence band are at the same momentum, namely $k = 0$; see Figure 5.15), there is a chance that electrons and holes recombine "radiatively" to release a light quantum (photon) of energy $hf = E_g = hc/\lambda$.

This is the basis of the junction injection laser, an essential element of fiber-optic communications, compact disk players, and "laser pointers" whose colors (now) range from red to blue. Thus, junction lasers now can be made from semiconductors having a range of bandgap energies E_g, although the most important applications still are in communications: fiber optics, in the infra-red range.

PN Junction Injection Laser

To make a junction laser, it is essential to put the junction region in a high-Q optical cavity (but also to let some fraction of the resonant light leak out to the optical fiber or to the screen on the wall). The Q of a resonator is defined as $f/\Delta f$, where Δf is the frequency width of the resonant peak at f and the residence time of the photon in the resonant cavity is $\tau = Q/f$.

One form of conventional optical cavity or resonator comprises two parallel mirrors spaced by a distance L (discussed in Section 5.3.2 in connection with the Casimir force). If the mirrors are metallic, the electric field at each mirror must be zero, and the resonance condition is $N(\lambda/2) = L$, where N is an integer (the violin string condition). Since $c = \lambda f$, this can be expressed as $f = Nc^*/2L$, and $c^* = c/n$ is the speed of light in the semiconductor, whose index of refraction is n. So the result of the optical cavity, which is usually formed by the accurately cleaved ends of a semiconductor bar of length L, is a set of resonant modes with frequency spacing $\Delta f = c/2nL$. (If the mirror is the cleaved end of a semiconductor of optical index n, then the reflection coefficient is

$$R = (n-1)^2/(n+1)^2, \tag{5.55}$$

which approaches unity as n gets large.) In the literature of optics these are called "Fabry–Perot" modes: analogs of harmonics on a violin string.

In the action of the single-mode laser, a single mode, indexed by $N = N_{sm}$, accumulates photons, of frequency $f = N_{sm}c/2nL$ (n the index of refraction). An example of a single optical mode, although taken from a different context, is shown in the first panel of Figure 4.2.

How does a single mode accumulate photons? The phenomenon involved is called "stimulated emission". An electron in the junction region at energy E_g overlaps a hole at the same location and energy zero. The electron can fall into the vacant position in the valence band (the hole) and release a photon of energy E_g. (This is an example of "population inversion", the result of "pumping"; in this case the flow of electrons and holes from the diode into the junction accomplishes the inversion.) The chance that this emission will occur is increased if a photon of energy E_g is already present (thus stimulating the emission) and the important concept is that the newly released (stimulated) photon will have *exactly the same quantum state* as the stimulating photon. So if the stimulating photon is the single optical mode shown in the first panel of Figure 4.2, which initially has one photon in it, then the result will be the single optical mode now occupied by two photons. (As mentioned in Section 5.3.2, each mode is like a quantum state of a simple harmonic oscillator (4.76), whose energy is $E_n = (n + ½)hf$; in the example n starts at 1, and after the stimulated emission, n becomes 2.) Thus, a snowball effect can occur, a richly populated mode gains population, and the final result may be that all photons go into exactly the same mode. This is the action leading to a single-mode laser.

The bandwidth Δf (distinct from the mode spacing Δf, above) of the emitted single-mode light is determined by the Q of the mode of index N_{sm}, as $\Delta f = f/Q$. The Q depends on the lifetime of the photon inside the cavity before it is absorbed or exits (through one of the partially reflecting mirrors).

The *directionality* of the laser output arises in the single optical mode. In the example of parallel mirrors of limited diameter, the photons of longest lifetime will travel exactly perpendicular to the mirrors.

In the development of injection lasers for many purposes, such as CD players, where battery life is an issue, the *critical current density* is a parameter to be minimized. Intuitively this means raising the Q of the optical cavity. Additional factors, however, are important in reducing the critical current density for laser action. One of the most obvious other factors is the efficiency η in converting electrical current into light photons. In the discussion of the PN junction diode, the value of η was zero! Raising the efficiency η of photon emission involves trapping the electrons and holes in the junction region, giving them more time to annihilate and reducing their chance to exit into the electrical leads.

Increasing Radiative Efficiency η

One approach to raising the radiative efficiency η is to incorporate quantum dots (QD) (Section 4.6.5) into the junction region (the whole assembly including quantum dots can still be placed in, or coupled to, a high-Q cavity). In an ideal case, referring to (4.68), the electrical leads with bias voltage applied will allow electrons to flow (tunnel) into the quantum dot levels labeled $E_{n,\text{electron}}$ and holes to flow (also by barrier tunneling) into quantum dot levels labeled $E_{n,\text{hole}}$.

In this scenario the incoming electron and hole particles may arrive in QD states $n \geq 1$, but will quickly fall into their lowest energy states, $n = 1$ for electrons and

$n = 1$ for holes, thus avoiding their loss to the external electrical circuit (thus raising η, as is desired for the laser). The radiative lifetime τ of these $n = 1$ states to recombine can be much longer, to narrow the frequency range (reduce the bandwidth) of the radiation caused by the recombination [the $n = 1$ state electron falls into the $n = 1$ hole state, giving off light of (external) wavelength described by (4.68)]. In all such cases the energy of the emitted photon exceeds the bandgap E_g of the host semiconductor (blue shift), precisely by the confinement energies of the electron and hole states in the dot of size L.

Finally, a useful conventional formula, relevant to Figure 5.17(d), with high doping levels, for the critical injection current density for the PN junction laser is

$$J = 8\pi e W(\alpha L - \ln R)/(L\lambda^2 \eta\tau). \tag{5.56}$$

Here L is the spacing of mirrors of reflectivity R (equation (5.55)), W is the junction (depletion layer) width (equation (5.45)), α is the absorption per meter of light of internal wavelength $\lambda = n\lambda_{ext} = nc/f$, η is the efficiency of electron-to-photon conversion, and τ is the lifetime for radiative electron–hole recombination in the semiconductor junction.

Single-nanowire Electrically Driven Laser

As a recent nanoscale example [12] of an injection laser, Figure 5.18(a) shows schematically a nanowire of CdS lying on a conducting Si substrate.

It is fortunately true that a variety of small-diameter perfect crystal wires (NW), distinct from carbon nanotubes, can be grown, with full cylindrical packing. These perfectly crystalline cylindrical objects typically form along simple crystalline directions, such as [100] for CdS and [111] for silicon. The curved cylindrical surfaces,

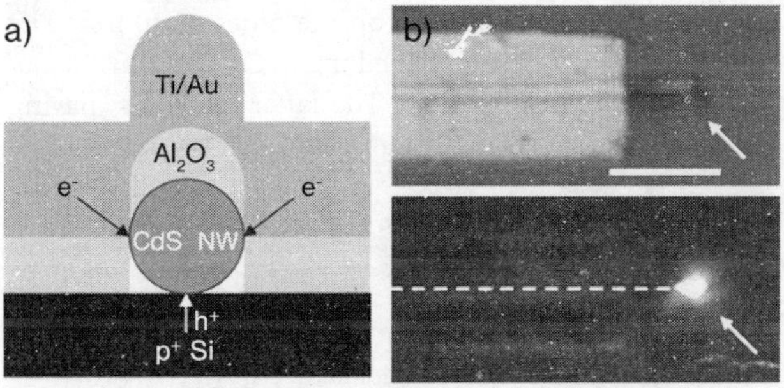

Figure 5.18 (a) A CdS nanowire (NW), coated above with thin alumina, and a Ti/Au electrode, lying on a heavily doped conductive P^+ Si substrate, acts as an injection laser as described in the text. Electrons flow into the nanowire from the Ti/Au electrode, and holes flow into the device from the substrate. (b) The device is seen in a top view in the upper panel, and is seen to be emitting light from its end in the lower panel. The scale bar in panel (b) is 5 µm. In this device the reflective cavity is provided by the discontinuity in refractive index at the accurately cleaved ends of the single-crystalline (001) CdS nanowire [12]. (Reprinted from Nature with permission from Macmillan Publishers, Ltd.)

composed of (curved) planes of fully bonded atoms, are saturated in their bonding; hence, the only growth direction is along the fiber. So the amazing fact is that nanowires of CdS of diameters in the 100 nm range but with lengths of 30 μm or more, and great perfection, have been grown [12].

The CdS nanowires of diameters in the 80–200 nm range were synthesized [12] at 880 °C by laser-assisted catalytic growth, using Au as catalyst. (A schematic of this pulsed-laser ablation technique, illustrated for the case of Si nanowires, is shown later in Figure 6.5.) The nanowire products, after collection on a cooled surface, were dispersed in ethanol and ultrasonically agitated ("sonicated") for 30–60 s, which, in breaking some of the wires, was found to produce a high yield of wires with cleaved ends. The cleaved ends are accurately perpendicular to the (001) growth direction of the nanowires, and suitable for forming the optical resonating cavity, typically of length $L \sim 20$ μm, with reflection coefficient R as described by (5.55).

In this case, Figure 5.18(a), holes are injected into the CdS nanowire from its contact with a p-type Si substrate, electrons from an evaporated Ti/Au metal contact on the top. The upper panel of Figure 5.18(b), with a 5 μm scale bar, shows the nanowire extending horizontally, flat on the highly doped p-Si substrate, with its right end uncovered by the Ti/Au (electron-injecting) electrode. The lower panel shows light emitted from the uncovered end of this CdS nanowire under the electrical injection. The critical current for laser action in the indicated device is ~280 μA. The II–VI compound CdS has a direct bandgap E_g near 2.5 eV and relative dielectric constant about 2.5. The external wavelength of bandgap radiation expected for CdS is $hc/E_g = 496$ nm, which is close to the wavelength of the laser emission peak, at 494 nm, as shown [12] in Figure 5.19.

It is stated that the spacing of the "Fabry–Perot" modes for this device (data not shown) is about 1.8 nm, corresponding from the equation $\Delta f = c/2nL$ (so that $\Delta\lambda = -\lambda^2/2nL$, with $c = f\lambda$) to L about 28 μm, assuming $n = 2.5$ and $\lambda = 494$ nm. This agrees with measurement of the length of the device, on the order of 30 μm. The authors suggest [12] that similar results could be obtained using GaN and InP nanowires, which can be grown also by the same method. The latter materials, having larger bandgaps, would provide shorter wavelength light, into the UV region (see also Figure 4.5).

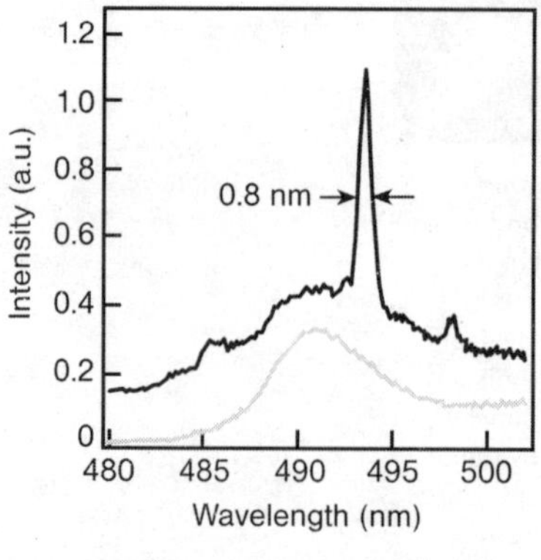

Figure 5.19 Light emission spectrum [12]of the CdS nanowire at 8 K and injection currents 200 μA (lower) and 280 μA (upper). The sharp peak in the upper trace at 494 nm indicates that the laser threshold has been reached, although it was not concluded by the authors that only a single mode was involved. (Reprinted from Nature with permission from Macmillan Publishers, Ltd.)

5.8
More about Ferromagnetism, the Nanophysical Basis of Disk Memory

In a ferromagnet, each atom has an electronic magnetic moment, and these moments are all aligned in the same direction. The alignment comes from the electrostatic exchange interaction, when J is positive, discussed above:

$$\mathscr{H}_e = -2J_e\, \mathbf{S}_1 \cdot \mathbf{S}_2. \tag{5.18}$$

The consequence of the spin alignment, forced by the electrostatic interaction, itself related to the symmetry of the wavefunction against exchange, is that the sample has a bulk magnetization, magnetic moment M per unit volume. It turns out that regions of alignment, called *domains*, appear, and the domain sizes adjust as the *domain walls* shift. If the walls are not easily moved, then the "coercive field" H_c is large, see Figure 5.20.

In zero applied field, although there is no preferred direction for the domain magnetization, the domains tend to remain oriented, leading to a "remanent magnetization". The limit of large applied field gives the saturation magnetization, when all domains are aligned parallel. As a result of the, sometimes complex, domain structure, the magnetization M of a sample has a characteristic dependence upon applied magnetic field, as sketched in Figure 5.20.

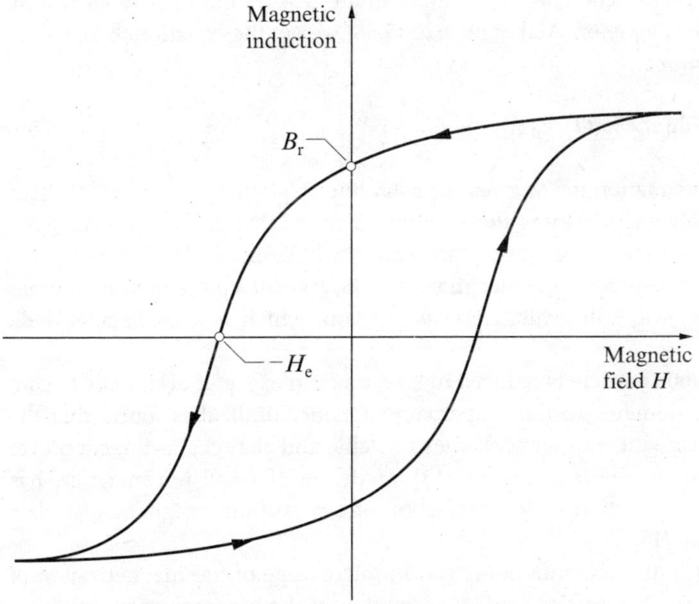

Figure 5.20 Bulk moment M vs. applied magnetic field H, for typical ferromagnet. Hysteretic behavior arises from motion of domain walls. A large remanent M at $H = 0$ is typical of a permanent magnet.

The critical temperature T_c for ferromagnetism is approximately related to the strength of the positive exchange interaction, J_E, by the following expression:

$$J_E = \frac{3k_B T_C}{2zS(S+1)}, \tag{5.57}$$

where z is the number of nearest neighbors, S is the effective spin quantum number of the group of electrons in the atom that constitute its magnetic moment, and k_B is Boltzmann's constant.

In thinking about this equation, it is important to realize that the driving force for the effect is the positive exchange interaction J_E, and that the ferromagnetic transition T_C is the consequence.

An understanding of the temperature dependence of the magnetization below T_C can be obtained by making an assumption that the exchange interaction can be simulated by a fictitious internal exchange field B_E. We assume that the saturation magnetization $M_S(T)$ (which occurs within domains) is proportional to this "magnetic field":

$$M_S(T) = B_E/\lambda, \tag{5.58}$$

where λ is a proportionality constant. A self-consistency equation can be obtained, which is solved to determine $M_S(T)$: The magnetization $M_S(T)$ at low temperatures for N spins (say $S = 1/2$) in a magnetic field (say B_E) is $M_S(T) = N\mu_B \tanh(\mu_B B_E/k_B T)$. This is the Brillouin function, which is an accurate description of *paramagnetism*, and μ_B is the Bohr magneton. Making use of (5.58) we get the equation to be solved to find *ferromagnetism*:

$$M_S(T) = N\mu_B \tanh[\mu_B \lambda M_S(T)/k_B T] . \tag{5.59}$$

Solution of this equation for M gives the solid line as plotted in Figure 5.21. This equation reasonably well balances the strength of the exchange interaction, represented by λ, and the thermal agitation represented by $k_B T$, for a *bulk* sample.

An appropriate question for nanotechnology is: how small a piece of iron, or other magnetic material, will still exhibit ferromagnetism, which is a cooperative bulk effect?

As a ferromagnetic particle is reduced in size, one can expect it eventually to contain only a single domain. At the "superparamagnetic" limit, it is found that the direction of the magnetic moment becomes unstable and changes its direction over time, depending upon the temperature [13]. Modeling of small ferromagnets has been reported, in connection with a prediction of a maximum magnetic disk data density of 36 Gb/in^2 [14].

Observations [15] in a transmission electron microscope of the magnetization of 10 nm magnetite Fe_3O_4 particles confirmed motion of the magnetization, but on a much slower time scale than the 5 ns predicted. It is reported that the superparamagnetic limiting size for $SmCo_5$ is 2 nm [15].

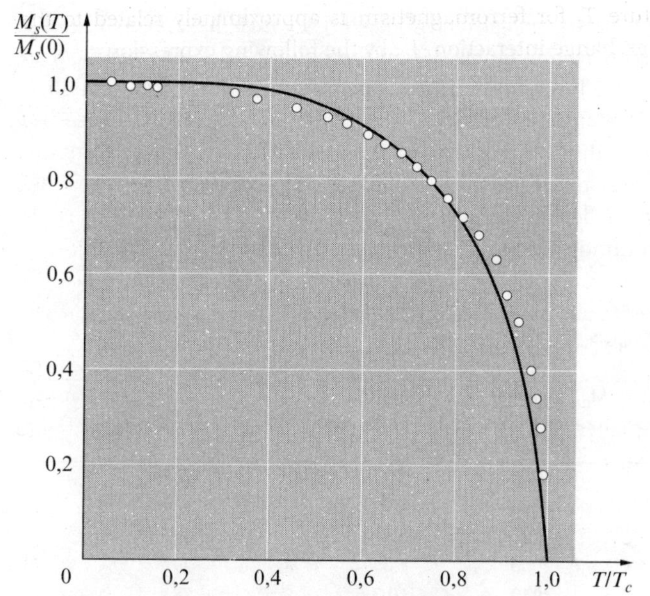

Figure 5.21 The saturation magnetic moment $M_S(T)$ of a typical ferromagnet vs. temperature, normalized to critical temperature T_c. Circles are data for ferromagnet nickel. Solid line comes from solution to equation (5.59).

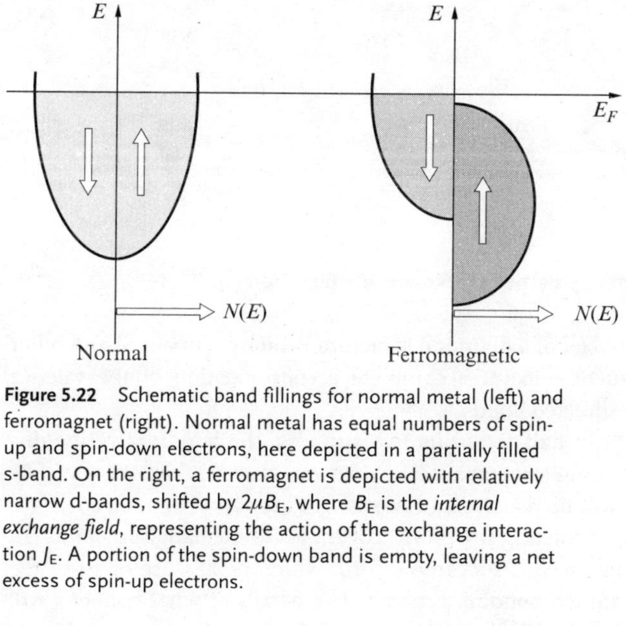

Figure 5.22 Schematic band fillings for normal metal (left) and ferromagnet (right). Normal metal has equal numbers of spin-up and spin-down electrons, here depicted in a partially filled s-band. On the right, a ferromagnet is depicted with relatively narrow d-bands, shifted by $2\mu B_E$, where B_E is the *internal exchange field*, representing the action of the exchange interaction J_E. A portion of the spin-down band is empty, leaving a net excess of spin-up electrons.

In Chapter 6 we will discuss *magnetotactic* bacteria, which have been found to have arrays of 100 nm scale ferromagnets within their single cells. The magnets are near the superparamagnetic limit, and it may be true that the linear array stabilizes even superparamagnetic magnetite particles into magnetic alignment. These bacteria, and the question of the limiting size of magnetic bits written on computer magnetic hard drives, are examples of the nanophysical phenomena of ferromagnetism appearing near the smallest limiting size.

Properties of some common ferromagnets are summarized in Table 5.3 [16].

Table 5.3 Properties of some common ferromagnets. The column "$n_B(0\,K)$" is expressed in multiples of the Bohr magneton, $\mu_B = 9.27 \times 10^{-24}$ J/T [16].

Substance	Magnetization M_g, in gauss Room temperature	0 K	$n_B(0\,K)$, performula unit	Curie temperature, in K
Fe	1707	1740	2.22	1043
Co	1400	1446	1.72	1388
Ni	485	510	0.606	627
Gd	–	2060	7.63	292
Dy	–	2920	10.2	88
MnAs	670	870	3.4	318
MnBi	620	680	3.52	630
MnSb	710	–	3.5	587
CrO_2	515	–	2.03	386
$MnOFe_2O_3$	410	–	5.0	573
$FeOFe_2O_3$	480	–	4.1	858
$NiOFe_2O_3$	270	–	2.4	858
$CuOFe_2O_3$	135	–	1.3	728
$MgOFe_2O_3$	110	–	1.1	713
Eu	–	1920	6.8	69
$Y_3Fe_5O_{12}$	130	200	5.0	560

5.9 Surfaces are Different; Schottky Barrier Thickness $W = [2\varepsilon\varepsilon_o V_B/eN_D]^{1/2}$

Figure 5.14 shows the unit cell of a diamond structure, which occurs in Si and other semiconductors. The structure features complete accommodation of the valence electrons of each atom in directed bonds.

If such a crystal is split in half, exposing two surfaces, the bonding will be disrupted. If the electron bonding structure locally does not change, then there will be "dangling bonds", which will be reactive and unstable. Passivation of such a surface is sometimes carried out by reacting it with hydrogen, to fill the dangling bonds. In other cases the surface may reconstruct into a form which makes use of all of the valence electrons in a modified bonding scheme. The details of what happens will depend upon in which direction the crystal was split. The properties of a (111) surface, as was illustrated in connection with Cu, will be different from those of (110) or (100) surfaces.

A frequent and important surface effect is an *electrical barrier layer* (Schottky barrier). A simple way of thinking of bending of the energy bands at the surface is that the position of the Fermi level relative to the bands may be different at the surface than it is in the interior.

For example, in N-type silicon the Fermi level in the bulk of the material is just below the conduction band edge. At the surface, however, the Fermi level may lie in the middle of bandgap. The surface typically will have a large density of states in the midgap energy range, and electrons from the adjoining N-type material will fall into these states, making the surface electrically negatively charged and leaving a positively charged depletion layer. The equilibrium situation that results is that the conduction band edge at the surface is raised by $eV_B \sim E_g/2$, leaving a depletion region between the surface and the bulk. The width of the Schottky barrier, a depletion region, is

$$W = [2\varepsilon\varepsilon_o V_B/eN_D]^{1/2}\,. \tag{5.60}$$

For donor concentrations N_D convenient for bulk device operation, the resulting W is so large that electrical current will not cross the Schottky barrier. So there is an inherent difficulty in making conductive Ohmic contacts to Silicon. One way to correct this difficulty is to make the surface layer heavily doped (N^+), so that, according to (5.60), the width W is small enough that electrons can readily tunnel through the barrier.

5.10
Ferroelectrics, Piezoelectrics and Pyroelectrics: Recent Applications to Advancing Nanotechnology

These "cooperative" effects are, in principle, similar to ferromagnetism, in that large numbers of particles behave identically, rather than independently. This can be described by an "order parameter", or coherent nonzero distortion: e.g., an average *electric* field in material, analogous to spin polarization P (equation (4.91)) in a ferromagnet.

Cooperative effects are rare in metals and semiconductors, but are sought out for applications. From the point of view of electron behavior, most frequently, the motion of individual carriers is understood by ignoring all the other carriers, or treating their effects, in an average way, as a correction such as the effective mass in metals and semiconductors. The most striking collective effects involving carriers are probably ferromagnets and superconductors. The classes of materials in the title, however, are basically insulators, which undergo cooperative behavior with regard to small displacements of *atomic positions*. If the uniformly displaced atoms are slightly charged, then the distortion can lead to the collective electric field, for the ferroelectric or the pyroelectric, for example.

Piezoelectric materials, being insulators, certainly do not have free carriers. The effects of interest involve small distortions of the crystalline lattice under applied electric fields. These distortions in polycrystalline samples turn out to be smoothly dependent on applied potentials, and allow the smooth movement on the angstrom (1 Å = 0.1 nm) scale of scanning electron microscope tips and similar objects. This is one of the experimental essentials to the success of the scanning tunneling microscope.

Piezoelectric Materials

A piezoelectric material can be characterized, e.g., by the fractional change in its x-dimension per unit electric field in the z-direction. As an example, for the useful and commercially available (polycrystalline) material known as PZT-5H, a quoted value of the coefficient $D_{31} = S_1/E_3$ = strain in 1-direction / electric field in 3-direction=−2.74 Å/V. Here the acronym PZT refers to "lead–zirconate–titanate").

"Strain" is a fractional distortion, so that for a bar of length $x = L$, the strain, $\Delta L/L$, is calculated as $D_{31}V/t$, if the electric field is applied across the thickness t in the z-direction. The resulting motion of the end of a bar of length L and thickness t, with applied voltage V across the t dimension, is $\Delta L = D_{31}VL/t$, so the "piezo constant" is $D_{31}L/t$. So for an $L = 1$ μm bar of thickness 0.1 μm, with voltage applied to electrodes across the narrow z-dimension, the "piezo constant" is −27.4 Å/V.

Although this example relates to the scanning tunneling microscope, the most frequent application of PZT materials is as generators of high frequency (~15 kHz) "ultrasound" for medical imaging in a liquid environment, for general agitation ("sonication") as an adjunct to cleaning procedures, and in underwater applications to generate SONAR waves, historically for ranging, i.e., measuring the distance to an underwater obstacle, such as the bottom of the lake (or, possibly, to a nearby submarine).

The use of piezoelectric transducers, mainly based on the PZT ceramics fed with strong AC (in the kHz range) voltages, is a highly developed technology, which allows generation of intense ultrasonic pressure waves in liquids. The pressure amplitudes in ultrasonically generated pressure waves in water, for example, can easily exceed atmospheric pressure, ≈ 15 lb/in^2. (An example of 5.5 bars is shown in Figure 5.25.)

Ultrasonic Initiation of Bubbles, by a Negative Pressure

A consequence of such a strong pressure wave amplitude, exceeding atmospheric pressure, is that the pressure becomes negative during a part of the cycle! Such a negative pressure, sustained by the surface tension of the liquid, nonetheless fosters occasional *bubble formation* in the liquid (especially if a seed or nucleus of the bubble is provided).

This phenomenon is referred to as "cavitation", a subject with some history. Cavitation in water was originally treated elegantly by Lord Rayleigh, ca. 1900, in connection with degradation of the propeller blades of British naval vessels, which occurred at high power (in speed trials of destroyers), conditions fostering water turbulence around the propeller blades. The propellers were observed to be eroded after time trials, and the Admiralty asked Lord Rayleigh to analyze this effect, which he did with considerable success. One can speculate that erosion of the metal blade would indicate locally high temperature.

Ferroelectrics and Pyroelectrics

Ferroelectricity arises as a coherent distortion of a partially ionic crystalline lattice, to give accumulations of (oppositely charged) charge densities on opposite sides of the insulating sample, under the influence of an applied external electric field E. Large surface charge densities can be obtained, to produce large electric fields E, perpendicular to the surfaces of such materials.

The basic law is $E = \sigma/2\varepsilon_o$, where σ is the charge density on the surface, which results, in the present case, from the ferroelectric distortion. In some cases there will be a transition temperature, say T_c, at which the internally driven distortion occurs. As the temperature is varied above and below this value, one can expect a surface charge density (and consequent surface electric field) that is strongly dependent on the precise value of the temperature. This is the situation for a pyroelectric crystal.

A Nanotechnological (Pyroelectric) Compact Source of Neutrons

A recent nanotechnological application of ferroelectricity (in its closely related pyroelectric form) [17] is a simple and compact source of neutrons. Neutrons, conventionally produced in fission reactors and high energy accelerators, are potentially of importance, e.g., in detecting unwanted fissionable (fissile) materials, perhaps hidden in a shipping container. A fissile material, such as ^{235}U or ^{239}Pu, when exposed to a neutron flux, responds characteristically with induced neutrons which might be detected to set off an alarm.

In the simple, harmless device demonstrated [17], no external high voltages are needed, in principle only a few AA batteries! In this case, heating the $LiTaO_3$ crystal from 240 K to 265 K, using a 2 W heater, is stated to increase the surface charge density s by 0.0037 C/m^2. This should correspond to a surface electric field

$$E = \sigma/2\varepsilon_o = 3.7\times10^{-3}/(2\times8.85\times10^{-12}) = 0.209\ \text{V/nm}.$$

The authors state [17] that in the device geometry (see Figure 5.23) this gives a potential of 100 kV. Neutrons (n) are unquestionably produced in this device by the nuclear fusion reaction

$$D^+ + D^+ \rightarrow {}^3He^{2+}\ (820\ \text{keV}) + n\ (2.45\ \text{MeV}),$$

where D is the deuteron, the nucleus of deuterium, ^2H. As shown in Figure 5.23, a nanoscale tungsten tip of radius 100 nm is connected via a post (length 2.3 mm) to a copper electrode encasing one surface of the lithium tantalate $LiTaO_3$ pyroelectric

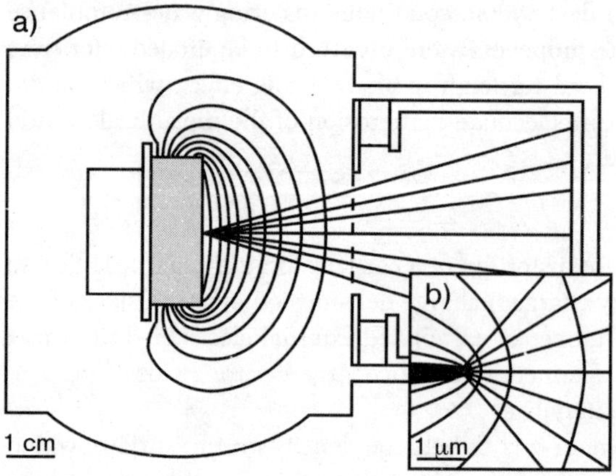

Figure 5.23 Sketch [17] of compact pyroelectric-based field-ionizing neutron generator. The pyroelectric crystal (center left) produces a strong dipole-like set of electric field lines, which appear as a function of temperature. The strongly (positive) biased (up to 80 kV) tip of diameter 100 nm (see inset) is attached to a copper electrode encasing the $LiTaO_3$ pyroelectric crystal, producing a distinct set of radial electric field lines which trace toward the (grounded) target, at the right side of the apparatus. The right surface of the pyroelectric crystal reaches a potential of +80 kV or so relative to its left surface, which is grounded. The ambient gas is D_2, dilute such that the mean free path exceeds the dimension of the chamber. The electric field near the tip (at voltage) locally ionizes the deuterium gas to produce D^+ ions, which then fall down the potential gradient and impact the (deuterated) target at 80 keV energy. This energy is enough to drive the well-documented $D + D = {}^3He + n$ reaction, verified by direct observation of the neutrons. (Reprinted from Nature with permission from Macmillan Publishers, Ltd.)

crystal. The other side of the crystal is grounded, as is the collector (counter-electrode) at the right side of the apparatus.

The positively biased tip extends from the positively charged copper electrode into a dilute deuterium D_2 gas. At the operating tungsten potential of 80 kV, the electric field E produced at the tip is sufficient to ionize deuterium! The resulting D^+ ions then fall down the 80 kV potential and impact D in the counter-electrode (a Mo disk coated with ErD_2). The observations confirm D–D fusion, based on the observation of the expected neutrons.

Electric Field Ionization of Deuterium (Hydrogen)

The electronic properties of deuterium are the same as those of hydrogen H, which differs only by the additional (uncharged) neutron mass in the nucleus. Thus, the simplest estimate of the required electric field E to ionize deuterium might be the field of the proton at the Bohr radius in H, which is $E = ke/a_o^2 = 514.5$ V/nm. Of course, this is an overestimate: a much smaller electric field, $E \approx 25$ V/nm, quickly ($\sim 10^{-15}$ s) removes the electron.

In an electric field E, the potential energy has a term $U(x) = -eEx$, so that, at a spacing $x^* = E_0/eE$, the electron will have the same energy as in its ground state. Here E_0 is the electronic binding energy, 13.6 eV, for deuterium as well as for hydrogen.

The electron can tunnel through the barrier (Section 4.6.4) which extends from $x = 0$ to x^*. In detail this is a difficult problem to solve, but a simplified treatment is possible. The earliest estimate of the lifetime of H against field ionization was given by Oppenheimer in 1928. Oppenheimer's notable estimate, one of the early validations of the theory of electron tunneling, is that for H in a field of 1000 V/m the lifetime τ is $(10^{10})^{10}$ s.

(For a simplified estimate of the *E*-field ionization of H, see Exercise 5.12, which will lead the interested reader to easily estimate that, for $E = 25$ V/nm, the field ionization lifetime of hydrogen and deuterium is on the order of 10^{-15} s, while, for $E = 2.5$ V/nm, τ is extremely long, estimated as 6×10^{33} s or 1.9×10^{26} y. This latter estimate, for 2.5 V/nm, is effectively an infinite lifetime, exceeding the age of the universe, ~ 13 billion years. Such extremely nonlinear dependences, arising in the exponential barrier tunneling factor, are characteristic of tunneling phenomena (4.64a). For example, lifetimes of radioactive elements against alpha-particle decay, well explained by a similarly straightforward tunneling model, range from 10^9 y to 10^{-3} s.)

The authors [17] indicate that $E = 25$ V/nm is sufficient for 100% field ionization, thus generating electrons and D^+ ions from D_2 gas. The positively charged D^+ particles accelerate to an energy of 80 keV and impact the deuterated counter-electrode to generate neutrons and ^3He by nuclear fusion, releasing about 2.45 MeV of energy.

With a tip potential $V(a) = V_0$, the electric field E for $r > a$ is $E = V_0 a/r^2$. Using values $V_0 = 80$ kV and $a = 100$ nm, this indicates that the required $E = 25$ V/nm is available out to $r = 566$ nm. The measurements in Figure 5.24(c) indicate a peak deuteron ion current of 4 nA, and the heating power to run this neutron generator is given as 2 W (see Exercise 5.13).

The fusion reaction of the two D^+ ions to form a ^3He nucleus ($^3He^{+2}$, containing two protons and one neutron), releasing the other neutron, requires ~ 80 keV ion energy. Once the threshold is reached, this reaction is exothermic, releasing particles with kinetic energies of 0.82 MeV and 2.45 MeV. This energy release exceeds by about 10^6 that of an analogous exothermic chemical reaction: burning hydrogen in oxygen to make water. (The spacing between two protons corresponding to potential energy 80 keV is about 18 fm (1 fm = 10^{-15} m = 10^{-6} nm). The nominal radius of a nucleus of particle number $A = Z + N$ is $R = A^{1/3} r_0$, where $r_0 \sim 1.2$ fm.

The present device [17] is not suited to capturing the energy release, but is intended only as a source of neutrons, which are otherwise quite hard to produce. It illustrates that it is relatively easy, using a nanoscale tip, to achieve very large electric fields. The same D–D fusion reaction in a hot gas (plasma) might require a temperature such that $k_B T = 80$ keV, which corresponds to 9.3×10^8 K! Again, this is an overestimate. The consensus is that 10^8 K, or about 10 keV per particle, is enough, given that the reactants at high density may be able to tunnel through a remaining electrostatic barrier against fusion (perhaps 18 fm in the case of the D–D reaction at 80 keV).

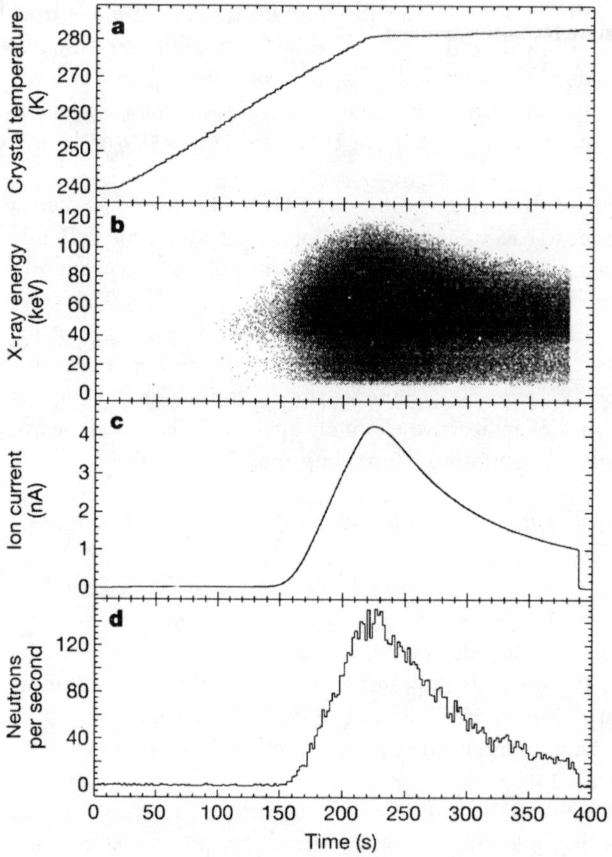

Figure 5.24 Documentation of one cycle of neutron generation [17]. (a) At time $t = 0$ a ramp of temperature from 240 to 280 K leads to crystal potential rising from zero to about 80 kV near $t = 230$ s. (b) The onset of x-rays which come as electrons released from the ionization events fall onto the positively charged copper plate encasing the tantalate pyroelectric crystal. The x-ray energies, observed up to about 100 keV, can only come from a tip potential of the same order, which produces a local electric field sufficient to strip electrons from the deuterium gas. (c) The ionic current, presumably the sum of electron current into the copper and positive ion current into the right-hand electrode, adding to 4 nA maximum. This number can be checked by elementary considerations involving the radius and surface area around the tip leading to certain ionization, the density of the gas at the stated pressure, and the number of deuterium molecules in random gas diffusion that would cross that surface per unit time. (d) The measured number of neutrons per second. The satisfactory coincidence of the peaking of the several indicators at about 230 s puts these results really beyond question. (Reprinted from Nature with permission from Macmillan Publishers, Ltd.)

An Unexpected High Temperature Nano-environment

How can one achieve the needed 10^8 K? The fusion reactions of the Sun, providing about 1 kW/m^2 of radiant power on Earth, are believed to occur at such temperatures, under high pressures and densities of reactant nuclei, such as D and $^3He^{+2}$. Attempts, on the terrestrial lab scale, to achieve nuclear fusion do not seem to be promising to generate large amounts of energy.

Collapse of Ultrasonically Produced Bubbles in Dense Liquids

Piezoelectric technology is capable of creating pressure waves $P = P_0 \cos(kx - \omega t)$, where P_0 may be as large as 5.5 bars, about 5.5×101.3 kPa, in liquids. If the pressure amplitude is 5 bars, then the pressure in the liquid will be negative over a portion of the cycle, and, given a nucleation center, bubbles will appear.

Methods have evolved to nucleate single bubbles, and with proper choice of the sound frequency, a single bubble may enlarge over many cycles, reaching an unstable maximum size. Then the bubble abruptly collapses, with remarkably high temperatures and pressures resulting from the rapid adiabatic compression of the gas by the dense (cool) liquid.

Rare gases are typically dissolved in the liquid, and form the bulk of the gas in the bubble as it collapses. This nanoscale laboratory environment has been verified [18] as producing $T \sim 15\,000$ K, with theoretical estimates up to 10^6 K [19]. It is well known that compression of air in the diesel engine allows the injected fuel to ignite without a spark plug. The effect described is similar. Although 15 000 K is far short of $T \sim 10^8$ K mentioned above, this situation is of great ongoing interest. The light emitted is called "sonoluminescence" and was first discovered in 1933.

Evidence of $T \sim 15\,000$ K in this case is indicated in Figure 5.25 [18] which shows optical emission spectra collected from a collapsing bubble containing argon gas. The liquid in this case is 0.85 H_2SO_4 in water. Analysis of the radiation provides a good estimate of the temperature of the outer portion of the bubble, also indicating that the gas is at least partially ionized.

Once nucleated, a bubble can grow over several cycles of the oscillating ultrasonic pressure field, and achieve a radius r_{max} on the order of 100 μm. The unstable bubble collapses abruptly and disappears with the production of light: "sonoluminescence". So the radius r of the ultrasonically induced bubble falls rapidly from r_{max} to a minimum value r_{min}. The bubble disappears, from an experimental point of view. The details of the bubble collapse are not clear, but the data in Figure 5.25 clearly indicate temperatures up to 15 000 K!

In a simple model, an ideal gas is compressed without exchanging heat with the outer liquid. In this case the adiabatic law:

$$PV^{\gamma} = \text{constant}, \tag{5.61}$$

combined with the ideal gas law $PV = RT$ leads to the relation

$$T \propto (1/V)^{\gamma-1}, \tag{5.62}$$

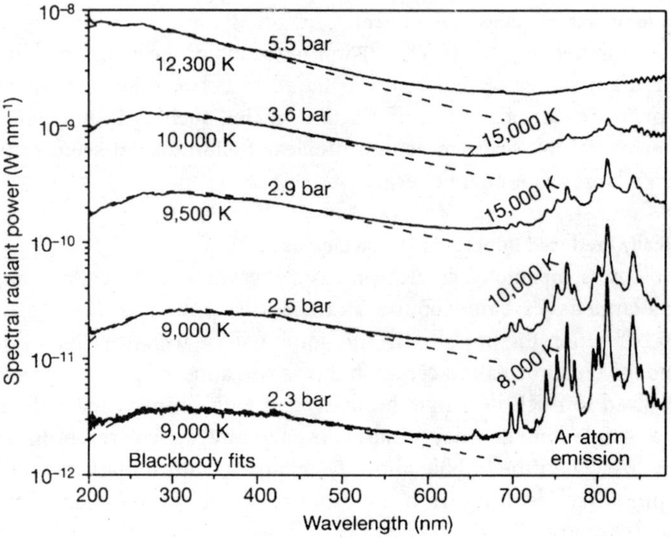

Figure 5.25 Light emission [18]: single bubble sonoluminescence spectra for argon gas above 85% H_2SO_4 in water at ultrasound pressure amplitudes ranging from 2.3 to 5.5 bar. The curves are smooth at small wavelength, and are fit by blackbody radiation curves (dashed) with temperatures T ranging from 9000 to 12 300 K. The λ = 700–900 nm spectra reveal identifiable emission lines of the argon atom. (These lines are known to arise from transitions among Ar atomic levels which lie 13 eV above the ground state.) Fits to these line spectra indicate that T rises from 8000 K to more than 15 000 K as the pressure amplitude P_0 rises from 2.3 to 5.5 bar. The emitted power densities here are relatively high: it was noted that by changing from H_2O to H_2SO_4 (higher density and lower vapor pressure) the power density increased by a factor of 2700 [18]. At the highest P_0 the Ar line spectra broaden and disappear, indicating short lifetimes of atomic levels, due to rapid ionization events by hot electrons from an inner plasma core of much higher temperature. The Ar atom levels identified in these spectra are too high in energy (>13 eV, or 156 000 K) to be consistent with the emission temperatures (12 300 to 15 000 K). The authors [18] suggest that there is an inner (opaque) plasma core (at perhaps 156 000 K) which showers the outer Ar with high energy electrons, exciting the observed Ar levels, known to lie more than 13 eV (156 000 K) above the ground state. This analysis suggests a temperature gradient within the collapsing bubble, so the plasma core is much hotter than 15 000 K at the outer region. The T of the outer region, near the liquid interface, is known to fall as the thermal conductivity of the gas is increased, by admixing Ne into Ar. (Reprinted from Nature with permission from Macmillan Publishers, Ltd.)

where $\gamma = c_P/c_V = 5/3$ for a monatomic gas. (For this reason, diesel engines, with compression ratios (V_{max}/V_{min}) in the range 12–24, need no spark plugs.) Expressed as a scaling law for temperature in terms of the bubble radius r, this adiabatic law gives

$$T \propto (1/r^3)^{\gamma-1} = 1/r^2. \tag{5.63}$$

If we apply this simplified analysis to the data in the figure, identifying the initial and final temperatures as 300 K and 15 000 K, the inferred compression ratio is

353.5, or $(r_{max}/r_{min}) = 7.07$. Starting at $r_{max} = 100$ μm, then $r_{min} = 14.1$ μm. This $r_{min} = 14$ μm is *much larger* than one might expect for the minimum radius of a collapsing bubble in a cool liquid! Indeed, the effect of surface tension should make a cool liquid surface *smooth*, even on a nm scale.

If, instead, the minimum radius of the collapsing bubble is taken as 10 nm, then, starting with 0.1 mm radius at 300 K, the relation gives the final temperature as 3×10^{10} K! This is an astonishing (and defensible) result, suggesting potential for fusion events if the gas included deuterium, for example. There is no guarantee that the bubble will collapse in a spherical fashion, but since the bubble does indeed disappear (in some sense $V \rightarrow 0$), the shape may not matter.

Rather than a mathematical criterion, $r = 0$, the minimum radius r_{min} may signal that the internal liquid/gas interface is rough (ill-defined on a scale $\Delta r = r_{min}$). One certainly expects, by the action of surface tension forces minimizing the surface area, that the surface of a cool liquid is smooth on a nm scale. Nanometer scale smoothness of cool liquid surfaces is confirmed by atomic force microscope (AFM) topographs (surface images) of amorphous materials, formed by freezing of liquids.

At the achieved and verified 15 000 K, the gas/surface interface certainly is 'rougher', and this may explain the (anomalously large) effective minimum volume and (anomalously small?) maximum temperatures reported. The inner liquid surface of the collapsing bubble will be bombarded with molecules, atoms and/or ions of energy $kT \sim 1.25$ eV (at 15 000 K), which will make the interface less smooth. The time scale would have to be considered to decide how long the surface remains "cool". (If this temperature were to extend at all into the "liquid", the affected region would obviously evaporate.) The time scale of bubble collapse is beyond the scope of discussion.

Following this line of reasoning, one might naively suggest (to achieve nuclear fusion in the bubble compression scenario) an extremely low mass gas (least disruptive to the gas/liquid interface) like H_2 or He (or D_2 or ^{3}He), dissolved in an extremely heavy liquid, such as WF_6, WCl_6, or even UF_6 (see Figure 5.2). Tungsten hexafluoride, used in the semiconductor industry (in spite of its toxicity; it easily forms HF, hydrofluoric acid), has a molecular weight 297.84 g/mol, density 3.43 g/cm^3, and boiling point 290 K. It was noted in [17] that changing the liquid from water to sulfuric acid (certainly of higher density) made the emitted light much brighter, which may be attributed to a lower vapor pressure or a higher density.

The authors [17] also noted, however, that the thermal conductivity of the gas was a factor, and that Ne, with a higher thermal velocity (and higher thermal conductivity) than Ar, gave lower temperature results from measurements such as those in Figure 5.25. The authors [17] also have given evidence that there is an inner (ionized plasma) core of the bubble, where the temperature, not evident from the light emission from the outer shell of the bubble, may be *higher* than 15 000 K.

This, now more completely understood, situation (leading to local pressure and temperature singularities) *clearly seems more open for exploitation as a potential energy source* (see also [20]). On fundamental grounds, this approach may be more promising for nuclear fusion than compression based on zero-mass particles from lasers.

References

[1] B. K. Tanner, *Introduction to the Physics of Electrons in Solids* (Cambridge, Cambridge, 1996), p. 186.
[2] F. London, Zeit. für Phys. **63**, 245 (1930).
[3] L. Pauling and E. B. Wilson, *Introduction to Quantum Mechanics with Applications to Chemistry* (Dover, New York, 1985) p. 387.
[4] K. S. Krane, *Modern Physics* (2nd Edition) (Wiley, New York, 1995). Figure 11.12, p. 344.
[5] E. A. Reitman, *Molecular Engineering of Nanosystems* (Springer, New York, 2001) p. 71.
[6] K. E. Drexler, *Nanosystems*, (Wiley, New York, 1992) pp. 64–65.
[7] H. G. B. Casimir, Proc. Kon. Ned. Akad. **51**, 793 (1948).
[8] Reprinted with permission from H. B. Chan, V. A. Aksyuk, R. N. Kleiman, D. J. Bishop, and F. Capasso, Science **291**, 1941–1944 (2001). Published online 8 February 2001 (10.1126/science.1057984). Copyright 2001 AAAS.
[9] B. Alberts, Nature **421**, 431 (2003).
[10] P. Ball, *Designing the Molecular World: Chemistry at the Frontier* (Princeton, Princeton, 1994), p. 151.
[11] R. L. de Kronig and W. G. Penney, Proc. Roy. Soc. Ser. A **130** 499 (1931).
[12] X. Dujan, Y. Huang, R. Agarwal, and C. M. Lieber, Nature **421**, 241 (2003).
[13] E. F. Kneller and F. E. Luborsky, J. Appl. Phys. **34**, 656 (1963).
[14] S. H. Charap, P. L. Lu, Y. He, IEEE Trans. Magn. **33**, 978 (1997).
[15] S. A. Majetich and Y. Jin, Science **284**, 470 (1999).
[16] C. Kittel, *Introduction to Solid State Physics* (Sixth Edition) (Wiley, New York, 1986). Table 2, p. 429.
[17] A. Naranjo, J. K. Gimzewski, and S. Putterman, Nature **434**, 1115 (2005).
[18] D. J. Flannigan and K. S. Suslick, Nature **434**, 52 (2005).
[19] R. Toegel and D. Lohse, J. Chem. Phys. **118**, 1963 (2003).
[20] R. P. Taleyarkhan, C. D. West, R. T. Lahey Jr., R. I. Nigmatulin, R. C. Block, and Y. Xu, Phys. Rev. Lett. **96**, 034301 (2006).

6
Self-assembled Nanostructures in Nature and Industry

Atoms and molecules are formed spontaneously when the proper ingredients (particles or atoms) and the proper conditions, such as temperature and pressure, are available.

Atoms on the whole have formed naturally. The nuclei of all atoms, especially those of many protons and neutrons, are believed to have formed in the interior of large stars, under conditions of extreme temperature and pressure. It is believed that the heavy atoms in our solar system have condensed from remains of exploded stars, called supernovas. The statement of Carl Sagan: "We are made of star-stuff" (in terms of the starting atoms) is well documented in decades of work of astronomers and astrophysicists. Self-assembly is definitely the preferred route to the formation of atoms.

Molecules, also, are mostly self-assembled and available in nature. (Certainly this statement is overwhelmingly true on a weight basis; consider the mass of the solar system, but there are now many key molecules that man has invented and produced even to the point of altering the ecosystem, e.g., DDT).

The entire array of biological molecules, from the point of view of science, has remarkably self-assembled. The task of the scientist has been to figure out how it happened! The learning has been rapid, especially after the genetic code represented by DNA was understood. Now there is a major industry devoted to modifying certain molecules of biology in the quest for new properties.

Many of the molecules of modern experience, however, have been synthesized for the first time by synthetic chemists or by enterprising chemical engineers making sensible substitutions. Butyl rubber, dynamite, sulfuric acid, organic dyes, many drugs, polystyrene and chlorofluorocarbons are examples of useful molecules not readily available in our environment. (The starting materials for many useful molecules, of course, are derived from nature, perhaps from the bark of a willow tree or the berry of a suitable plant.) Many other molecules, such as those found in fuels such kerosene and gasoline are fairly easily made by fractionating heavy crude oils from natural deposits on which our energy economy strongly depends.

The task of the synthetic chemist is to provide the proper conditions so that the desired molecule will form. It may take many steps in sequence before the final product can be formed with a large yield. But the formation is still self-assembly, from our point of view.

Nanophysics and Nanotechnology: An Introduction to Modern Concepts in Nanoscience. Second Edition.
Edward L. Wolf

ISBN: 3-527-40651-4

As elements of nanotechnology, molecules have some desirable characteristics. They are small enough, in many cases, to be unique and identical and free from errors, as are atoms. Quality control can be absolute in cases of small molecules such as water, diatomics, and even for relatively large molecules of high symmetry, such as the C_{60} Buckyball. (For all but the most esoteric practical purposes one can overlook the isotopic mass variability that does exist in most atoms and molecules.) This occurs in even the smallest atoms; atomic hydrogen comes in masses of 1,2,3, and helium comes in masses of 3 and 4, for example. The astounding success of biology as a nanotechnology has come with no need for isotope separation.

On the other hand, thermal excitations can lead to changes in shape of molecules on a rapid time scale. The thermal vibrations in cases of diatomic molecules and in the vibration of atom positions in extended solid are well understood and predictable in terms of thermal expansion and heat capacity, for example.

In contrast, devices built in the semiconductor technology are all different, in their complete atomic specifications, and only the essential parameters for the operation of the device are the realistic targets for quality control considerations. Devices of the most rudimentary sorts built up by assembly using scanning microscope tips are very challenging to build, and even more challenging to build in identical versions. The scanning tunneling microscope, as used in such cases as shown in Figure 3.8, does offer the opportunity to inspect in detail for errors and to correct them.

6.1
Carbon Atom $^{12}_{6}C\ 1s^2 2p^4$ (0.07 nm)

The carbon atom is the basic unit of organic chemistry, biological molecules, notably DNA, the code for all terrestrial life. The six protons and six neutrons of the $^{12}_{6}C$ nucleus, self-assembled either in the big bang or in the interior of a star (~10^6K). It is believed that supernova explosions have ejected nuclei and electrons into space, and these have gradually condensed in the atomic, molecular, liquid and solid matter of planets such as those in our solar system. The six electrons of the structure $1s^2 2s^2 2p^2$ can be expected to uniquely self-assemble about the $^{12}_{6}C$ nucleus on a very short time scale at almost any terrestrial temperature <10^3K. The radius of the 2p wavefunction is of the order of 0.1 nm. The outer 3s and 2p wavefunctions, amenable via hybridization to tetrahedral and other symmetries, easily bond into molecular and periodic solid structures, notably (metallic) graphite and (insulating) diamond, and, or course, organic chemistry and biology.

There is great flexibility in the way carbon bonds with itself and with other elements. The flexibility must be inherent in the wavefunctions as enumerated in Table 4.1, for these are well known to be the correct basis for all covalent bonding of carbon. In more detail, s and p_x combine for the sp bond of angle 180°, as in acetylene, C_2H_2; s, p_x, and p_y combine to form the trigonal sp^2 bond of angle 120°, as in ethylene C_2H_4; and s, p_x, p_y and p_z combine for the tetrahedral sp^3 bond of angle 109° in methane, CH_4, and diamond.

6.2 Methane CH_4, Ethane C_2H_6, and Octane C_8H_{18}

Methane is a covalent tetrahedral molecule involving the $2s2p^3$ electrons of carbon. The wavefunctions are listed in Table 4.1. In forming the bonds pointing to the corners of the tetrahedron, linear combinations of the 2s and 2p wavefunctions are selected. This is called hybridization of the underlying wavefunctions. The result is the filled $1s^2$ shell and the four orbitals pointing in tetrahedral directions, which are composed of linear combinations of 2s and 2p wavefunctions.

Ethane C_2H_6 involves the same carbon electrons. The carbons bond to each other and also to three hydrogens, again in the tetrahedral directions.

Octane C_8H_{18} has one linear structure and also a rolled-up structure. The bonding in the linear form can be thought of as ethane, with six additional $-CH_2$-groups inserted between the original two carbons. This structure is not at all rigid. Molecular simulations of octane at a temperature of 400 K are shown in Figure 6.1 after Drexler [1]. The time intervals between the succeeding conformations are about 10 ps. The energy estimates come from computer modeling, as carried out by Drexler [1]. The great variability in conformation of this molecule under thermal conditions qualifies the earlier statements to the effect that molecules, once formed, are unique.

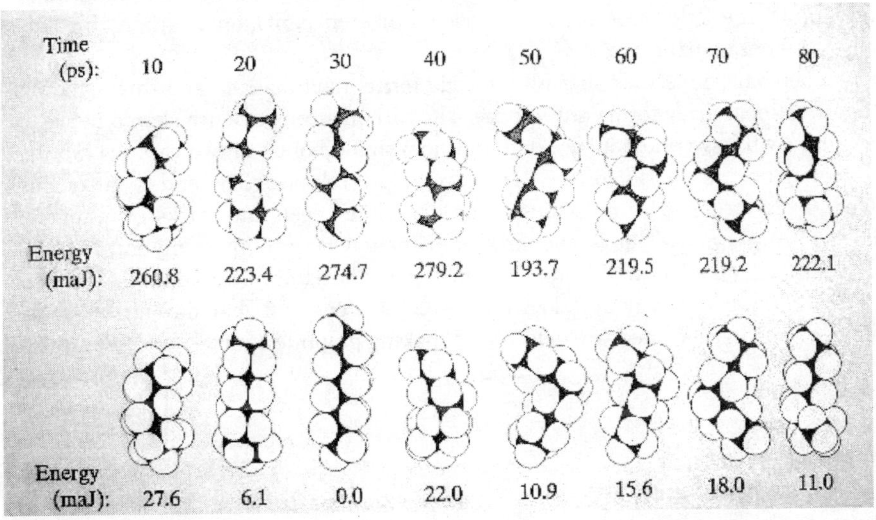

Figure 6.1 Modeled thermal conformation changes of octane C_8H_{18} at 400 K.
1 maJ = 10^{-21} J = 6.25 meV. From [1].

6.3
Ethylene C_2H_4, Benzene C_6H_6, and Acetylene C_2H_2

The carbon molecules considered so far have involved only single bonds, based simply on the covalent bonding found in molecular hydrogen that was discussed in Chapter 5, acting on wavefunctions as listed in Table 4.1. Carbon and other atoms also sometimes form double and triple covalent bonds.

The linear acetylene molecule C_2H_2 is formed by carbon atoms which each share three of their four valence electrons with each other, a structure called a triple bond. Thus, six valence electrons are concentrated in the region between the two carbons.

Ethylene C_2H_4 involves a double bond between the carbons, so each carbon devotes two electrons to its neighboring carbon and two electrons to hydrogens.

Benzene C_6H_6 involves a hexagonal ring of six carbon atoms. Each carbon is bonded to two carbon neighbors on the ring and to a hydrogen atom. The bonding between carbons is single on one side and double on the other, so that each carbon makes use of four valence electrons. The positions in the hexagonal ring at which the double- and single-carbon bonds appear, can be exchanged without changing the structure. This suggests that the carbon electrons involved in carbon–carbon bonding in benzene are delocalized and can move around the ring. There is evidence that this is the case from the large diamagnetic susceptibility of benzene. The six-fold carbon rings of benzene appear in graphite and its relatives, the carbon nanotubes and C_{60}. In the latter case there also five-membered pentagonal carbon rings, as well as the hexagonal rings.

Two benzene molecules sharing one side forms naphthalene, and three sharing sides in a linear array forms anthracene. The carbon atoms that are shared between two rings will have one double bond and now single bonds to two carbons. So the hydrogens on the interior of the 2D networks are released. The end point of this planar array of benzene molecules is the graphene sheet, and stacks of graphene sheets form graphite. This is one form of elemental carbon, involving only hexagonal rings with double and single carbon–carbon bonding. The bonding between sheets in graphite is weak and does not involve any covalent shared-electron bonds. The electrons in each sheet are delocalized making graphite a metal.

6.4
C_{60} Buckyball (~0.5 nm)

These molecules are empty spherical shells, containing exactly 60 carbon atoms in five- and six-membered rings. It is essentially a graphene sheet closed onto itself, with pentagons added to allow curvature. The pentagonal and hexagonal benzene rings are located in a similar fashion to pentagonal and hexagonal panels in a soccer ball.

These molecules are very stable and fully tie up the four valence electrons of each carbon atom. A simple model of this molecule is shown in Figure 6.2 [2].

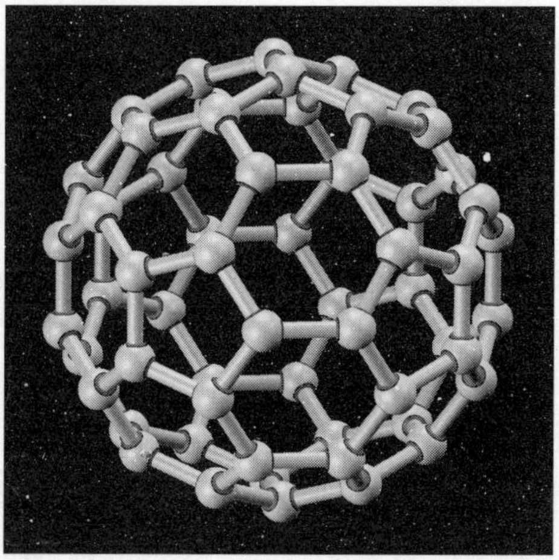

Figure 6.2 Carbon 60 molecule, in perspective model view. Each carbon atom participates in one double bond and in two single bonds. All valence electrons in each carbon are bonded. Each hexagonal benzene ring shares sides with three hexagonal and three pentagonal rings [2].

These molecules are present in tiny amounts in nature. They spontaneously form in oxygen-free atmospheres containing carbon atoms. There are reports of several other "magic numbered" molecules, notably C_{70}. It has been found possible to condense C_{60} molecules onto plane surfaces and to make superconducting compounds by adding stoichiometric amounts of alkali metals.

6.5 C_∞ Nanotube (~0.5 nm)

Carbon nanotubes are cylindrical shells made, in concept, by rolling graphene sheets, choosing an axis vector, and a radius, and closing them. The diameters are in the range of 0.5 nm, similar to the C_{60} molecule, but the length can be micrometers. Once the radius and pitch are established, the tube will grow by adding carbon to its length, since the carbon bonds around the cylinder are fully saturated and do not offer an opportunity for additional carbon atoms to attach. The choice of rolling axis relative to the hexagonal network of the graphene sheet (the pitch), and the radius of the closed cylinder, allows for different types of single-walled nanotubes, which vary from insulating to conducting. There are also varieties of multi-walled nanotubes. These materials are very light and very strong. They self-assemble under various controlled conditions: by carbon arc discharge, laser ablation, chemical vapor deposition based on hydrocarbon gases, and laser assisted catalytic growth. The smallest diameter nanotubes, 0.5 nm diameter, have been grown inside zeolite cavities.

Figure 6.3 after Baughman [3] shows, in Panels A, B, and C, single wall nanotubes of three different types. Panel D shows a scanning probe microscope picture of a chiral single-wall nanotube, and Panel C shows a TEM image of a nine-wall nanotube.

The electrical conductivity of properly oriented (armchair) nanotubes is extremely high, with the observation that carriers can pass through the micrometer length of

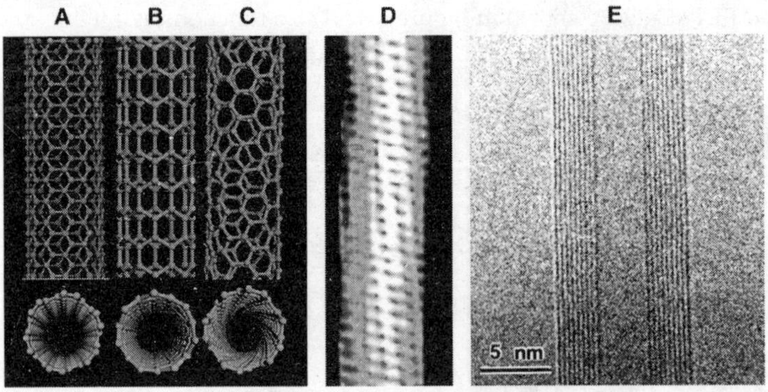

Figure 6.3 Images of carbon nanotubes [3]. (A), (B), and (C), respectively are armchair (metallic), zigzag (small bandgap) and chiral (semiconducting) nanotubes. The twist of the chiral nanotube is clearly evident in the lower panel (C), a perspective view along the tube axis. Panel (D) shows an STM image of a 1.3 nm diameter chiral nanotube. Panel (E) shows a TEM image of a nine-walled nanotube, a concentric cylindrically nested assembly, in which the binding between the adjoining nested tubes is very weak.

available tubes with zero scattering, and zero heat dissipation. It has been reported [4] that a 1.4 nm diameter single-wall nanotube transmits 6 μA, corresponding to > 10^8 A/cm^2 (see Figure 6.3). Nanotubes frequently appear to be perfectly periodic over their lengths, on the scale of μm. The only source of electrical scattering is thermal vibration of the atoms. Vibrations have small amplitude in a rigidly bonded structure. In consequence, the electrical mobility is exceptionally high, which makes the units desirable for the active regions of high-speed electron devices.

Carbon nanotubes have also extremely high thermal conductivity, measured to be 3000 W/mK. This value is comparable to the best thermal conductors, diamond and basal plane of graphite, 2000 W/mK. The high thermal conductivity is significant for nanotechnology, because removing heat is one of the central challenges in making electronic devices smaller. Their tensile strength and Young's moduli are high, comparable to the strongest known materials (silicon carbide nanorods) and between 20 and 60 times stronger than steel wire [3].

Figure 6.4 [4] shows a 1.4 nm diameter single-wall nanotube assembled into a *field-effect transistor* (FET). The nanotube itself is regarded as self-assembled, but its incorporation into the electrical device using the methods of semiconductor technology is not yet self-assembly.

A challenge to the use of nanotubes in such detailed electrical applications is, indeed, how to get the right kind of nanotube (radius, pitch, wall thickness) into exactly the right location to do the task. Sorting of nanotubes can be approached on the basis of their electrical conductivity, highest for the metallic armchair tubes [see Fig. 6.3 (A)]. Another general approach is to grow the tube from a catalyst placed in an optimum position.

The nanotube based FET structure shown in Figure 6.4 [4] also functions as a light source. Electrons and holes are provided in the central portion of the nanotube by the separate source and drain contacts. Some electrons and holes recombine radiatively, radiating bandgap light. The bandgap of the nanotube (evidently a semiconductor) of about $E_g = 0.75$ eV, gives light emission of wavelength $\lambda = hc/E_g$ about 1650 nm. This is in the infrared, but not so different from the wavelengths involved in fiber optic communications. The length of the active region of the carbon nanotube, modeled as being between 400 and 800 nm, is large enough that this structure does not behave as a quantum dot. That is, the wavelength of the emitted light is determined by the nanotube's bandgap and semiconductivity (related to its pitch angle), and not by the length of the active portion of the nanotube.

The authors [4] report that the devices were fabricated by dispersing nanotubes in solvent, spreading the solvent on the lightly thermally oxidized surface (150 nm of gate oxide) of the P^+ Si gate electrode. After the solvent had evaporated, the nanotubes used in device fabrication were selected in a microscope. The source and drain and other supporting structures then had to be deposited using photolithography and methods of silicon technology.

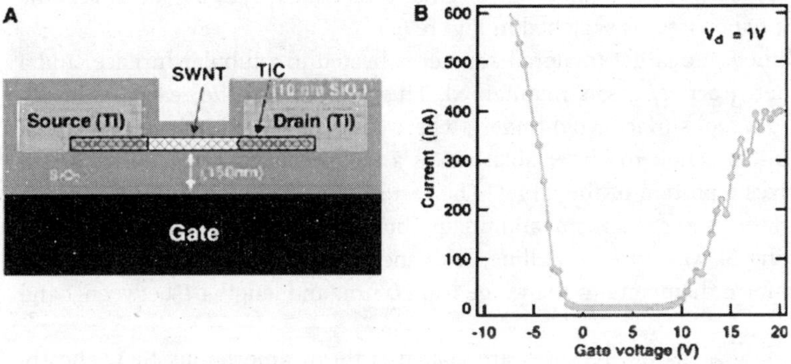

Figure 6.4 Incorporation of 1.4 nm diameter single-wall nanotube (SWNT) as active element of field-effect transistor [4]. Panel (A) shows 1.4 nm diameter SWNT placed 150 nm above Gate (P^+ Si), spaced by 150 nm thermally grown SiO_2 dielectric barrier layer. Source and drain contacts to the nanotube are thermally annealed Ti, which establish covalent bonding to the C nanotube by the formation (during annealing) of titanium carbide. Electrically, the Ti-TiC-C contacts are highly transmissive Schottky barriers. The whole device is covered with 10 nm of SiO_2. Panel (B) Current–gate voltage characteristics at fixed source–drain voltage. Current is small except in positive and negative bias ranges when the nanotube becomes metallic in hole or in electron conduction. Saturation current of 600 nA corresponds to current density of $> 10^8$ A/cm^2.

6.5.1
Si Nanowire (~5 nm)

CdS nanowires of diameters in the 80–200 nm range, said to be synthesized at 880 °C by laser-assisted catalytic growth, using Au as catalyst, were shown in Figure

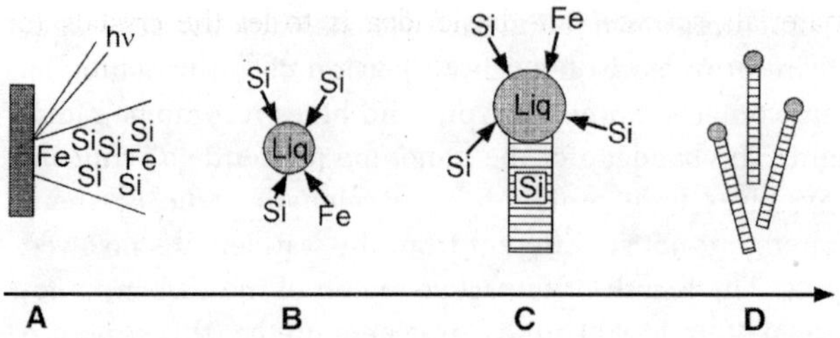

Figure 6.5 A laser ablation method for growing free-standing semiconductor nanowires of controlled diameters by metal nanocluster catalyzed growth [5]. (A) Laser ablation with photons of energy *hν* of the $Si_{1-x}Fe_x$ target creates a dense, hot vapor of Si and Fe species. (B) The hot vapor condenses into small clusters as the Si and Fe species cool through collisions with the buffer gas. The furnace temperature is controlled to maintain the Si–Fe nanoclusters in a liquid state. (C) Nanowire growth begins after the liquid becomes supersaturated in Si and continues as long as the Si–Fe nanoclusters remain in a liquid state and the Si reactant is available. (D) Growth terminates when the nanowire passes out of the hot reaction zone in the carrier gas flow onto a cold finger and the Si–Fe nanoclusters solidify. (Reprinted with permission from Science, copyright AAAS.)

5.18. The same fabrication method was originally described [5] as a general scheme for making Si nanowires, as sketched in Figure 6.5.

In this method, the target (material source) is heated in a tubular furnace, and a flow of an inert (carrier) gas is maintained. This flow carries the reactants downstream toward a cool surface (cold finger), where they condense, and are finally collected. The initial action in "laser ablation" is a focused laser beam pulse, which locally vaporizes a portion of the target. The resulting atoms and ions react among themselves in the inert gas stream, and finally condense on the cold finger.

The resulting Si wires are crystalline, with the [111] direction along the growth axis, with uniform diameters in the range 6 to 20 nm, and lengths L between 1 and 30 μm.

The crystalline Si nanowire cores are coated with an amorphous SiO_x sheath, related to residual oxygen in the buffer gas. It is stated that the Si/SiO_x interface is atomically sharp, and that the oxide can be removed using HF. The authors [5] demonstrate the method also for Ge nanowires, and suggest that the method is suitable also for SiC, GaAs, BN, and Bi_2Te_3. Nanowires (NW) are also known as "whiskers", sometimes observed to grow out from crystalline samples. It was reported that CdS nanowires have a [100] growth direction.

6.6
InAs Quantum Dot (~5 nm)

Quantum dots have been mentioned in Chapters 1 and 4, in connection with fluorescent markers used in biological research. These are nanometer-sized single crystals grown in solution, taking care to give a uniform size. This type of self-assembly, to produce large numbers of nearly equal-sized products, is an accomplishment of modern chemistry and materials science. The basic idea is to let the crystals (or amorphous spheres, in the case of polystyrene beads) grow until the solution is depleted of dissolved atoms feeding the growth. When the supply stops the crystal stops growing, at a well defined size. In this scenario, there is also a need to control the number of nucleation sites.

Quantum dots in the application mentioned in Chapters 1 and 4 are useful because their controlled size, L, in turn controls the wavelength of light that is emitted when the crystal is illuminated by shorter wavelength light, sufficient to generate electron–hole pairs across the bandgap of the semiconductor. These quantum dots (QDs) are grown in solution, then coated in elegant ways, first, to protect the semiconductor crystal from an aqueous environment, and then, further coated to attach the fluorescent marker to the desired type of tissue.

Semiconductor (InAs) quantum dots (QDs) have found a rather different application in certain types of lasers. Lasers (Light Amplification by Stimulated Emission of Radiation) are widely used as optimum light sources in telecommunications, as well as in consumer products such as CD players. This QD application in telecommunications, again, is based on the *tunability* of the emission spectrum with the size, L.

However, these quantum dots are incorporated into a laser hetero-junction, in which electrons are injected from one side of the junction and holes are injected from the other side. The quantum dot is thus *electrically* "pumped" into an unstable electronic state from which it will radiate a photon, exactly as in the previous application. This is an example of an "injection laser", which is industrially important. But the question is "How can one grow quantum dots inside a P–N heterojunction"?

The answer is in the realm of *epitaxial film growth* on a single crystal substrate in a molecular beam epitaxy (MBE) machine, which is based on an ultra-high-vacuum (UHV) chamber. One example [6] is provided by InAs quantum dots, which can be made to self-assemble within a strained InGaAs quantum-well structure, grown epitaxially on (100) GaAs.

Epitaxial growth means that the material that is deposited (under ultra-high vacuum conditions, requiring suitable deposition rates and suitable substrate temperature) continues to grow on the same crystalline lattice as its substrate. Epitaxy may be preserved even if the chemical composition is changed, at the expense of some strain in the grown layer. For example, adding some Indium (In) to the Gallium (Ga) beam in growing GaAs on a GaAs substrate can lead to a perfect (but slightly strained) crystalline layer of IndiumGalliumArsenide (InGaAs), if the In concentration is not too large.

The effect of this Indium (In) addition to the growing layer, electrically, is to locally reduce the bandgap, E_g. Alternating layers of GaAs and (heteroepitaxial) InGaAs create a *square wave modulation of the bandgap, and this is called a quantum well structure*. The electrons will segregate into the layers having smaller bandgap (lowered conduction band edge), and these electrons are now in 2D bands (Chapter 4) because the thickness of the InGaAs layers is small. (This scheme raises the mobility, since the carriers collect in regions away from the donor ion charges, which would scatter them, and reduce their mobility.)

The quantum dot aspect appears if the In concentration is raised, increasing the local strain. Finally, the epitaxy will no longer be tolerated, and locally there will be *inclusions of InAs* (QDs). These are the "self-assembled InAs quantum dots", whose size (and emission spectra) are controlled by the details of the deposition conditions in this industrial process.

This is an advanced use of MBE technology to produce quantum dots (local deviations from epitaxy), which, in this case, make lasers vastly more efficient! Why? Because QDs have a sharp narrow range of light emission, much like atoms [6].

6.7
AgBr Nanocrystal (0.1–2 μm)

Figure 6.6 Crystals contained in photographic emulsion are typically Silver Bromide (Image courtesy © Eastman Kodak Company. KODAK is a trademark.)

Crystals of AgBr and AgI are historically the basis for conventional silver halide photography. Figure 6.6 shows an array of such crystals in a tabular grain photographic emulsion. The range of sizes of such crystals is of the order of 0.1 μm – 2 μm. The process by which light absorption sensitizes such a crystal to allow it subsequently to be developed to a silver image (negative) is known in its outline but not in detail. It is likely to involve ruptures in the bonding of the AgBr crystal which lead to clusters of two or more Ag ions, which constitute a latent image, susceptible to the development process, in which relatively massive clusters of metallic silver are catalyzed by the latent image site.

Silver halide photography has historically accounted for a substantial fraction of the commercial sales of silver metal. The proper conditions for the AgBr crystals to self-assemble, and, certainly, the variety of subsequent steps in the category of "sensitization" to extend the spectral range of sensitivity, are known in the industry but not extensively in the open literature.

6.8
Fe_3O_4 Magnetite and Fe_3S_4 Greigite Nanoparticles in Magnetotactic Bacteria

Magnetotactic bacteria such as *Magnetospirillum magnetotacticum* are known to contain linear arrays of antiferromagnetic nanocrystals of Fe_3O_4 (magnetite) or Fe_3S_4 (greigite). The arrays are oriented parallel to the long direction of the bacteria. Both minerals are antiferromagnetic, which means that there are two magnetic sublattices of opposed but unequal magnetization, leading to a net magnetization in the direction of the stronger moment.

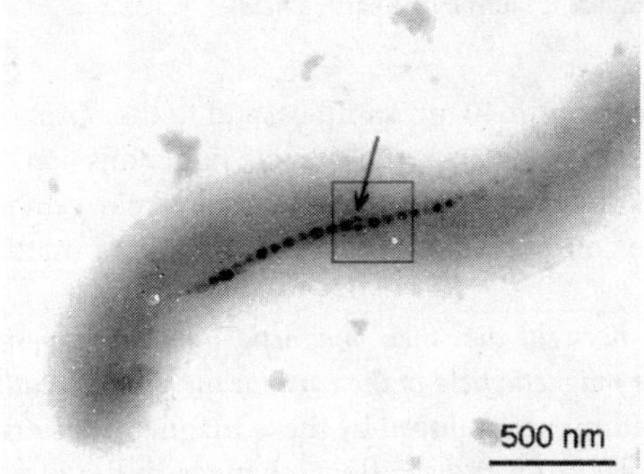

Figure 6.7 Image of magneto-tactic bacteria, showing linear array of magnetic crystals. The nanocrystals are greigite, an iron sulfide, dimensions on the order of 40 nm. These crystals are 40 nm and are yet each large enough to be antiferromagnets. Arrow in box locates a bi-crystal defect in the magnetic chain, see Figure 6.8 [7].

A typical bright field TEM observation [7] of a single cell of *M. magnetotacticum* (Figure 6.7) reveals a "magnetite chain 1200 nm long containing 22 crystals that have average length and separation of about 45 nm and about 9.5 nm, respectively".

The bacteria were cultured cells of Magnetospirillum magnetotacticum strain MS-1 and were deposited onto carbon-coated grids for use in the transmission electron microscope. The image was obtained at 200 keV with a field-emission-gun TEM. It was deduced that the linear array of the tiny magnets had a larger effect on the direction of the overall magnetization than did any magnetic anisotropy within individual grains. It is known in these systems that the crystals are combinations of (111) octahedron and (100) cube forms, and that the (111) magnetic easy axes of the crystals are primarily parallel to the chain axis [7].

Magnetic studies indicate that each crystal forms a single magnetic domain, and that the magnetization M in the magnetite is consistent with the bulk value, 0.603 T. The magnetic moment of one typical bacterium at room temperature was measured to be 5×10^{-16} Am^2. Measurements of the coercive field have given values in the range 300 Oersted to 450 Oersted. There are indications that smaller nanocrystals in the chains are sometimes superparamagnetic, rather than antiferromagnetic, and that their magnetization is stabilized by the magnetic field from larger neighboring antiferromagnetic nanocrystals in the chain. This may apply to the bi-crystal defect arrowed in the figure. These magnetic minerals evidently grow in the cell from iron-containing chemicals dissolved in the fluid inside the bacterium.

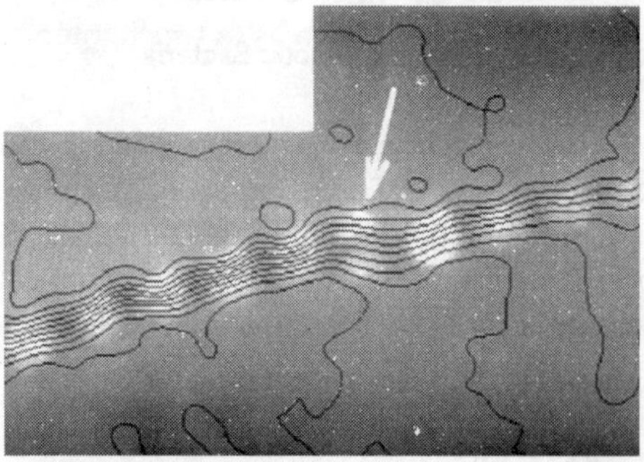

Figure 6.8 Magnetic field mapping [7] of central boxed region of magnetic bacterium image, shown in Figure 6.7 (see box and arrow). More precisely [7], this image is an off-axis electron hologram obtained in field free conditions. The apparent magnetic field lines thread along the array of antiferromagnetic nanocrystals, including the bi-crystal defect shown by the arrow, proving that the array acts as an extended bar magnet (believed to orient the bacterium in the earth's field).

These observations, of antiferromagnetism in 40 nm samples, add to the observations on the band gap, size-quantized excitation states, and tunable light emission of semiconductor quantum dots; as evidence that even very small samples may exhibit sophisticated phenomena for which the conceptual machinery of condensed matter physics is fruitful and necessary.

It is further believed that when these bacteria die, their magnetic moments remain intact and retain the orientation set by the magnetic field of the earth at the time of death. Over eons of time, extensive layers of sediment produced by these magnetic bacteria have accumulated. In taking core samples from the sea floor, changes in the direction of their magnetization as a function of depth have been recorded. It is believed that these data provide a record over long periods of time revealing the history of the orientation of the earth's magnetic field.

It is also known that magnetic crystals below 100 nm in size, occur in organisms in many biological phyla. Since these nanocrystalline magnets are apparently widespread in such primitive forms of life, one suspects that they may be present, at least in vestigial forms, in higher forms of life. If one does not know where to look in a large organism for such a tiny features, they may be hard to locate!

6.9
Self-assembled Monolayers on Au and Other Smooth Surfaces

Gold is a nonreactive metal, but sulfur-containing groups called thiols do have a tendency to weakly bond to a clean gold surface. Organic chain molecules, such as octanes, terminated in thiol groups, have been found to self-assemble on gold surfaces. They exhibit ordered arrays defined by their interactions with the surface, and with adjacent organic chain molecules.

The type of ordering that occurs in a given situation, when the thermal energy is low enough, is that which provides the lowest energy. For a two-dimensional array of spheres on a structureless surface, this is likely to be hexagonal close packing, so that each member of the array has the most benefit from the attraction it feels from its neighbors.

If there is a different kind of order in the substrate on which the array is deposited, for example, cubic ordering, then there will be a competition as to which interaction has the most to offer. If the spheres in the example have a strong attraction to the underlying cubic structure, then the adsorbed spheres may take on a cubic order to match. In cases of such competition, a remarkable degree of complexity can occur. Some of this has been documented in STM studies. It is even possible, in the case of two competing types of order, that domains of each type may coexist.

A good example [8], of an ordered monolayer, emphasized the locking of the monolayer structure to the structure of the underlying gold surface. This is not always the case. In the next example, there is no relation between the monolayer and the surface on which it rests.

Another example of an ordered self-assembled monolayer, shown in Figure 6.9 [9], is a hexagonal close-packed array of C_{60} molecules on an alkylthiol organic surface. The alkylthiol layer is similar to the layer in the previous example.

These authors emphasize that the hcp close packed layer, in which the C_{60} nearest-neighbor distance is 1.0 nm, is *unrelated* in its orientation to the rectangular lattice of the alkylthiols, which has dimensions 0.9994 nm by 0.8655 nm.

The measurements are made at 5 K in a cryogenic scanning tunneling microscope. Panel (a) shows a 3.5 nm × 35 nm scan of the field of C_{60} molecules. In this image, the structure of the Buckyball molecules can clearly be seen. The characteristic features are three hexagonal benzene rings which are clearly resolved by the STM image at 5 K. Further, and striking, is the fact that these three hexagonal features are seen to be locked in identical positions in all of the C_{60} molecules.

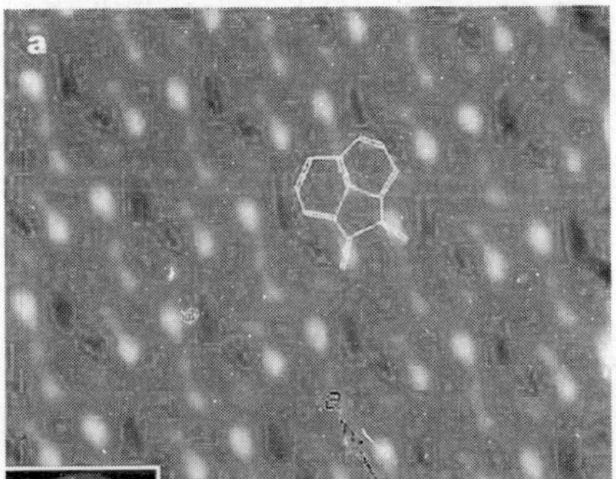

Figure 6.9 Self-assembled monolayer of identically oriented C_{60} molecules observed by cryogenic scanning tunneling microscope [9] Panel (a) reveals C_{60} array, individual molecules are identified by three resolved hexagonal rings (see white overlay in upper central field). Inset shows theory modeled image based on bond orientation in white overlay and axis identification acb in lower right. Panel (b) shows bond model of C_{60}: (left) oriented as it appears in image and (right) as it would appear if *c* axis is perpendicular to the monolayer.

This gives an appreciation that molecules are absolutely identical, and in this case, identically oriented. The inset to the panel (a) shows a theoretical model of the STM image expected if the single and double bonds are oriented as shown (in white) in the center of the field. The agreement is seen to be excellent.

The bond models of C_{60} seen in panel (b) are oriented to match the images in panel (a).

The authors mention that in measurements at 77 K and at room temperature the images of the individual C_{60} are featureless. From this they infer that the molecules are in some state of rotation at 77 K and higher, but have locked into position at 5 K.

References

[1] From *Nanosystems: Molecular Machinery, Manufacturing & Computation* by K. Eric Drexler. Copyright © 1992 by Wiley Publishing, Inc. All rights reserved. Reproduced here by permission of the publisher.

[2] Courtesy Joseph W. Lauher, Chemistry Department, SUNY Stony Brook.

[3] Panels A, B, C and E reprinted with permission from R. H. Baughman, A. A. Zakhidov and W. A. de Heer, Science **297**, 787 (2002). Copyright 2002 AAAS. Panel D reprinted with permission from Nature: J.W.G. Wildoer *et al.*, Nature **391**, 59 (1998). Copyright 1998, Macmillan Publishers Ltd. and with permission from C. Dekker, Delft University of Technology.

[4] Reprinted with permission from J. A. Misewich, R. Martel, Ph. Avouris, J. C. Tsang, S. Heinze, and J. Tersoff, Science **300**, 783–786 (2003). Copyright 2003 AAAS.

[5] A. M. Morales and C. M. Lieber, Science **279**, 208 (1998).

[6] G. T. Liu, A. Stintz, H. Li, K. J. Malloy, and L. F. Lester, Electron. Lett. **35**, 1163 (1999).

[7] Reprinted with permission from R. E. Dunin-Borkowski, M. R. McCartney, R. B. Frankel, D. A. Bazylinski, M. Posfal, and P. R. Buseck, Science **282**, 1868–1870 (1998). Copyright 1998 AAAS.

[8] G. E. Poirier, Chem. Rev. **97**, 1117 (1997).

[9] Reprinted with permission from Nature: J. G. Hou, Y. Jinlong, W. Haiqian, L. Qunxiang, Z. Changgan, Y. Lanfeng, W. Bing, D. M. Chen, and Z. Qingshi, Nature **409**, 304 (2001). Copyright 2001, Macmillan Publishers Ltd.

7
Physics-based Experimental Approaches to Nanofabrication and Nanotechnology

The previous chapters have outlined the rules of nanophysics that apply to nanometer-sized objects. These rules describe chemical matter and biological matter. The previous chapters have also given some examples of nanometer-scale machines and devices achieved by nature and by modern science and technology.

It appears, to date, that the most advanced machines and devices on the nanometer scale are those produced in nature. The basic reason for this assessment is that nature seems uniquely to have developed an information base and locally-available instructions for assembly and replication of complex nanometer scale machines and devices.

On the other hand, the accomplishments of modern science and technology in two particular areas: the silicon and computer technology, and the scanning probe microscope approaches, exceed those available in nature.

Computation at gigahertz rates is not available in nature. Nor does nature allow individual atoms to be placed on selected sites to represent bits of information. That these accomplishments have come from essentially physics-related science and technology is notable, and encouraging to further efforts based on nano-physics.

On the other hand, it is not unlikely that the most immediately productive approaches to biotechnology may, indeed, be based on modifying natural (biological, genetic) processes to achieve practical goals, in the same spirit that fluorine substitutions for hydrogen have produced fluids, such as freon, more suitable for refrigeration. Genetically engineered corn is available, for example, which effectively generates its own insecticide. This is a contemporary accomplishment of biotechnology.

Of the many promising nanophysical approaches to nanotechnology, at present, a leading pair are: 1) Extensions of the hugely successful silicon/computer technology, and 2): Extensions of the methods of scanning probe microscopy. The scanning probe microscopy approaches, indeed, are closest, in a literal sense, to the imagined "molecular assembler" devices. There have been many examples, in the figures of this book, of the contemporary *silicon technology*, and of *scanning probe technology*,

These two areas have stimulated development of a large set of fabrication and measurement tools, which, in total or individually, may be useful as starting points for the invention of new forms of nanotechnology.

The continued advance of the silicon technology, it is argued, may lead to computers with capacity greatly exceeding the capacity of the human brain. A huge achieve-

Nanophysics and Nanotechnology: An Introduction to Modern Concepts in Nanoscience. Second Edition.
Edward L. Wolf

ISBN: 3-527-40651-4

ment (which has largely compensated its creators). *Is it also a threat?* When will these hugely intelligent computers and computer networks insist on a life of their own?

On the scanning probe technology side, advances may conceivably lead to realization of the "self-replicating molecular assembler", an imagined universal tool able to build "anything", quickly, on an atom-by-atom basis. What about this? If it is possible, how can one be sure that the "assemblers" might not run amok and use up all the chemicals in the world as we know it? Annihilate human life?

7.1
Silicon Technology: the INTEL-IBM Approach to Nanotechnology

Silicon microelectronics technology is a large topic. Only the barest outline of this material can be covered here. Several figures have already illustrated ingenious adaptations of silicon technology methods to the fabrication of hybrid structures; see, for example, Figure 6.4.

7.1.1
Patterning, Masks, and Photolithography

An essential element in silicon technology is preparing a sequence of patterns, miniaturizing the patterns, and transferring the miniaturized patterns sequentially onto the silicon chip to define the appropriate layers of metal (patterned into wires), insulation, and P and N regions needed for fabrication of logic and memory cells.

Patterning Deposition Masks

The patterning process starts with an image, for example, a detailed wiring diagram. This pattern has to be transformed from black and white (historically on drafting paper), to open and closed areas on a metal mask. The metal mask, which may be in the form of metal layers deposited on glass, would be open where the wire is to be placed. The size scale of the pattern may be 0.1 m on the drafting paper, but has to be reduced, say, to 1 μm on the contact mask. The full technology deals with single crystal wafers of, say, 6 inches diameter, containing many identical chips. There is a step-and-repeat aspect in the real process that we will pass over.

To utilize the mask, an initial layer of photoresist polymer, such as polymethylmethacrylate (PMMA), is uniformly applied to the chip surface. Then the mask is brought close to the coated surface, and light is exposed onto the photoresist through the mask. In this case the photoresist polymer would be of the type that is weakened by light exposure, so that the resist can be removed chemically, leaving the regions of the chip to be metallized ready for deposit of a metal layer, while the rest of the surface is protected by the photoresist. The whole surface is then coated with the metal for the wiring. Finally, in a chemical step, the remaining photoresist is removed, leaving the chip surface with a deposited wiring network.

Masking Layers to Limit Etching

Masking layers are used for other purposes than to define a deposit. Masking is also used to delineate areas of the surface that are to be removed by chemical etching or by dry etching. Dry etching is done in a plasma which may include reactive molecules such as CHF_3. (In many cases, these masking layers are also initially patterned by the basic deposition patterning outlined above.)

Masks to protect a surface are sometimes simply grown silicon dioxide SiO_2. Heating the silicon in oxygen or steam grows silicon dioxide layers of a wide range of thickness, up to many μm.

Alternatively, chemical vapor deposition (CVD) may be used to deposit a masking layer, such as Si_3N_4. In a CVD reactor the sample surface is exposed to chemical vapors, or carrier gases, which contain the chemical ingredients of the desired deposit. With control of the temperature of the surface and the flow rates and compositions of the carrier gases, CVD is a way to rapidly grow layers of a variety of compositions. For example, carrier gases silane and hydrogen at 630 °C can be used to deposit polycrystalline silicon (polysilicon), which may also be doped N-type with addition of a phosphorus-bearing chemical vapor to the reactor. Phosphosilicate glass can be deposited by CVD methods.

Metal layers are also commonly used as masks against etching, for example, gold and chromium. Extremely thin layers of Cr or Ti are also used in advance of a thicker layer of a conducting metal, to improve adhesion of the electrode to the silicon.

7.1.2 Etching Silicon

Wet Etches

Etching away silicon is accomplished by wet etches, which include acids HF (which removes oxide quickly and Si slowly), nitric and acetic acids. These are isotropic etchants.

An isotropic etch, when exposed to a surface protected by a planar oxide or metal masking layer, will have a certain tendency to undercut the masking layer, producing in a limiting case a spherical cavity. The shape of the etch pit can be controlled, of course, by limiting the etching time.

Still in the category of wet etches, the alkali KOH is an example of an anisotropic etchant, which removes material from certain crystallographic planes faster than from others. KOH can be used on Si to produce triangular etch pits exposing (111) planes on a (100) surface.

Dry Etches

Dry etches, protected by masking layers such as oxides or metals, are accomplished in a vacuum chamber or gas reactor. In this case energetic ions, produced in a plasma discharge, hit the surface and remove material. (Removal of material under bombardment by ions is called sputtering, and is also adapted as an important method for thin-film deposition.) The ions may be non-reactive, for example, argon ions, or reactive, such as CHF_3 ions. These etch processes are somewhat directional, because the momentum carried by the ions in the plasma is predominantly perpendicular to the surface.

Directional dry etching can be carried out using an ion-beam gun, which is an accessory to a vacuum chamber, and generates a directed beam of ions, which can be reactive or non-reactive. In this way, deep trenches with vertical walls can be produced using the conventional metal- or oxide-etch masks.

7.1.3
Defining Highly Conducting Electrode Regions

Transistors, bipolar and FET, require precisely defined localized regions of N- and P-type conductivity. An NPN transistor might be formed by diffusing two closely spaced N regions into a P-type surface. A field effect transistor (FET) requires highly conductive source and drain contacts to be closely located and connected by the channel, which is a high-mobility conductive region. The conductivity of the channel of an FET is modulated by an electric field provided by voltage bias on the gate electrode, which is insulated from the channel by a thin layer of oxide (the gate oxide).

The traditional means of producing N or N^+ and P or P^+ regions is by thermal diffusion from a dopant-containing layer positioned appropriately. (Doping can also be introduced by ion-implantation, see below.) The chemical dopant is contained in a glass, so that when the heating is carried out, the dopant does not simply evaporate. Typical dopants for N-type silicon are phosphorus and arsenic; for P-type silicon, boron is typical. Silicon is a robust (although very easily oxidized) material and temperatures above 600 °C are needed to produce significant diffusion. At these temperatures photoresist layers cannot be used, but oxide or metal masking layers are still appropriate.

Ion implantation is another method for introducing impurities and is described in the next section.

7.1.4
Methods of Deposition of Metal and Insulating Films

Evaporation

Metal deposition is traditionally accomplished by evaporation. Evaporation of a soft metal like Al can be accomplished by heating the metal in an electrically heated boat of Ta until it melts and until its vapor pressure becomes high enough to eject energetic atoms of Al toward the surface to be coated. This process requires a high vacuum system, typically evacuated with an oil diffusion pump, an ion pump, or a cryopump.

For more refractory metals, like Cr or Ti, evaporation is done using an electron-beam gun (e-gun). In an e-gun an electron beam is focused on the surface of the metal charge, locally heating it to a very high temperature, leading to evaporation as in the case of Al. Evaporation, including e-gun evaporation usually cannot be used to deposit compounds, because the different elements in the compound typically have different vapor pressures.

Sputtering

Sputtering is another basic technique for making deposits of almost any metal. In sputtering a discharge of non-reactive ions, such as argon, is created, which fall on the target and break loose atoms (and atom clusters) which are collected on the surface to be coated. Sputtering can also be used to make deposits of compounds. Different sputtering rates for the different elements of a compound may require a target composition somewhat different from that of the desired deposit, but this method for depositing compounds is much more suitable than evaporation.

Chemical Vapor Deposition (CVD)

Chemical vapor deposition is also a general method for depositing metals as well as compounds. In a CVD reactor, carrier gases, containing the elements of the desired compound, flow over the surface to be coated. This surface is heated to a suitable temperature to allow decomposition of the carrier gas and to allow mobility of the deposited atoms or molecules on the surface. Mobility of deposited atoms helps to produce a highly ordered crystalline deposit, desired for improved electrical properties. In some cases laser light may be used as an assist (LA-) in the decomposition of the carrier gas. Processes called LACVD are used to produce nanowires, for example.

Laser Ablation

Laser ablation is a general method for depositing even a complicated compound from a target to a nearby surface in vacuum. Focused laser pulses fall upon the target, completely vaporizing a small region of material. The ablated material is collected on the nearby surface, and has essentially the same composition as the target. This is a slow and expensive process by its nature, but finds application in high-value products where a complicated compound, such as a high-temperature superconductor (HTS) with many different elements, must be deposited with accurate stoichiometry. Incorporation of HTS films as wavelength filters in cellular telephone transmitter output, can multiply the number of mobile telephone subscribers who can be served by a given cellular phone transmitting station, for example.

Molecular Beam Epitaxy

Finally, molecular beam epitaxy (MBE) is an elegant process for depositing atomically perfect (epitaxial) layers of compound semiconductors, such as GaAs and InGaAs. This is an expensive and slow process, but is still industrially competitive in cases where the product is valuable.

In MBE, separate ovens with careful temperature regulation are provided for each element in the desired deposit. Temperature control is needed to control the vapor pressure and hence the deposition rate. The rates of each element have to precisely match the stoichiometry. For InGaAs, e.g., there will be three ovens. Deposition rate monitors and shutters will be available for each oven. With these elaborate precautions, in ultra high vacuum, epitaxial layers and heteroepitaxial layers can be deposited.

For example, multiple quantum well devices are produced using MBE. These methods were described in Chapter 6 in connection with "quantum-dots in a quantum well" lasers, valuable for the improvements and economies they may offer in optical-fiber communications or even in lasers for portable CD players, where battery life may be an issue.

Ion Implantation

High-energy ion beams can be used to implant chemical impurities such as donor and acceptors. An ion implanter is an electrostatic accelerator capable of giving 50–100 keV energy to ions, which allows them to penetrate deeply, on a micrometer scale, into a single crystal of silicon. The violent ion implantation produces undesirable damage to the host crystal, which would drastically lower the mobility for current flow, but annealing (heating) can remove the implantation damage. The temperature required for annealing (removing) implantation damage is lower than that required to diffuse impurities into the silicon from a surface layer of, e.g., phosphosilicate glass. For a large-scale contemporary silicon device facility, ion implantation is a standard tool.

Ion implantation can be used to create a *buried* layer of impurity, including the buried oxide layer (BOX) used in prototype FET transistors on silicon. In this case oxygen ions are highly accelerated and a heavy dose is buried at micrometer depths into the silicon crystal. After annealing, the result can be a nearly perfect silicon layer (Silicon on Insulator, or SOI) of perhaps 4 – 8 nm depth above a thick SiO_2 layer, above the single crystal Si wafer.

7.2
Lateral Resolution (Linewidths) Limited by Wavelength of Light, now 65 nm

7.2.1
Optical and X-ray Lithography

The widths of the finest (narrowest) wires that can be put on a chip, in optical lithography, are limited by optical diffraction. The light that exposes the photoresist comes through an aperture defined by the metal walls of the mask. Light, as a wave, has a tendency to bend around a corner, and this means that the sharpness of the image of the mask on the photoresist, is limited by λ, the light wavelength.

At present, this limits wires to being at least 65 nm. There is no easy way around this limitation, because higher energy (shorter wavelength) light sources are inconvenient and expensive. In addition, the higher energy light, which can extend to the x-ray range, then also has a tendency to penetrate the conventional metal mask. Metal wires, if they were made much narrower than 65 nm, still retain their essential characteristics as conductors, although the resistance per unit length increases. This incremental effect will increase the amount of heat per unit length, on a constant current basis.

65 nm is a lot bigger than the 1.4 nm diameter nanotube of Figure 6.3, which carried a current of 0.6 μA. On the other hand, there is no known prospect at present for creating the correct number of nanotubes of the right radius and pitch, and to place them appropriately on a 6 inch diameter wafer.

These issues, among others, definitely limit the advance of the traditional "INTEL-IBM" silicon technology. This performance is conventionally compared, in the industry, to the numerology of "Moore's Law" (see Figure 1.2). The current performance and projections for semiconductor technology are summarized on a timely basis at "The International Technology Roadmap for Semiconductors" [1], available on the Web at public.itrs.net/.

7.2.2
Electron-beam Lithography

Photolithography is the conventional (hugely successful) process for make computer chips *in bulk*. Finer lines can be produced by an alternative process, electron beam lithography, but this process is not compatible with the mass production available with photolithography, where the masks are simultaneously illuminated with light, exposing many chips at once.

In the e-beam process, an electron beam is focused to a point, as in an electron microscope, to individually expose the photoresist. The diameter of an electron beam can be as small as the resolution of an electron microscope, approaching 0.1 nm. However, the electron beam has to be scanned, as in a TV picture tube, to illuminate every area of photoresist. *This is a slow process.* Consequently, e-beam lithography is used in research, but not in production, in applications to produce features of dimensions smaller than 180 nm.

7.3
Sacrificial Layers, Suspended Bridges, Single-electron Transistors

We have already seen, in connection with our discussion of the Casimir force, in Chapter 5, a structure in which a flat plate was freely suspended above a silicon surface by two torsion fibers. This structure (see Figures 5.5 and 5.6) is an example of one in which a "sacrificial layer" (in this case silicon dioxide) was initially grown.

A deposited *polysilicon layer*, suitably patterned, was grown above the sacrificial layer, also providing supports to the underlying crystal. Then the SiO_2 layer was etched away, leaving the suspended "paddle" free to rotate in response to electric and Casimir forces. The method is seen in Figure 7.1 [2].

Referring to Figure 7.1, panel (1) shows e-beam exposure of a PMMA (polymethylmethacrylate) resist, which has been uniformly applied above the three layers comprized by the base silicon, the sacrificial oxide layer, and a polysilicon deposited layer. Panel (2) shows a patterned metal etch mask on the polysilicon layer. Panel (3) shows liftoff of the unexposed resist layer, followed in panel (4) by dry etch to delineate the suspended bridge. In panel (5) a wet etch, such as HF, undercuts the oxide beneath the suspended layer, leaving it freely suspended. Finally, in panel (6), metal electrodes are applied.

This scheme is used in the class of MEMS (NEMS) devices Micro (Nano)Electro-MechanicalSystems, which include the accelerometers used in automobile collision

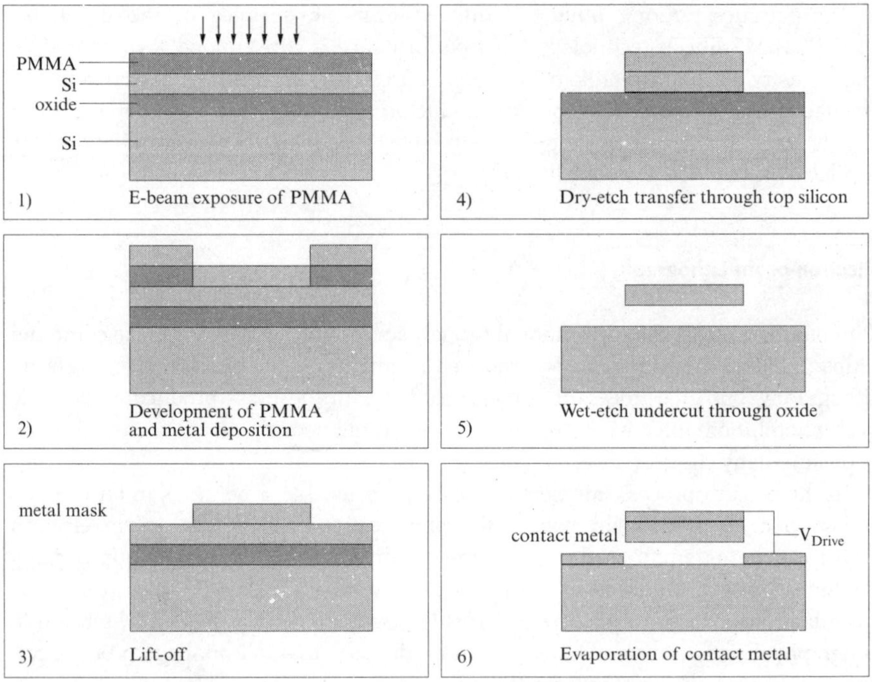

Figure 7.1 Steps in formation of a suspended silicon plate above a silicon chip [after 2].

sensors. It is also used in the recent "Millipede" prototype data storage device, where more than a thousand individual atomic force microscope sensors are produced, using the silicon photolithographic process illustrated in Figure 7.1, on a single chip [3].

Another example of this elegant version of the silicon process is illustrated in Figure 7.2, in which a single-electron-transistor (SET) is used to sense the 0.1 GHz vibrations of a suspended nanometer-scale beam [4,5].

The structure in Figure 7.2 panel (b) is patterned and etched from a GaAs-GaAlAs single crystal heterostructure. The vibrating beam is a doubly clamped single crystal of GaAs with dimensions 3 μm × 250 nm wide × 200 nm thick, with a resonant frequency of 116 MHz. The beam (the gate electrode of the SET) was patterned [5] using electron beam lithography and a combination of reactive ion etching (to make the vertical cuts) and dilute HF etching (to remove the sacrificial GaAlAs layer originally under the beam). The beam, which serves as the gate electrode, is capacitively coupled to the Al island, the closest approach being 250 nm, with a gate-island capacitance of 0.13 fF.

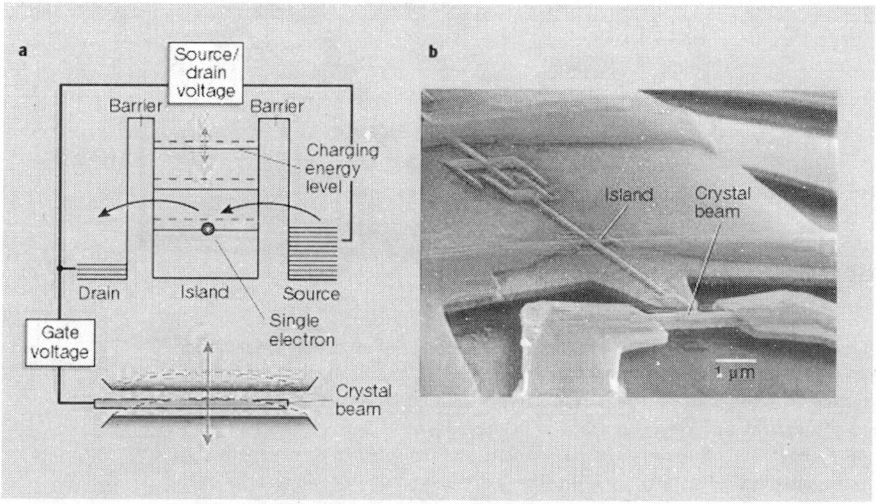

Figure 7.2 Single-electron-transistor used to sense vibration of freely suspended crystal beam [4,5]. In panel (b) a "crystal beam" is suspended and free to vibrate at 116 MHz. It is metallized and acts as gate electrode in the FET transistor, denoted "island". The "island" is connected to source and drain electrodes (right and left) by Al-AlOx-Al tunnel barrier contacts. At fixed gate bias and fixed source-drain voltage, motion of the gate electrode induces charge on the island, and sensitively controls the source-drain current. Panel (a): Schematic diagram of electrical operation of the device. A motion of the bar by 2 fm (about the radius of an atomic nucleus) can be detected with 1 Hz bandwidth at 30 mK.

7.4
What is the Future of Silicon Computer Technology?

The ongoing improvement of silicon computer chips continues, with innovative new device designs appearing frequently. In addition to the International Technology Roadmap for Silicon [1], a further analysis of the trends is given by Meindl *et al.* [6]. One of the recent trends has been to build the basic field effect transistor, FET, or MOSFET (Metal Oxide Silicon Field Effect Transistor), on a silicon chip with a "buried oxide" layer (BOX). The effect of the BOX is intended to reduce the thickness of the active channel, which carries the current when the device is in the "ON" (conductive) condition. One advantage of the reduced channel thickness is to reduce capacitance between the source and drain. Reducing this capacitance allows the device to be switched more rapidly, with less energy wasted per switching cycle.

Figure 7.3 shows a *prototype* field-effect transistor described by IBM [7] as the "world's smallest working transistor". It is an example of "silicon on insulator" technology, with a silicon thickness of 4–8 nm. The silicon channel length, L_{Gate}, is 6 nm. Some features of this device are produced lithographically using light of 248 nm wavelength.

It seems clear that the gate, because of its extremely short length, can only be produced by e-beam lithography, an unconventional and slow method. Further, there is

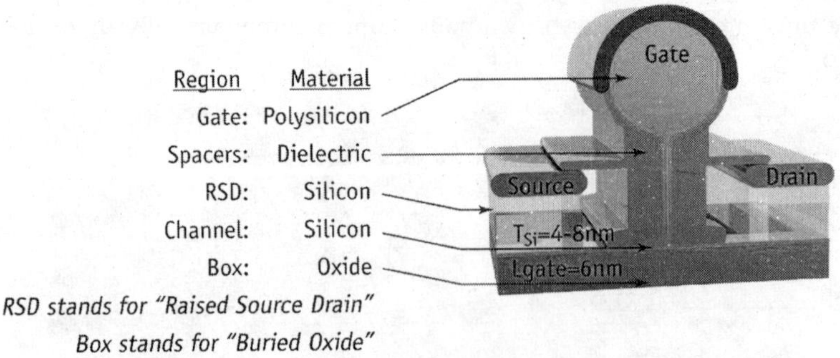

Figure 7.3 Schematic indication of MOSFET *prototype* of extremely small dimensions, based on the conventional silicon technology [7]. Looking at this Figure, starting from the bottom: The bottom layer is "Buried Oxide" (Box), and at the top of this thick oxide layer, in a small thickness T_{Si} of 4 – 8 nm, is the active silicon channel. The other notably small dimension in this device is the length L_{Gate} of the gate electrode, stated as 6 nm. This extremely small dimension implies the use of e-beam lithography, certainly a complicating factor.

no mention of the gate oxide, which is one of the most difficult features of the basic MOSFET design in connection with maintaining the trend as set by Moore's Law. The problem that occurs with scaling is that the gate oxide becomes so thin that it is no longer a barrier, as required in the framework of an FET device. Barriers so thin as to allow nanophysical tunneling, inevitably the consequence of reducing the size scale, are incompatible with conventional FET operation.

An innovative feature in this design is the thinning of the active silicon channel by means of the buried oxide layer, or "Box". One way to form the Box is to heavily implant oxygen ions deeply into a silicon crystal, followed by annealing. The details (of the implantation and anneal) control the channel thickness T_{Si} and determine whether the active channel is depleted or partially depleted.

It appears that the buried oxide method is preferred over the more obvious approach of depositing the thin active silicon layer onto a sapphire chip. (Sapphire is a good substrate because of its extremely high thermal conductivity.) The quality, including mobility and reproducibility, of the resulting thinned silicon layer is presumably higher in the Box approach.

This energy-saving approach (and the publicized use of copper, rather than aluminum wiring, for its modest improvement in conductivity) is a hint that the heat generated in the chip is recognized as a limiting factor in the scaling of the silicon technology.

7.5
Heat Dissipation and the RSFQ Technology

The generation of heat in silicon chips is a problem at present, and is a problem that will get more difficult the more devices there are on a chip [1]. Large installations, as well as the uncomfortably warm contemporary laptop computers, already exhibit this problem.

There is a complete, *but problematic*, potential solution to this heating problem, suitable for large installations where expensive air conditioning is already required. The potential solution is the *Rapid Single Flux Quantum (RSFQ) superconducting technology*. Although the full details of this technology, purely based on nanophysical phenomena of superconductivity, are beyond the scope of this book, it is a wonderful example of the notion that the new laws of nanophysics may offer concepts for different, and possibly superior, devices.

An example, the performance of an RSFQ counting circuit [8] is shown in Figure 7.4.

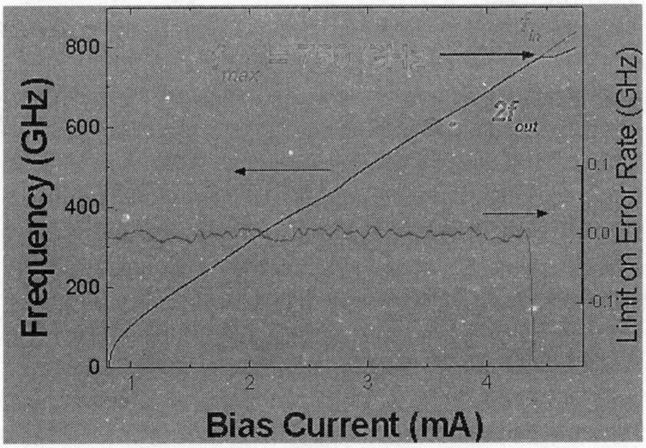

Figure 7.4 Frequency divider [8] operates at 750 GHz, using RSFQ superconducting junctions. This device produces half-frequency output. Twice the output frequency, $2f_{out}$ is shown superimposed on f_{in}, both quantities referenced to left-hand scale. Bias current (abscissa) in mA does not signify heat, because the resistance is zero [8].

The problem, for the RSFQ technology, is that refrigeration is needed to maintain RSFQ devices at the temperature of liquid helium. Liquid helium is a coolant widely used in Magnetic Resonance Imaging (MRI) medical installations. In addition to saving energy, the RSFQ technology allows *extremely fast clock speeds*, as seen in Figure 7.4.

The heat dissipation that occurs during the RSFQ switching events is described [8,9] as on the order of 10^{-18} J/bit, leading to an overall power dissipation estimated as 10^{-5} that of an equivalent silicon device. These devices are fast, as can be seen by the graph. These RSFQ technology devices have been considered for analog-to-digital converters, e.g., in missiles where *fast* data acquisition is urgently desired and refrigeration is already in place to cool infrared detectors.

The RSFQ technology is relatively simple, tolerating larger linewidths, but suffers from a primitive (underfunded) infrastructure for fabrication and testing. The active devices are niobium-aluminum, niobium (tri-layer) [10,11] Josephson tunneling junctions, which must operate well below the superconducting transition temperature of niobium, 9.2 K. Compact low power refrigerators that approach this temperature are becoming available.

The energy consumption and the cooling power (note that these quantities are additive, in terms of operating costs) needed for a large computer in this technology are less, by large factors, than for an equivalent silicon (CMOS) machine. The savings in power and floor-space are significant. It can be argued that the size of the overall machine would be much smaller because the design would not have to have the fins, open channels, fans, and air conditioners that are normal in large silicon computer/server-farm installations, simply to keep the devices from dangerously overheating, and to carry away the heat.

A benchmark in large-scale computing is the Petaflop computer, which carries out 10^{15} floating point operations per second. A typical modern Pentium-chip laptop computer operating at 1.3 GHz has a power supply rated at 54 Watts. A rough conversion of 1.3 GHz clock speed to 1 GFlop, would imply that a million laptop computers, if suitably coupled, might constitute a parallel architecture Petaflop computer. (This practical approach to making supercomputers has been implemented, as mentioned in the press, including use of Pentium computers and also SONY PlaystationII devices).

On this basis, a Petaflop computer would consume 54 MW of power, or 1.7×10^{15} *J/year. At a conventional rate of 10 cents per KWh, the annual power operating cost would be $47 M,* not to mention the cost of the air-conditioning.

Binyuk and Likharev [9] offer the following projected comparison of the specifications for a Petaflop computer (10^{15} operations/s) in RSFQ technology, operating at 5 K, vs. silicon CMOS technology operating somewhat above room temperature:

In CMOS, one might have [9] $10^4 - 10^5$ advanced CMOS silicon chips, total power about 10 MW. Neglecting the air-conditioning needs, the operating cost would be around $8.7 M/y. The linear size of the computer might be 30 m, leading to a relativity time delay $L/c = 300$ ns. Such a large signal-propagation time is a drawback for a fast computer. This is a huge installation, on the scale of those in existence at major stock exchanges.

In RSFQ, one might have [9] 500 RSFQ logic chips and 2000 fast superconductor memory chips, at an estimated power at 5 K of about 1000 W. The linear size of the computer CPU might be 1 m, corresponding to a speed-of-light time delay of 20 ns. To achieve the reliable cooling of this 1 m "core" to about 5 K, a closed-cycle (recycling) liquid helium refrigerator would be required, at a total power of about 300 KW. Again neglecting air-conditioning, the annual cost is estimated as $263 000 per year.

This is a noticeably lower figure than $8.7 M/y estimated for the CMOS facility, to which one should add much larger costs in floor space and air-conditioning.

The ever-increasing energy cost of the pervasive silicon information technology was hinted at in the media discussion of the US power blackout of August 14, 2003. There were statements to the effect that the US needs an (electric) power grid that "matches the needs of information technology". "Server farms", typically housed in urban buildings with windows blocked in, that have appeared in information technology, are users of vast amounts of electrical power.

This costly power shows up as heat in the environment, cost on the balance sheet, and even represents a potential security threat, in its dependence on large amounts of (demonstrably insecure) electrical power.

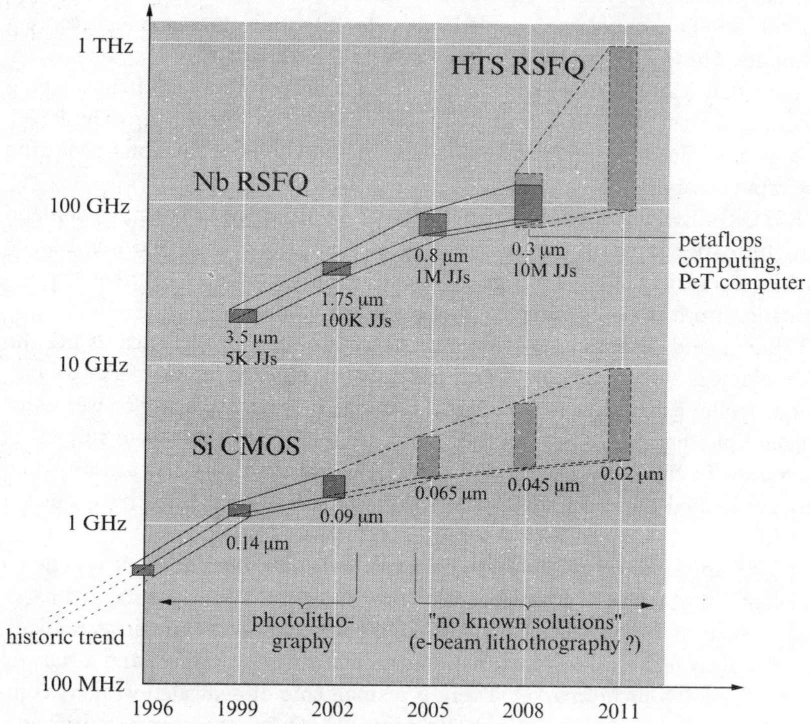

Figure 7.5 Estimated clock rates in RSFQ (upper) and conventional Si CMOS (lower) projected to the year 2011 [after 9].

Figure 7.5, adapted from Binyuk *et al.* [9], shows comparative projections of clock rate in RSFQ and CMOS computer technology. In the upper portion, "Nb RSFQ" refers to technology based on Nb-Al-Nb trilayer [10,11] Josephson junctions (JJ's), which operate below 9.2 K. In the upper right, "petaflops computation" and "PeT Computer" ("Personal Teraflop Computer", see [9]), with 10 million JJ's on a single chip, are predicated on an assumed 0.3 μm Nb technology. Note that 0.3 μm exceeds today's silicon technology linewidths, and is likely achievable with funding. In the far upper right, "HTS RSFQ", is a rough guess of RSFQ performance if the high-temperature superconductors could be brought into play.

The HTS superconductors are notoriously difficult to incorporate into electronic device technology, except in the simplest single-layer applications, such as high-Q radio-frequency filters for cellular telephone transmitting stations. For HTS superconductors the required refrigeration to 77 K is less costly.

In the lower portions of Figure 7.5 relating to the silicon technology [1], note that the linewidths by 2003, have already come down to the limit of photolithography. The very uncertain projections labeled "no known solutions" reflect that no mass production technology has appeared to replace the photolithography. These projections are based on the Silicon Technology Roadmap [1].

Other large issues relevant to Figure 7.5 are the greatly reduced power consumption in the RSFQ technology, coupled with its extremely inconvenient cooling requirements. These issues have been discussed in the text above.

It is clear that RSFQ technology will never find its way into wristwatches, laptop computers and most other basic computing venues for the silicon chip. The RSFQ technology may find commercial application in high-end installations requiring massive rapid computation.

The RSFQ technology offers a chance for much reduced power consumption and better performance in computing technology, at the major cost of learning how to make and use a variety of good refrigerators and making a large adjustment in mental orientation from silicon to superconductors like niobium.

To challenge and surpass a pervasive and successful technology, such as the silicon technology, is seldom achieved, and certainly not quickly. Jet airliners now outnumber propeller driven planes, and digital cameras may be displacing conventional silver halide photography. On the other hand, in spite of the complete success of nuclear power in the US Navy for submarines and aircraft carriers, private enterprise in the United States regulatory environment seems unable to make nuclear power work.

The RSFQ approach [9] offers an opportunity for innovators and entrepreneurs who are conversant with the advanced, extensive, and almost wholly untapped nanophysical science of superconductivity. To launch this radical computer technology, which is certainly technically sound; requires a largely new workforce and a start-up approach aimed at a niche market. There is a small core of dedicated workers competent in superconducting technology, including RSFQ, in laboratories at universities and a few leading defense contractors.

7.6
Scanning Probe (Machine) Methods: One Atom at a Time

The scanning tunneling microscope (STM) and the atomic force microscope (AFM) are similar single-point probes of a surface. Both have a sharp tip which can be moved in an x-y raster scan over a 2D surface and both have a servo loop which raises the height of the tip to maintain the current (or the surface-tip force) to be constant, thereby providing a topograph of the surface. Either instrument can be imagined as capable of carrying an atom or molecule to a point on the surface and leaving it to provide one step toward fabricating a surface structure. Either can be imagined as a starting point for the "machine assembly" of a nanostructure, atom by atom. The essential ideas of the two instruments are presented in Figure 7.6 [12].

The STM is capable of atomic resolution, as is evident from many images, including those in Figure 7.7. The basic idea is that the tunneling occurs when the wavefunction of an atom on the tip overlaps the wavefunction of an atom on the surface. In ideal cases these wavefunctions decay exponentially with characteristic distances on the order of the Bohr radius, around 0.1 nm. With such a short decay length, a single tip atom may carry almost all of the observed current. If the wavefunction on that atom is, e.g., a directed bonding orbital, the spatial resolution may be remarkably high, as is sometimes observed.

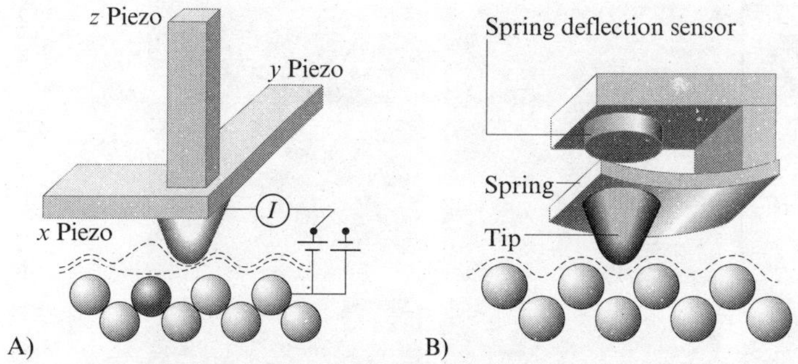

Figure 7.6 Schematic diagrams of (A) scanning tunneling microscope (STM) and (B) atomic force microscope (AFM). For STM the tunnel current *I* is maintained constant by tip height, controlled by *z*-piezo. For AFM the force (spring deflection) is maintained constant [after 12].

The maximum scanning and sampling rates are determined in large part by the resonant frequencies of the support structure, the tip and the *xyz* piezo-electric element on which it is mounted. The upper range of such frequencies in the best instruments is about 1 MHz.

The AFM (see panel (B) in Figure 7.6) is more complicated. Sensing of the force between tip and surface is by deflection of the cantilever on which the tip is mounted. The sensing in modern AFM instruments is by deflection of a light beam, focused on the upper surface of the cantilever; or by changes in the resistance of the specially designed cantilever with deflection (piezoresistance). The earliest instruments sensed the deflection of the cantilever by changes in the tunneling current at fixed bias between the tip-holder and an upper spring-deflection-sensor electrode (see panel (B)).

The force between the tip and sample is attractive at large spacing (van der Waals regime) and repulsive at small spacing (exclusion principle overlap). Atomic imaging has been reported with the AFM, but is more difficult than in the STM.

7.7
Scanning Tunneling Microscope (STM) as Prototype Molecular Assembler

7.7.1
Moving Au Atoms, Making Surface Molecules

A recent state-of-the-art STM study, which combines atomic resolution imaging and nanofabrication using an STM, is indicated in Figures 7.7, 7.8 and 7.9. [13].

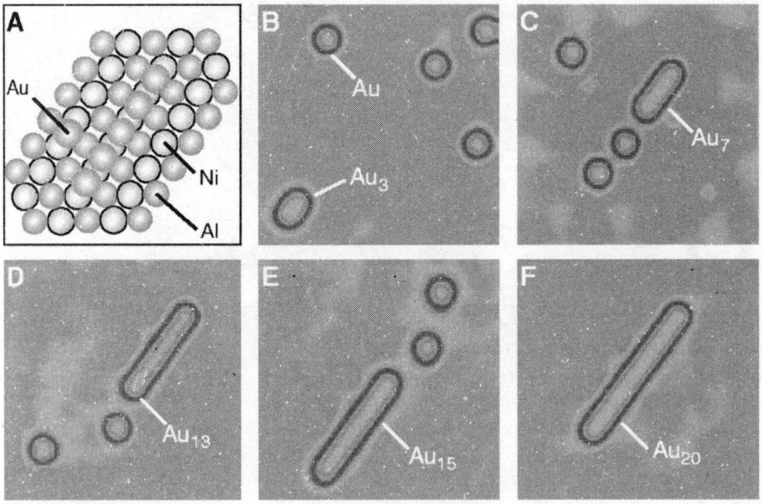

Figure 7.7 STM fabrication and atomic imaging of gold Au atoms and molecules Au_n on a surface of (110) NiAl [13]. The NiAl (110) surface has dimpled channels matching the 0.3 nm spacing of Au atoms in Au molecules. A 20-atom Au_{20} wire is formed and its density of states matched to a 1D band model (see Chapter 4).

Figure 7.7, panel (A) schematically shows the (110) surface of the NiAl crystal on which Au atoms were randomly deposited. (B) 9.5 nm by 9.5 nm image shows 4 Au atoms and an Au_3 cluster observed in STM at 12 K, using the constant current mode at 1 nA. The sample is at positive 2.1 V bias, which means that electrons from the tip tunnel to empty states in the Au 2.1 eV above the Fermi level. The STM tip is used to pull the atoms into line in panels (C,D,E,F), leading to the 20-atom nanowire.

Pulling the Au atoms is accomplished by decreasing the tunneling resistance to 150 KOhms, compared to 1 GOhm for observation. The higher electric field, at reduced resistance, induces a larger electric dipole in the Au atom, which is then attracted to the tip, where the field is strongest. Pulling is a balance of forces between the Au atom and the tip and surface. The Au atom is neither attached to, nor lifted above, the surface by the tip.

Figures 7.8 and 7.9 show tunneling spectroscopy measurements and their analysis along the 20-atom nanowire. These measurements, and their analysis, are useful in preparing for a further step in nanofabrication [14], in which these authors react the Au nanowire with a CO molecule. The bonding between the Au wire and the CO depends upon the electronic structure of the Au wire, which is analyzed here.

In the theory of electron tunneling, the conductance, dI/dV, is a measure of the density of states per unit energy [15]. Positive bias voltage here signifies the energy relative to the Fermi energy of the empty state in the wire that is being filled by the tunneling electron.

The dI/dV–V curves taken near the middle of the wire show a large peak at 0.78 eV. This peak is easiest to think of as the singularity in the density of states at the bottom of a 1D band, as described in Chapter 4. It also represents the $n = 1$ state of the electron trapped in the 1D potential well of length L. The bottom of the empty band is determined as $E_o = 0.68$ eV above the Fermi energy. The details of the dI/dV–V curve reflect the fact that the system is a 1D electron trap of length L. Thus, the wavefunctions of the states that the electrons can tunnel into are of the form $\sin(n\pi x/L)$, where x is measured from one end of the wire. It is determined that the principal values of n involved (see Figure 7.9, panel (A)) are 1, 4, 5, 6, and 8. The width of the levels is about 0.35 eV, corresponding to a lifetime of about 10^{-15} s.

It is seen that the states overlap in energy, so that in the density of states measurement at 0.78 V (Figure 7.9, panel (C), upper curve), the electrons fall partially into

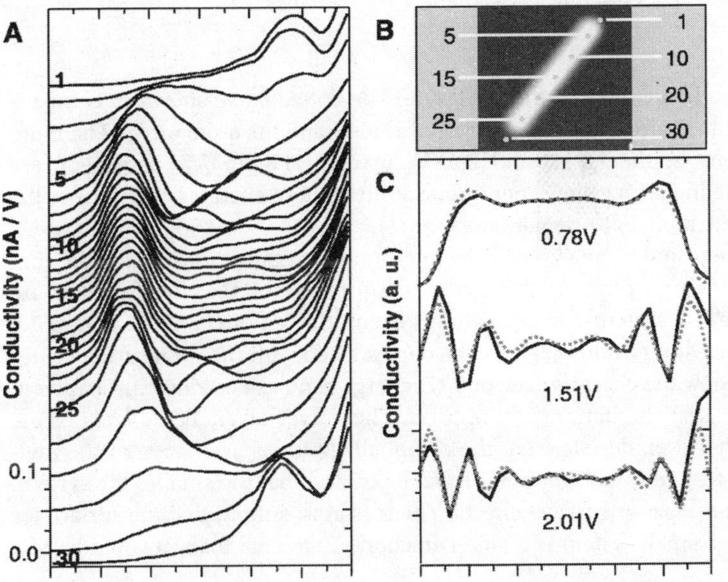

Figure 7.8 STM measurement of DOS vs. bias V and position along the Au_{20} wire. V denotes energy of empty state where tunneling occurs, and dI/dV (nA/V) measures DOS. Panel (A): curves of dI/dV at locations seen in panel (B). Panel (C) shows DOS vs. x along the wire, taken at three different energies [13].

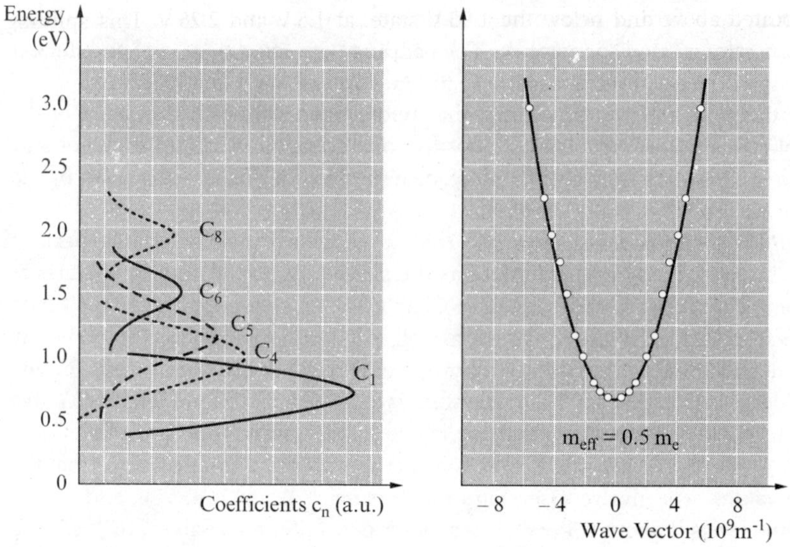

Figure 7.9 Determination of effective mass of electrons in the one-dimensional nanowire [after 13]. The effective for electrons in the nanowire is determined to be about 0.5, from the *E*–*k* curve shown in Figure 7.9, panel (B).

the $n = 1$ state and partially into the $n = 4$ state. The dotted curve shown in comparison to the solid measured curve is based on this idea, and fits quite nicely. The more rapid oscillations in the 1.51 V and 2.01 V curves of Figure 7.7, panel (C), are explained by the higher n values appropriate at these these energies, and the oscillatory nature of the probability density $\sin^2(n\pi x/L)$.

This experiment and its analysis can be interpreted as again showing that the concepts of solid state nanophysics, such as the energy band, and the effective mass, are useful even for tiny systems. In a 20-atom system, the set of curves in Figure 7.8 panel (A) is interesting in this regard. The curves taken with the tip near the center of the sample show the characteristic of a 1D energy band as described in Chapter 4, with a huge peak at the bottom of the band. The curves taken at the ends of the 20-atom wire, however, deviate from the essentially bulk behavior seen in the middle. This illustrates an important point in nanophysics, that the smaller the system, the larger are the edge- or surface- effects. The reason is, simply, that the surface (or end or edge) of a small system is a larger fraction of the total than it would be of a large system.

One aspect of the situation not shown in these figures is observation of the bonding and antibonding states of the Au_2 molecule [13]. The single gold atom (not shown) shows a large density-of-states peak at 1.95 V, which is understood as a prominent empty state in the atomic density of states of an Au atom on the NiAl surface. However, in the Au_2 molecule, as in the H_2 molecule studied in Chapter 4, this state splits into two states. For the Au_2 molecule, the two states [13] are symmetrically located above and below the 1.95 V state, at 1.5 V and 2.25 V. This splitting indicates a strong interaction between the adjacent atoms at their spacing of about 0.289nm, spacing similar to the spacing of Au atoms in bulk gold.

From the point of view of an STM as a prototype "Molecular Assembler", this work is also instructive. The 20-atom gold wire is assembled by the tip, *after the atoms have been randomly deposited* on the carefully prepared surface at a very low temperature. The low temperature is necessary to keep the atoms from randomly diffusing about, because their trapping into the dimples of the NiAl surface is weak. The tip does not *carry* the atoms into the nanostructure to be assembled, but merely nudges existing atoms into position. The process illustrated in Figure 7.7 can be assumed to be extremely slow.

7.7.2
Assembling Organic Molecules with an STM

Chemical reactions have been carried out on single crystal surfaces of copper by Hla *et al.* [16], as illustrated in Figure 7.10 [17]. The upper and lower panels show molecules lined up along a monoatomic step on the surface, which offers stronger bonding than does a flat surface. The starting molecules are iodine-substituted benzene rings, namely iodobenzene. In the first inducement by the tip, a voltage pulse breaks off the iodine, leaving a phenyl ring. The tip is then used to pull the iodine out of the way, and then to pull two phenyl rings close together. The second inducement provides energy to allow the two adjacent phenyl rings to fuse and form $C_{12}H_{10}$, the desired product. Finally, in panel (f), pulling on one end of the biphenyl is demonstrated to move the whole molecule.

This experiment shows a new use of the STM tip, in adding energy to either dissociate a molecule into parts or to help two molecules to react, by providing energy to overcome a barrier to the reaction. In this experiment a specially prepared Cu surface was used, one cut at a slight angle to form monoatomic steps (shown as shaded horizontal strips in the upper and lower sections of Figure 7.10). This step-edge location offers stronger binding to an atom or molecule, because two sides of the atom are affected, rather than one, on a single flat surface.

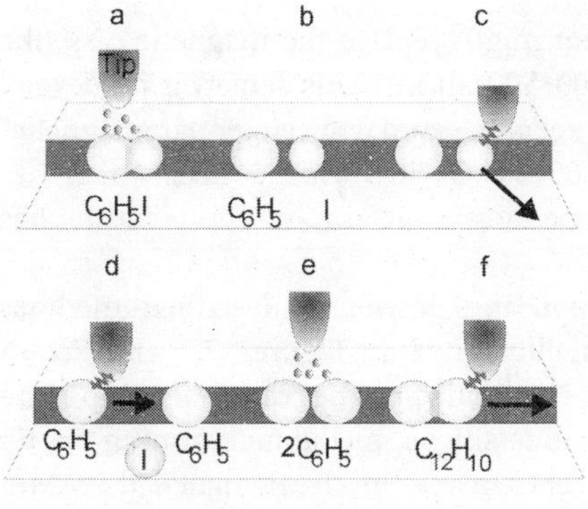

Figure 7.10 Schematic of single molecule STM organic chemistry on a copper surface [16,17]. The reaction is the copper-catalyzed transformation of iodobenzene C_6H_5I to make $C_{12}H_{10}$. This reaction is known as the Ullman reaction, here carried out at 20 K, with the usual thermal activation energy replaced by pulses of energy from the tip. The first pulse is applied in panel (a) to dissociate the I, leaving a phenyl group C_6H_5. The second pulse is applied in panel (e) to cause two adjacent phenyl molecules to bind. In panels c, d, and f the tip is used to move a molecule.

7.8
Atomic Force Microscope (AFM) Arrays

The atomic force microscope is useful for making marks on surfaces, which can be described as nanolithography, as well as for measuring topography of surfaces that need not be conductive. An example of an AFM image is Figure 4.5 revealing several InP nanowires on a flat surface. The AFM tip can be used to push and drag objects like nanowires or nanotubes around once they have been deposited on a flat surface. The AFM has been adapted to work in liquid environments in biological research. As a prototype for the "molecular assembler" the AFM tip is quite similar to the STM tip.

In neither case does the single tip provide an advanced and controllable facility for picking up, manipulating, or specifically releasing an atom or molecule. To assemble a structure using LEGO toys, the hand, under a watchful eye, selects a part, grasps that piece, orients the piece, inserts the piece into the desired site, and then releases the piece. *These functions are essentially unavailable with STM and AFM tips.* The smallest radius that can be expected for an AFM tip, mass produced in silicon technology, is likely to be 20 nm [3].

Additional functions that an STM/AFM tip system can provide include a combination of current and force sensing, as would be available from a piezoresistive, electrically conductive cantilever. An added ability [3] is for tip heating and tip temperature sensing. In [3] the tip temperature was used as a means of reading digital information imprinted into the surface. However, there is no indication in the present literature suggesting that there will ever be a means of grasping and releasing an atom or molecule from an STM or AFM tip.

7.8.1
Cantilever Arrays by Photolithography

The Millipede project [3], carried out by IBM at its Zurich Laboratory, provides a 32×32 array (1024 tips) contained in an area of $0.3\,\text{mm} \times 0.3\,\text{mm}$, and capable of writing and reading 10 Gb of information with access times on the order of 25 ms. The information is recorded by thermal-mechanical indentation of 40 nm size pits in a re-writable thin PMMA polymer film.

Conceived of as a data storage device that might replace the magnetic disk, the Millipede achieves data storage density of 400–500 Gb/in^2. This density is far beyond present disk memory and also beyond the reliably estimated "super-paramagnetic" limit for magnetic disk storage, around 60–100 Gb/in^2. This device, which has access times similar to disk memory and operates in air at room temperature; has been built, tested, and quite fully described.

The technology of Millipede builds upon the previously described methods for creating suspended beams or bridges, illustrated in Figures 7.1 and 7.2. A 32×32 field of partially detached cantilevers is provided on the "cantilever chip". The cantilevers are constructed entirely of crystalline silicon, including scanning tip, tip heater, and temperature sensor. The cantilever chip and the closely matched PMMA-coated "recording chip" are fabricated separately, and are subsequently mounted together using a delicate x,y,z drive which moves *all cantilevers in unison* relative to the recording chip. The cantilevers are not individually moved in x,y, or z. The writing and sensing of information by each cantilever are accomplished, respectively, with the tip heated under indentation force, and, for reading, by the individual temperature sensor.

The dimension of a cantilever cell is 92 μm × 92 μm. Each cantilever has two legs, which join at the tip-tip-heater assembly. The legs are doped silicon: 50 μm long, 10 μm wide and 0.5 μm thick. The resonant frequency of each cantilever is 200 kHz and the force constant is 1 N/m. Each of the 1024 lithographically patterned Si tips is 1.7 μm in height, triangular in shape with an apex sharpened to a 20 nm radius using oxidation sharpening [3]. The tip heater time constant is a few μs, which is expected to give a rate of 100 kHz for reading and writing. The tips are heated to 500–700 °C for writing (making an indentation in the PMMA polymer), and to around 350 °C for thermally sensing the presence of a data point (indentation). Provisions are available for erasing data points.

7.8.2
Nanofabrication with an AFM

The Millipede device described above is intended as a memory device, but we can think of its writing function as a step of nanofabrication or molecular assembly.

Ignoring all questions about the facility of a single 20 nm radius tip to do the traditional functions in mechanical assembly, let us focus on how fast the assembly could possibly be carried out.

The Millipede stores 10 Gb = 10^{10} bits with an access time of 25 ms. We can optimistically translate that as a rate of 4×10^{11} steps/second, which is a great overestimate. (It is a great overestimate because each tip can access only 1, certainly not all, of its own $10^{10}/1032 = 10^7$ sites in 25 ms.) Suppose the goal is to fabricate a diamond structure of mass 12 g. The molar mass of carbon is 12, so the product will contain Avogadro's number, 6.02×10^{23}, of atoms. Even at the overstated rate of 4×10^{11} steps/second, the required time will be 1.5×10^{12} seconds. *This is a very long time, about 48 000 years.*

To fabricate the 12 gram carbon structure in 1 s would require 1.5×10^{12} Millipede devices working at once, again overlooking other absurdities of this idea.

For a second estimate, assume a single assembler tip at the highest possible rate, about 1 GHz. *At that rate, the time for a single tip to process 12 grams of carbon would be 6×10^{14} seconds, 19 million years.*

These rates are so slow, using even wildly optimistic estimates on tip rates and capabilities, as to make clear that *bulk matter cannot be assembled in an atom-by-atom mode* except by huge numbers of "molecular assemblers" working in parallel. The assemblers themselves must be very small and probably have to grow in numbers by reproducing themselves, as do the cells in biology.

This is the point in the argument for molecular assembly where the further assertion is made that the assemblers self-replicate. Following an illustrative analysis provided by Smalley [18], assume that the "molecular assembler" can reproduce itself at the same rate of 10^9 atoms per second. Let's further assume that the "molecular assembler" comprises only 10^9 atoms. On this assumption it will ideally take one second to make one copy of itself; if each copy reproduces itself in turn, then in 60 seconds the number of "molecular assemblers" will be 2^{60}, or about 10^{18}! (This is pure fantasy, but it is similar to the fantasy upon which the nanotechnology myth of "gray goo" rests.) If all of these huge numbers of assemblers subsequently work at the same rate, 10^9/s, then carbon can be assembled at a rate of 10^{27} atoms/s! For comparison, this is $10^{27}/(6.02\times10^{23}) = 1661$ Moles/s =20 Kg/s for molar mass 12 g for carbon. This is a large and industrially relevant rate, but it is *impossible*, since it is based on the flawed assumption of the molecular assembler.

This is not at all to say that arrays of AFM tips are useless for data storage, quite the contrary. It is clear that the Millipede device exceeds the projected limiting performance of any possible magnetic disk memory. The dollar value of disk memory in today's commerce is vast. Thus, a new memory technology is anticipated, based on nanophysics and nanotechnogy.

7.8.3 Imaging a Single Electron Spin by a Magnetic-resonance AFM

The MRI (magnetic resonance imaging) apparatus of medical application requires approximately 10^{12} nuclear spins or about 10^7 electron spins. A primary limiting factor is detection of the rotating magnetic field of the ensemble of resonant spins (see equations (4.85)–(4.90)). This occurs by the Faraday induction effect from time-varying magnetic flux cutting a macroscopic receiver coil, producing an oscillating voltage (see equation (4.20a)).

The "Magnetic Resonance Force Microscope" (MRFM) uses a vastly more sensitive detection scheme, indicated [19] in Figure 7.11, capable of detecting (with considerable effort) the magnetic resonance of a single electron spin. The detection is based on the atomic force microscope (Section 7.6) modified with a magnetic tip (see equations (4.85)–(4.94)).

The 150 nm-wide SmCo magnet is attached to the end of the AFM cantilever, driven into oscillation of fixed amplitude x_{peak} (16 nm) at its resonant frequency f_c (about 5.5 kHz) along the x-direction (see Figure 7.11). The spring constant k of the cantilever (see Section 2.1) is about 0.11 mN/m, and its tip is about 125 nm above the sample surface. The sum of the tip's dipole magnetic field and the static magnetic field ($B_z \approx 30$ mT) add up to about 106 mT, corresponding, for resonance ($h\nu = 2\mu_B B$), to an applied microwave resonant magnetic field supplied at $\nu = 2.96$ GHz for an electron spin. Thus the "resonant slice" is a bowl-shaped surface that extends about 250 nm below the tip into the sample. For a spin slightly displaced from the tip's center position in the x-direction, resonance (with resulting forces) occurs twice per cycle of tip oscillation, providing a cyclic force on the tip. This force is in the range of 10^{-18} N, if the magnetic field gradient $G = \partial B/\partial x$ is in the range of one oersted (10^{-4} T) per nm. The x-component of the cyclic force on the tip can be modeled as a small change in the cantilever force constant, causing a

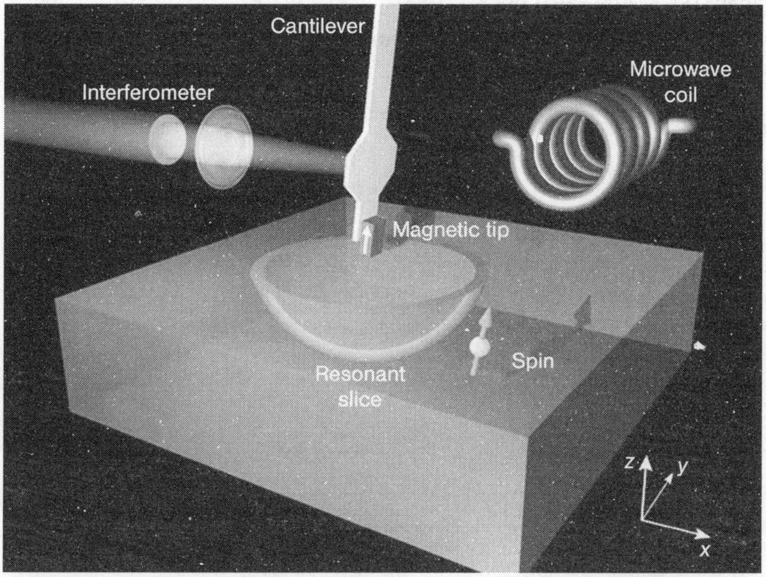

Figure 7.11 A schematic image of the AFM-type Magnetic Resonance Force Microscope (MRFM) device used to image a single electron spin [19]. The 150 nm-wide SmCo magnet attached to the tip oscillates at fixed amplitude (16 nm) at $f_c = 5.5$ kHz in the *x*-direction about 125 nm above the sample surface. The sum of the tip's dipole magnetic field and the static magnetic field ($B_z \approx 30$ mT) add up to about 106 mT, corresponding to an applied microwave resonant magnetic field supplied at $\nu = 2.96$ GHz. The "resonant slice" is a bowl-shaped surface that extends about 250 nm below the tip. For a spin slightly displaced from the tip's center position in the *x*-direction, the spin is flipped twice per cycle of tip oscillation, providing a cyclic force on the tip. The *x*-component of the cyclic force on the tip can be modeled as a small change in the cantilever force constant, causing a small change in its resonant frequency, which is detected by means of a reflected light beam focused on the tip. The device is mounted in a small vacuum chamber within the bore of a superconducting magnet and is operated at 1.6 K. (Reprinted from Nature with permission from Macmillan Publishers, Ltd.)

small change in its resonant frequency f_c of motion, which is detected in a synchronous fashion, by means of a reflected light beam focused on the tip. According to [19], the frequency shift is

$$\Delta f_c = \pm 2 f_c G \mu_B / (\pi k x_{peak}), \qquad (7.1)$$

where $G = \partial B/\partial x$ is the field gradient, $\mu_B = 9.3 \times 10^{-24}$ J/T is the magnetic moment of the electron, k is the force constant of the cantilever, and x_{peak} is the amplitude of the cantilever motion. The change of sign is related to conditions of magnetic resonance. For conditions attained [19] this frequency shift is about 3.7 mHz. This tiny frequency shift requires sophisticated use of synchronous detection (lock-in detector) and time averaging to be observed.

The device is mounted in a small vacuum chamber within the bore of a superconducting magnet and is operated at 1.6 K. To observe a single electron spin is quite a

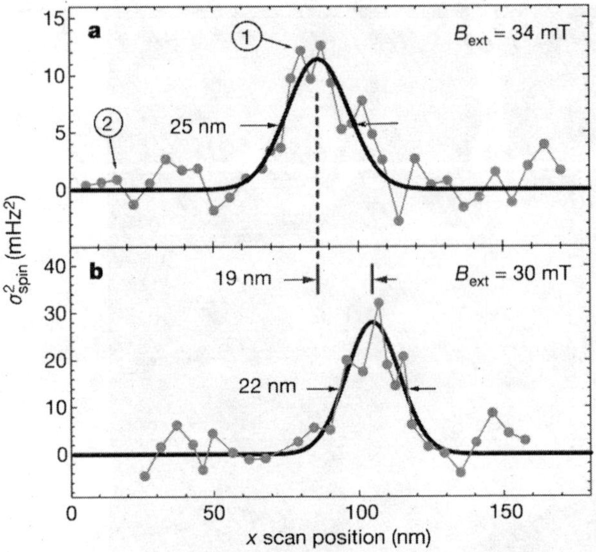

Figure 7.12 Data that support the observation of a *single electron spin* with the Magnetic Resonance Force Microscope (MRFM), an AFM-derived scanning device [19]. These plots show the spin signal (the shift in the cantilever oscillation frequency) as the sample was scanned slowly in the *x*-direction. The 19 nm shift in apparent position of the spin between upper and lower panels in the figure was caused by changing the applied static magnetic field from 34 mT to 30 mT. The spatial resolution in detecting a single electron magnetic moment is seen to be on the order of 25 nm. (Reprinted from Nature with permission from Macmillan Publishers, Ltd.)

feat, but this device is not very practical. The most obvious means of improving the signal size seems [19] to be by using a stronger tip magnet to increase the field gradient G, and, additionally, to further reduce the spring constant k.

7.9
Fundamental Questions: Rates, Accuracy and More

We have looked at the most advanced results from the leading scanning probe devices. These devices have produced new understanding of the behavior of matter at the atomic scale. *As far as possible use as "molecular assemblers" on an atom-by-atom basis for bulk matter, there is nothing that we have seen that remotely suggests that this is possible.*

The single tip, however refined it may possibly become in extra facility for grasping, orienting, and releasing particles, simply cannot act rapidly enough to produce a measurably large sample in a useful period of time.

Arrays of tips, projected [3] in the Zurich IBM Lab to rise to a million tips on a chip, do not come close to filling the rate discrepancy that is inherent in the large size of Avogadro's number.

References

[1] *The International Technology Roadmap for Silicon*, on the Web at www.public.itrs.net/.
[2] H. G. Craighead, Science **290**, 1532 (2000).
[3] P. Vettiger, M. Despont, U. Drechsler, U. Durig, W. Haberle, M. I. Lutwyche, H. E. Rothuizen, R. Stutz, R. Widmer, and G. K. Binnig, IBM J. Res. Develop. **44**, 323 (2000).
[4] Reprinted with permission from Nature: M. Blencowe, Nature **424**, 262 (2003). Copyright 2003 Macmillan Publishers Ltd.
[5] R. G. Knobel and A. N. Cleland, Nature **424**, 291 (2003).
[6] J. D. Meindl, Q. Chen, and J. A. Davis, Science **293**, 2044 (2001).
[7] Courtesy: IBM Research, T. J. Watson Research Center. Unauthorized use not permitted.
[8] Reprinted with permission from W. Chen, A. V. Rylakov, V. Patel, J. E. Lukens and K. K. Likharev, Appl. Phys. Lett. **73**, 2817 (1998). Copyright 1998, American Institute of Physics.
[9] P. Bunyk, K. Likharev, and D. Zinoviev, Intl. J. of High Speed Electronics and Systems **11**, 257 (2001).
[10] M. Gurvitch, M. A. Washington, and H. A. Huggins, Appl. Phys. Lett. **42**, 472 (1983).
[11] E. L. Wolf, *Principles of Electron Tunneling Spectroscopy*, (Oxford, New York, 1989) pp. 215–230.
[12] P. K. Hansma, V. B. Elings, O. Marti, and C. E. Bracker, Science **242**, 209 (1988).
[13] Reprinted with permission from N. Nilius, T. M. Wallis, and W. Ho, Science **297**, 1853 (2002), published online 22 August 2000 (10.1126/science.1075242). Copyright 2002 AAAS.
[14] N. Nilius, T. M. Wallis, and W. Ho, Phys. Rev. Lett. **90**, 186102 (2002).
[15] E. L. Wolf, op. cit., p. 317.
[16] P. F. Schewe and B. Stein, Physics News Update (AIP) Number 503, September 22, 2000.
[17] S. W. Hla, L. Bartels, G. Meyer, and K. H. Rieder, Phys. Rev. Lett. **85**, 2777 (2000).
[18] R. E. Smalley, Scientific American **285**, #3, 76 (2001).
[19] D. Rugar, R. Budakian, H. J. Mamin, and B. W. Chui, Nature **430**, 329 (2004).

8
Quantum Technologies Based on Magnetism, Electron and Nuclear Spin, and Superconductivity

Some of the most exciting possibilities for new electronic devices, computer technologies and communications technologies are based on the rules of nanophysics as we have developed them, and especially on magnetism, based on the spin property of electrons and of atomic nuclei. The electron spin has already been assumed in equation (5.6). In that discussion it was clear that electron spin was an important aspect in forming the covalent bond, and that the same symmetry-driven electrostatic exchange interaction (5.18) could lead to the ferromagnetic state. In devices today there are cases where individual electron motions depend upon the state of the spin. So we begin this chapter with a more detailed look at the discovery of electron spin and how it can affect the motion of individual carriers. The original experimental evidence for electron spin (the Stern–Gerlach experiment) will be mentioned below. (For earlier discussion of related topics, see equations (4.85)–(4.94), and Sections 5.4.1 and 7.8.3.)

Here we first look in more detail at the basic spin property of electrons: spin $S = ½$, with $m_s = \pm ½$ defined in equations (5.6) and (5.7). We focus below on how this affects electron motion in solids, especially ferromagnetic solids. To repeat, it is useful to separate the space part $\phi(x)$ and the spin part χ of the wavefunction, as given earlier by

$$\psi = \phi(x)\chi. \tag{5.6}$$

For a single electron $\chi = \uparrow$ (for $m_s = ½$) or $\chi = \downarrow$ (for $m_s = -½$). In much of the literature the quantum states for spin ½ are written as $|\uparrow\rangle$ and $|\downarrow\rangle$, and we will adopt this notation in this chapter.

The single spin ½ is an example of a "two-level system" in quantum mechanics. In some other cases, while there may be more than two quantum states, it may be possible to focus on only two states, ignoring the others.

Spin as an Element of "Quantum Computing"

One of the most interesting new ideas in information technology is the "quantum computer". The basis for "quantum computing" is the "qubit", the generalization of the binary bit (with just two values), to the linear combination quantum state

Nanophysics and Nanotechnology: An Introduction to Modern Concepts in Nanoscience. Second Edition.
Edward L. Wolf

ISBN: 3-527-40651-4

$$\chi = a|\uparrow> + b|\downarrow>, \tag{8.1a}$$

with the values of a and b representing information. There is a normalization condition,

$$a^2 + b^2 = 1, \tag{8.1b}$$

so that the information may be seen as equivalent to the real numbers occupying the unit circle, or exp(iΘ) with $0 < \Theta < 2\pi$.

In quantum mechanics, if ψ_1 and ψ_2, say, are solution functions to Schrodinger's equation, then any linear combination $\psi_Q = a\psi_1 + b\psi_2$ of these functions will equally solve the equation.

This concept of linear combinations is verified experimentally for microscopic systems, for example, the spin ½ in the magnetic resonance experiment. The spin can be prepared in any linear combination of spin up and spin down, and the result can be quantified as an angle, 90° for the equal combination of spin up and spin down. The "qubit" is precisely the angle Θ or, equivalently, the values of a and b. The apparatus detects the rotating (oscillating) magnetic field that is generated by a large number of spins. (The expectation value that is detected is an analog of (4.48), see also (5.13).)

There are several reasons why computing based on the "qubit" is *potentially* more powerful than that based on binary bits. The spin ½ is only one of several possibilities for the qubit. In a second case, e.g., which will be discussed in Chapter 9,

$$\psi_Q = a|L> + b|R>, \ a^2 + b^2 = 1. \tag{8.2}$$

Here $|L>$ and $|R>$, respectively, represent a single electron on the left side or the right side of a double potential well. This situation is analogous to the electron in a hydrogen molecule ion H_2^+.

The first potential advantage of a quantum computer is that fewer "bits" will be needed to provide the same performance, since the qubit contains more information. Potentially, today's computer chip with ~10^8 transistors may be replaced with a "quantum computer" chip needing many fewer "qubits" to do the same job, at lower power consumption.

Second, the quantum computer can efficiently solve problems that binary computers essentially cannot solve. The factoring of large prime numbers may seem esoteric, but lies at the heart of the encryption schemes used by banks and governments. One may say that the present "security" lies in the weakness of present binary computers, which is an incentive for increased understanding of quantum computation. The result is that effort is being expended on this (basically interesting) problem.

An authoritative review [1] of the state of "quantum information and computation" has been given by Bennett and DiVincenzo, of the IBM Yorktown Laboratory.

One of the fundamental weaknesses of the qubit is its susceptibility to "decohere", to have the precise values a and b disrupted. The time over which the qubit is valid (coherence time) is terminated by energy loss by the two-level quantum system, $\chi = a|\uparrow> + b|\downarrow>$.

So the successful qubit will be isolated from its environment, guarded from gain or loss of energy. (This also suggests a low power requirement.) The *nuclear spin* is particularly well isolated from the environment by being enclosed within successive shells of electrons. So the nuclear spin, from the beginning, has been a contender for the quantum qubit.

Isolation is especially hard to achieve in a solid state context such as silicon, which, for reasons of compatibility with existing semiconductor technology, may otherwise be favored. Thus the nuclear spin (magnetic moment), and its interactions with electron magnetic moments, are important subjects for the student of nanotechnology; see equations (4.85) to (4.90).

Reading the final value of a qubit is not as simple as reading the output of a binary bit. Within our understanding of quantum mechanics, the individual system $\chi = a|\uparrow> + b|\downarrow>$, when measured, will produce either state $|\uparrow>$, with probability a^2, or $|\downarrow>$, with probability b^2. (It is essential to realize that *measurement* has only two possible outcomes: it is only from the frequency of observation of these that the values of a and b in the superposition can be inferred.) If one deals with single qubits, there seems no alternative to making many successive equivalent tests, so that statistics on the number of $|\uparrow>$ vs. the number of $|\downarrow>$ outcomes can establish the values of a and b to required accuracy.

Another approach to reading the state $\theta(a,b)$ of the qubit is to monitor the expectation value (equations (4.48) and (5.13)) of some variable in the superposition state. For sensitivity reasons, this may not be possible for a single qubit, but may require a large simultaneous number of equivalent qubits. This approach has been demonstrated by calculations in a quantum computer based on nuclear magnetic resonance (NMR), as referenced in [1]. Here the number of quantum systems is closer to Avogadro's number, 6.02×10^{23}, so that the readout (taking the macroscopic average of Avogadro's number of answers, seen on the needle deflection) is much easier.

The challenges to the nanotechnologist are in finding experimental realizations of these projected qubits, to find ways in which they can be made to interact or "entangle" as is needed to perform calculations, making the coherence time as long as possible, and finally to read the coefficients to the desired precision.

Ferromagnetic effects already play a large role in computer technology, particularly in the areas of disk memory and nonvolatile random access memory. For these reasons, a more complete understanding of ferromagnetism is desirable, and will be approached below.

The reasons for the alignment of atomic magnetic moments in iron and other ferromagnetic substances, which underlies the operation of the magnetic disk memory, was explained, in Section 5.8 above, to arise from a basic symmetry of identical electron particles under exchange of positions. In the case of the H_2 molecule, the exchange constant J_e, in equation (5.18), is *negative*, leading to a binding energy of the molecule of 4.7 eV, with antiparallel alignment of spins; in contrast, in the ferromagnetic case, J_e is *positive*, leading to a large (electrostatic) energy, $E_x = -2J_e\mathbf{S}_1 \cdot \mathbf{S}_1$, favoring *parallel* alignment of electron spins. This E_x can be a huge effect, whose size may be better appreciated by imagining an effective "exchange" magnetic field. The interaction energy observed in H_2, which may be representative of ferromagnetic cases, –4.7 eV, if represented as a magnetic field (namely 4.7 eV = $-\mu \cdot \mathbf{B}^*$), gives $B^* = 8.12 \times 10^4$ T, as an effective field forcing spins to align parallel.

Finally, *superconductivity* is a fully quantum-mechanical phenomenon, whose theoretical and experimental aspects are well understood in the scientific community. The application of this effect has been limited by the needed low temperatures. Even so, there are several reasons to expect a greater of importance of superconductivity in future. First, it offers a complete solution to the energy dissipation problem in the present computer technology: the superconductor is essentially non-dissipative. Secondly, the superconductor is a large sample in a coherent quantum-mechanical state, so that it offers opportunities for qubits with long coherence times. An extremely fast superconducting computer technology, known as RSFQ (Rapid Single Flux Quantum, see Sections 7.5 and 8.13) is finding applications, for example, in digitizing fast analog signals. Finally, the area of superconducting applications stands to be advanced by better materials science applied to the existing class of "high temperature superconductors" and also possibly by the discovery of new superconductors whose transition temperatures are closer to room temperature. There are chances for major advances in several aspects of superconductivity and its technology, which makes it suitable to cover these topics in more detail below.

8.1 The Stern–Gerlach Experiment: Observation of Spin ½ Angular Momentum of the Electron

In the discussion of the hydrogen atom, the spin of the electron was assumed (equation (4.85)), and in equations (4.87)–(4.90) the corresponding magnetic moment was described. Here we return to the fundamental observation which established the electron spin, the Stern–Gerlach experiment, in preparation for more detailed discussion of spin effects below.

The experiment was designed to detect magnetic moments in atoms. Following (4.88), one expects $\mu = (e/2m)L$ for an electron with orbital angular momentum L in an atom. For such a magnetic moment μ, in the presence of a magnetic field B the interaction energy is

$$U = -\boldsymbol{\mu} \cdot \mathbf{B}. \tag{4.90}$$

This has the effect of aligning the magnetic moment parallel to B, but exerts no force on the magnetic moment if B is constant in space. The separation of atoms with different orientations of μ is possible if there is a gradient to B. Suppose B is in the z-direction, and that $\mathrm{d}B_z/\mathrm{d}z$ is positive. Then a force $F = -\mathrm{d}U/\mathrm{d}z$ will exist,

$$F_z = \mu_z \mathrm{d}B_z/\mathrm{d}z. \tag{8.3}$$

Beams of one-electron atoms (hydrogen and also alkali metal atoms like K, Na, Rb, plus Ag, which have only one outer electron), surprisingly, were split up and deflected in an inhomogeneous magnetic field.

Since the $n = 1$ (1s) ground state of the outer electron (equation (4.80)) has no orbital angular momentum L, and thus has no magnetic moment associated with it, these deflections of neutral ground state atoms were puzzling. A second puzzling aspect was that the atomic beams of neutral atoms were split into two beams, rather than three beams as might have been expected. In the Schrodinger treatment of the one-electron atom (see equation (4.75)), the orbital angular momentum quantum number l is restricted to values between 0 and $n - 1$, where n is the principal quantum number.

So $n = 1$ requires $l = 0$, and hydrogen atoms in the $n = 1$ state have no angular momentum L. In the case of $n = 2$ the value $l = 1$ is possible, and we have seen that the angular momentum can take $2l + 1$ different orientations (see Figure 4.10), which would lead to three beams for $l = 1$ in the Stern–Gerlach experiment. The only way to get two orientations is to set $l = s = ½$. This effect was recognized as distinct from the known orbital angular momentum case, and thus had to be a new feature of the electron.

So it came to be recognized that an electron in an atom can have two types of angular momentum, orbital and spin, and these are represented by L and S; the sum is referred to as J. It is similarly the case that neutrons and protons have spin ½, and the symbol I is used for the spin angular momentum of the nucleus of an atom. The magnetic moment of the nucleon (nuclear magneton μ_N) is smaller than the Bohr magneton μ_B by about 1/1838, reflecting the mass ratio of the electron to the nucleon.

8.2 Two Nuclear Spin Effects: MRI (Magnetic Resonance Imaging) and the "21.1 cm Line"

Protons (and neutrons) have spin ½, with nuclear magnetic moments on the scale of μ_N which, based on $\mu = (e/2m)L$, are expected to be smaller by the ratio of the electron to proton mass, thus $\mu_N = (1/1838)\mu_B$. The energy states of a proton spin magnetic moment in a magnetic field B are known experimentally to be $E = \pm 2.79285\mu_N B$. (The corresponding number for the neutron is −1.913, equally strange but observed to be true.) For a proton (there are two protons in each H_2O, and in living organisms the locations of the H_2O essentially describe the structure) at $B = 1$ T this energy difference is $\Delta E = 0.176$ μeV, which corresponds to a resonant frequency for absorption of 42.7 MHz. The imaging aspect of MRI is based on a scan (3-d raster scan) of x, y, z locations in the specimen where exactly $B = 1$ T (giving the proton resonance) is detected.

The mysterious value $\mu_z = 2.79285\mu_N$ for the moment of the proton (and the similarly strange value $\mu_z = -1.913\mu_N$ for the neutron) is believed [2] to come from complicated motions of the three underlying quarks (particles of charges ±⅓e, ±⅔e) which constitute the proton and neutron.

Electron spin ½ and nuclear spin ½ are both candidates for a qubit in quantum computing. The reason basically is that spins (unlike a macroscopic object) can be prepared and observed in quantum states that are *linear combinations* of the basis states (spin up and spin down). Thus, a qubit can take intermediate values between zero and one. A concrete example of a qubit is also a state basic to the operation of the MRI apparatus. One takes a spin-up (lower energy state) proton (in water) to a linear combination of down and up by the application of a "$\pi/2$ pulse". This produces a freely rotating magnetization at the resonant frequency (42.7 MHz, in the example, for $B = 1$ T), which is detected (equations (4.48) and (5.13)) by the receiver coil of the MRI apparatus, and tells the computer whether or not there are protons (water) at this location in the patient.

One of the complications in realizing an electron spin qubit in a semiconductor context, now actively pursued, is that the nuclei of the semiconductor atoms also usually have spins and magnetic moments, which can interact with (and possibly degrade) the electron spin qubit. For this reason, the interaction of electron and nuclear spins is of interest for the nanotechnology of quantum computing.

An important electron spin–nuclear spin interaction occurs in the hydrogen atom, and accounts for observed radiation of wavelength 21.1 cm from H atoms in outer space.

It turns out that this (clearly observed) effect is not easy to understand, in that the most obvious interaction, a (magnetic spin) dipole–dipole interaction (analogous to that outlined in (5.19) and (5.20) for the electric case) exactly cancels out in the hydrogen atom situation.

Still, it is instructive to estimate the magnetic dipole field of the electron magnetic moment at the position of the nucleus (proton) of the H atom. This is roughly $B \approx (\mu_o/4\pi)3g\mu_B(a_o)^{-3} \approx 20$ T, where $(\mu_o/4\pi) = 10^{-7}$, g for the electron spin is 2.0023, $\mu_B = 9.274 \times 10^{-24}$ J/T, and $a = 0.05291$ nm. Continuing a rough analysis, the proton magnetic moment will have spin-up and spin-down energies split by this magnetic field from the electron. (If the effect did not cancel by symmetry, $B = 20$ T would give a splitting about 20 times that mentioned in the MRI case above, the spin splitting should be about 20 times that above, thus a resonance frequency 854 MHz, or a radiation wavelength $\lambda = c/f = (3 \times 10^8)/(850 \times 10^6) = 0.353$ m $= 35.3$ cm.) This analysis can only be approximate, because we assumed (incorrectly, using a sometimes useful approximation) a *fixed* orbit of radius a_o, while we have seen in equation (4.80) that the wavefunction for the electron is $\psi_{100} = (\pi)^{-1/2}\exp(-r/a_o)$, with a range of radii centered on the Bohr radius, and zero angular momentum. However, the dipole field at the nucleus from the electron averages to zero, since the electron in the 1s state with equal probability samples all 4π possible angles on the unit sphere, leading to exact canceling of the dipole field.

The observed interaction (used in the qubits in quantum computing) comes to mind when we note (4.80) that $\psi_{100} = (\pi)^{-1/2}\exp(-r/a_o)$ is largest at $r = 0$, which would put the electron at the exact location of the proton! This is the situation that is described by the "contact hyperfine" interaction denoted by

$$\mathcal{H}_{IS} = (8\pi/3)\gamma_e\gamma_n\hbar^2\,|\psi_{100}|^2\,\mathbf{I}\cdot\mathbf{S}. \tag{8.4}$$

In this expression $\gamma_e \gamma_n$ is the product of the gyromagnetic ratios for the electron and nucleon, such that, e.g., $\gamma_e B = \omega_e$. The symbol I is used for the spin of the nucleus, S for the spin of the electron.

An intuitive explanation of the contact interaction is that the electron samples the "near field" of the nuclear magnetic dipole. Imagining the nucleus to be an array of current loops, B inside that distribution will be oriented uniformly, and the 1s electron spin moment will find a net effect.

To get the final answer for the energy of the contact interaction $E = -A^* \mathbf{I} \cdot \mathbf{S}$, the expectation value, analogous to equation (5.13), is calculated. Following such a procedure, in agreement with observations, the radiation frequency is 1420.4 MHz for H corresponding to $\lambda = 21.1$ cm and energy $hf = 5.86$ neV. These effects precisely arise from flipping the electron spin in the presence of the fixed nuclear spin, as at the start of this section. If the interaction is described as an effective magnetic field produced by the nucleus, it has value 50.8 mT for the 1s electron in H. The effective contact hyperfine field is bigger in other cases; for example for the 2s electron in fluorine F, the effective field is 1.7 T.

21.1 cm radiation is observed from outer space, enabling astrophysicists to map the location of hydrogen atoms in the universe! This implies that excitation energy $hf = 5.86$ neV is available in outer space, corresponding to an effective temperature $T^* = hf/k_B = 0.067$ K. The "black body" microwave electromagnetic spectrum observed from outer space, traced to the "big bang", peaks at a higher temperature, 2.7 K. Thus, hydrogen atoms in space can be expected to have equal populations with electron and proton spins parallel and antiparallel (split by 9.86 neV), consistent with observation of the 21.1 cm radiation.

The interaction between electron and nuclear spins is called the "hyperfine interaction". It is described in quantum mechanics by the relation, which includes a static magnetic field, B_0,

$$\mathcal{H} = g_e \mu_B B_0 S_z + g_N \mu_N B_0 I_z - A\mathbf{S} \cdot \mathbf{I}, \tag{8.5}$$

where $g = 2$ for the electron, μ_B is the Bohr magneton, $S_z = \pm 1/2$, g_N describes the nuclear spin I, μ_N is the nuclear magneton, $I_z = \pm 1/2$, and A is the constant for the hyperfine interaction in the system at hand. This is an example of a "spin Hamiltonian" operator representing the energy states of a set of coupled spins. The number of spins whose interaction is represented in this way can be more than two. The expectation energies of the electron–nuclear spin interaction can be obtained from the spin Hamiltonian operator $\mathcal{H}$ given above following the formula $<E> = \int \psi^* \mathcal{H} \psi dq$. The quantum numbers of the wavefunction will in general give distinct values for the interaction energies.

8.3 Electron Spin ½ as a Qubit for a Quantum Computer: Quantum Superposition, Coherence

Nuclear and electron spins are candidates for a qubit in quantum computing, for reasons given in the introduction to this chapter.

As a concrete example of a proton qubit state, the medical Magnetic Resonance Imaging (MRI) apparatus takes majority population of spin-down protons (in water) to a linear combination of down and up, by the application of a "$\pi/2$ pulse". This generates the $a = b$ superposition state, and in this macroscopic case leads to a magnetization freely rotating and tilted at 90° with respect to the magnetic field, at the resonant frequency (42.7 MHz, at 1 T magnetic field, in the example). What is meant by a "$\pi/2$ pulse", in the magnetic resonance language, is momentary application of a resonant rotating magnetic field, applied just long enough to tilt the (rotating) magnetic moment of the spin from vertical down into the direction perpendicular to the applied field B.

The MRI apparatus has a radio-frequency (RF) transmitter which fills the apparatus (and specimen) with a static ~1 T magnetic field and radiation at the resonant frequency $f = 2\mu_{\text{proton}}B/h$ (42.7 MHz, at $B = 1$ T). The *co-rotating component* of the resonant RF magnetic field,

$$\mathbf{B}_1 = \mathbf{B}_{10}\cos(\omega_0 t) = H_{10}\,[\exp(\mathrm{i}\omega_0 t) + \exp(-\mathrm{i}\omega_0 t)]2,$$

where $\omega_0 = 2\pi f$, exerts a constant torque on the rotating magnetic moments, which tilts the resultant nuclear magnetic moment into the plane perpendicular to B. The 90° ("$\pi/2$") pulse cuts off just as the spin magnetic moment is tilted to 90°, creating a macroscopic "qubit" (i.e., a large number N^* of identical qubits). Then the spin population (macroscopic magnetization $\mathbf{M} = N^*\boldsymbol{\mu}_N$) rotates (precesses) around the applied static magnetic field. Since the internal magnetic field in a magnetized region is $\mathbf{B} = \mu_o\mathbf{M}$, $\mathbf{B} = \mu_o N^*\boldsymbol{\mu}_N$ is the resultant rotating magnetic field vector, whose flux (i.e., the expectation value, in (4.48) and (5.13)) is detected, via Faraday's law, $V = -\mathrm{d}\Phi_m/\mathrm{d}t$, giving a time-dependent signal voltage in the pickup coil of the receiver. (The decay in time of this freely rotating magnetization (after the exciting field $\mathbf{B}_1 = \mathbf{B}_{10}\cos(\omega_0 t)$ is turned off) allows determination of the constant T_1, which would be called the "coherence time" in the qubit language.) Measuring this induced RF voltage is the basis for the MRI estimate of the proton (water) density in the sample at that location.

It is true, in this example, that if the state of any *particular spin* were measured, the answer would be spin up half the time and spin down half the time.

How long does this freely rotating qubit state last? (In the present semiconductor technology, the DRAM (dynamic random access memory) cells have a short storage time, and the information is refreshed periodically, requiring an energy input.) The freely rotating coherent spin state (qubit) produced by the 90° pulse is degraded by two processes. First, this is an excited state, and it may lose energy and return to the lowest energy state (pure spin up). The time for this return to the lowest energy is called T_1 in the NMR literature. A second possibility for degrading the macroscopic qubit is that the local resonant field at the position of the spin may be slightly different at different locations. Or, focusing on a single electron or nuclear spin qubit, the local resonant field may change in time if a nearby nuclear spin flips, via interaction with the local nuclear spins. The time for de-phasing of the rotating magnetization to occur is T_2 in the literature of magnetic resonance.

In the commercial MRI medical apparatus, diagnostic schemes are used to measure T_1 and T_2. Images may even be presented which are based on measurements of these times, which are possibly indicative of the local environments of the H_2O protons. Given a spin magnetization in the freely rotating state, the basic observation in the receiver coil is the induced voltage at ω_0 based on $V = -d\Phi_m/dt$. This RF voltage will fall to zero after a time T_1, because in that time the spin will have found its lowest energy state. The observation in the receiver coil may fall to zero on a shorter time scale, if the various moments are rotating at different frequencies (local B fields), or if the local field changes (maybe by change of the arrangement of nearby nuclear spins) during the time of observation. Thus, if $T_2 << T_1$, the signal may disappear before the spins are all in the lowest state. In this case, application of a "π pulse" after the onset of the free induction decay can *reverse the dispersion* of spin phases produced by disorder, producing a "spin echo". Thus, the commercial MRI machines map the locations of protons (the basic MRI image), and also provide images for the local spin relaxation times T_1 and dephasing times T_2. The latter allow discrimination of the local chemical/biological H_2O environments.

The use of coupled electron and nuclear spins to make a multi-qubit quantum computer based on Scanning Tunneling Microscope (STM) readout has recently been proposed [3], which depicts the energy levels for one electron interacting with a spin-½ nucleus (^{125}Te) located at the surface of a silicon crystal (this is known as the A center in Si). It is assumed that a large magnetic field of 10 T is applied to the system. Berman et al. [3] describe this system $S = I = ½$, $g_e = 2$, $g_N = 0.882$, B_0 a high field 10 T pointed in the z-direction, and $A/(2\pi\hbar) = 3.5$ GHz, in the spin-Hamiltonian expression:

$$\mathcal{H} = g_e\mu_B B_0 S_z + g_N\mu_N B_0 I_z - A\mathbf{S}\cdot\mathbf{I}. \tag{8.6a}$$

Considering this expression and Figure 8.1, one can see that the frequency of the electron resonance changes depending upon the orientation of the nuclear spin I. This is summarized as

$$f = f_{e0} = g_e\mu_B B_0/(2\pi\hbar) + A/(4\pi\hbar) \quad \text{(nuclear spin up)}, \tag{8.6b}$$

$$f = f_{e1} = g_e\mu_B B_0/(2\pi\hbar) - A/(4\pi\hbar) \quad \text{(nuclear spin down)}. \tag{8.6c}$$

The hyperfine splitting is $A/(2\pi\hbar) = 3.5$ GHz for the ^{125}Te donor center. The important point is that the state of the nuclear spin can be determined from the frequency of the electron's spin resonance. The frequency of the electron spin precession is proposed to be measured for each Te qubit by imaging in STM and measur-

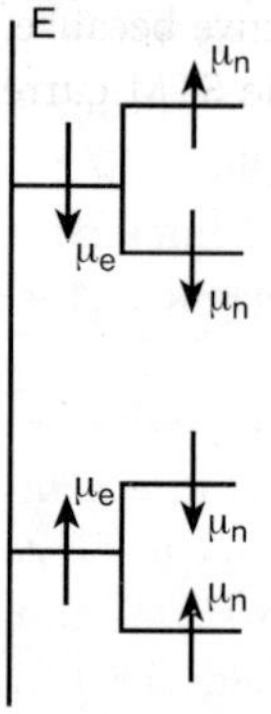

Figure 8.1 Energy levels for electron (μ_e) and nuclear (μ_n) magnetic moments in a high magnetic field B, after [3]. (Reprinted from Physical Review Letters with permission from APS.)

ing the frequency of the expected modulation of the STM tunnel current, identically as given above. The motion of the electron spin has been determined experimentally to produce a frequency modulation on the tunneling current, and this modulation frequency can be measured using a spectrum analyzer on the STM current [4].

Berman et al. [3] propose a linear array of 250 ^{125}Te I = ½ qubits, at accurate spacing a = 5 nm along the x-direction, along the surface of the Si crystal. They propose to make the B field strength slowly varying along the x-direction, so that for the kth location there will be distinguishable values for the kth nuclear resonance frequency and for the kth electron resonance frequency. With spacing a = 5 nm, and field gradient $dB/dx = 10^5$ T/m, the change of electron resonance frequency between adjacent sites is 14 MHz. For the nuclear resonant frequencies this difference is 6.75 kHz. The linear array of ^{125}Te I = ½ qubits will be addressed by a variable frequency rotating magnetic field $B_\perp$, perpendicular to the static B field, to allow spin-up to spin-down transitions (corresponding to a π pulse) on any desired single nucleus in the chain. It is thus possible to prepare the linear qubit array with each nucleus in its ground state; the STM measurement at the kth location of the electron spin frequency will identify (by the large hyperfine splitting 3.5 GHz, noted above) those sites where the nucleus is in the excited state. Then a π pulse can be applied at the kth nuclear frequency to put that nucleus into its ground state, which will persist on a long time scale because of the noted isolation of the nuclei from the outer environment. Because of the large field B = 10 T and low temperature T = 1 K, the electron spins will all be in their ground states, but the kth electron spin is also assumed to be controllable by electromagnetic pulses at the electron resonance frequency.

Finally, coupling between the qubits [3] is afforded by the interaction of the kth nucleus with the dipole fields of the electrons on the neighboring sites $k - 1$ and $k + 1$. The change in kth nuclear frequency with change in electron spin state on neighboring sites is estimated as about 200 Hz. (This dipole–dipole interaction was mentioned above in preliminary discussion of the hyperfine interaction.)

This proposal [3] is very clear. The desirable isolation of information is provided by the basic isolation of nuclear spins from the outside world.

From an experimental point of view this scheme may be unattractive because of the needed 10 T B field, and the difficulty of spectrum analyzing the STM current [4] while maintaining atomic resolution with the scanning tunneling microscope (STM). Methods of precisely implanting impurities such as ^{125}Te are being actively investigated [5] in connection with charge qubits in the silicon context, which will be mentioned in Chapter 9.

The electron or magnetic spin is a nanoscopic object producing a qubit in the presence of an applied magnetic field. The possibility of extremely dense information storage and of dense logic devices is clearly present.

Spin is not the only possible basis for quantum qubits in computation technology. A second example (there are many more) is the "charge qubit". One form of charge qubit corresponds to the location of an electron tunneling between two equivalent potential wells, as for example in the hydrogen molecule ion or the hydrogen molecule. In this context it is useful that the shallow donor centers in Si provide close analogs, and are well known in the Si technology. These possibilities will be addressed in Chapter 9.

8.4 Hard and Soft Ferromagnets

A ferromagnet has an internal magnetization $M(T)$ given by solutions to equation (5.59) (there, for simplicity, restricted to spin-½ systems) and the typical temperature dependence $M(T)$ is shown in Figure 5.21. Recall that ferromagnetism arises from the exchange interaction equation (5.18) in cases where the constant $J_E < 0$. It is accompanied by a shift of energy levels for the spin-up (majority) vs. spin-down (minority) electrons, as sketched in Figure 5.22. These energy shifts are large and have verified technological consequences.

The important thing to note is that the exchange energy shift, leading either to molecular bonding or to ferromagnetism, can be large, dominating the thermal energy $E = k_B T$.

Figure 5.22(b) shows that the total number of majority spins (by definition) exceeds the total number of minority spins, but the *density per unit energy* $N(E)$ of spin-up states at the Fermi energy, $N_\uparrow(E_F)$, can also be changed at the ferromagnetic transition temperature. In this case the change is of the *opposite sense:* (in Fig. 5.22(b)) $N_\uparrow(E_F) = n^+ << N_\downarrow(E_F) = n^-$. The change in the density of states in a ferromagnet with respect to spin direction at the Fermi energy is the basis for the "tunnel valve", which is a magnetic field sensor used in magnetic disk drives.

At low temperatures, $T << T_C$, M approaches its saturation value, $N\mu_B$, where N is the number of spins per unit volume. This means that all magnetic moments are aligned, so that the saturation magnetic field is $\mu_o M = \mu_o N\mu_B$. Table 5.3 gives typical values for the saturation internal field, e.g. 0.171 T, 0.14 T, and 0.0485 T, respectively, for Fe, Co, and Ni at 300 K. (Note that these values are smaller than the 1 T available in an MRI apparatus, by means of a superconducting electromagnet operated at 4.2 K.) The saturation field $\mu_o M$ exists within a magnetic domain. A given sample of

Fe for example will typically have a large number of magnetic domains whose magnetization directions will differ. The effect of an external field on such a multi-domain system is to cause alignment of the domains, leading to the typical hysteretic magnetization vs. applied field curve, as shown in Figure 5.20. The physics involved in the formation of domains is reduction of the total energy, and since the external B field carries an energy density $\mu_o B^2/2$, the domains arrange themselves to reduce the external magnetic field. An applied B field will enlarge domains whose magnetizations lie parallel, and the hysteresis results from internal energy barriers to the motion of the boundary between domains, called "domain walls". Referring to Figure 5.20, the remanent magnetization M_r or remanent magnetic field $B_r = \mu_o M_r$ is large in ferromagnets that have a large coercive field B_c. These metals have a domain wall structure that resists change, and when the maximal alignment is forced on the system by a large applied field, a large fraction of that magnetization is retained when the external field is removed. Such materials, called "hard" ferromagnets, are used for permanent magnets, and Co is an example of such a material.

On the other hand, if the domain walls easily move, the saturation magnetization (equivalent to an extremely large local B field) can be reached with a small applied field, as suggested earlier by

$$B = B_{\text{app}}(1 + \chi_{\text{m}}) = \kappa_{\text{m}} B_{\text{app}}. \tag{4.94}$$

Here $(1 + \chi_{\text{m}}) = \kappa_{\text{m}}$ is called the magnetic permeability. This equation describes the initial slope of the plot of Figure 5.20. In these cases, hysteresis is not noticeable, and the important aspect is a line of large slope M vs. B_{app} running through the origin. Large values of κ_{m} exceeding 10^4 are obtained in alloys of Ni and Fe (e.g. Permalloy, $Ni_{0.8}Fe_{0.2}$). If we take B_{sat} = 485 G (for Ni at 300 K, see Table 5.3) and $\kappa_{\text{m}} = 10^4$, then according to (4.94) the necessary B_{app} = 0.049 G. Such a magnet is called a "soft magnet", and Permalloy or a similar alloy is used in the "spin valve" and "tunnel valve" magnetic field sensors in hard disk drives, as mentioned in Section 1.5. Coercive fields of the order of 1.0 G have been observed in tunnel valve devices, discussed below.

8.5
The Origins of GMR (Giant Magnetoresistance): Spin-dependent Scattering of Electrons

Ferromagnetic metals have mobile electrons ("carriers") that are released from the d states of the atoms and also mobile electrons that are released from the s states of the same atoms. The conductivity of the metal is the sum of the conductivities of these two types (although usually the conductivity of the s electrons is dominant, because of the lower effective mass, see equation (5.44)). The two types of carriers are suggested in Figure 5.22, with the left panel showing free-electron-like bands that describe s electrons, as in Cu (see also Figure 5.10), while the right panel of Figure 5.22 shows "d bands" which cover a relatively narrow energy range (shifted by the exchange energy). Such sharply defined bands are typical of 3d electrons,

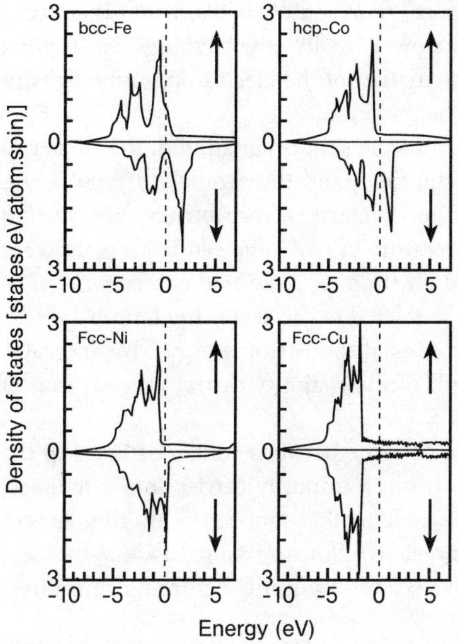

Figure 8.2 Spin-polarized densities of states for elemental metals Fe, hcp-Co, Ni, and Cu.

such as those of Fe, Ni, and Co. More realistically, carriers in a ferromagnetic metal lie in overlapping s and d bands, as shown in Figure 8.2.

Figure 8.2 shows the spin-polarized densities of states for the elemental metals Fe, hcp-Co, Ni, and Cu. In the lower right panel for fcc-Cu, it is seen that the 3d bands of Cu are below the Fermi level, and that the Fermi level lies in the s band. Looking at the lower left panel for fcc-Ni, it is seen that a small portion of the spin-down 3d band is pushed above E_F (i.e. becomes unoccupied), leading to the statement that the net moment of $0.54\mu_B$ for Ni arises from "0.54 hole in the $3d_\downarrow$ band" (which contains five electrons, corresponding to 10 electrons total in the 3d electron shell of the Ni atom). Note also from Figure 8.2 (lower left panel for Ni) that the spin-down (minority) density of states at E_F, n^-, greatly exceeds the spin-up (majority) density of states, n^+.

The electronic conductivity of ferromagnets is nearly entirely due to electrons from the overlapping s bands. The contribution to the conductivity of the d-band electrons is much smaller because of their higher effective mass m^*, which enters the denominator of the expression for the electronic conductivity, $\sigma = Ne^2\tau/m^*$ (see Section 5.4.1). Here τ represents the mean free time of the carrier between "scattering" collisions that change its direction of motion, thus affecting the conductivity and resistivity. The definition of the effective mass m^* (5.44) is inversely proportional to the curvature of the $E(k)$ relation, mentioned in connection with Figure 5.15. The $E(k)$ relation for s-band electrons typically has a high curvature, like that shown in Figure 5.15 (right panel) for electrons at $k = 0$ in GaAs. The $E(k)$ relation for d electrons typically has a lower curvature, representing a band that is broader on an E vs. k plot and more restricted (narrower) in its energy range. So in what follows it will be understood that, in the calculation of the electronic conductivity of ferromagnets, only the s-band carriers will be included.

A second effect on the resistivity comes from the spin orientation of the carrier in relation to the direction of the ferromagnetic field, and the symbols ρ^+ and ρ^- are used for the resistivity of majority and minority carriers, respectively (see Section 5.4.1). It is found that $\rho^+ < \rho^-$, and in the literature of GMR the symbol α is used for the ratio $\rho^-/\rho^+ > 1$. The resistivity, $\rho^+ = 1/\sigma^+$, with $\sigma^+ = Ne^2\tau^+/m^*$ can be written as $\rho^+ = m^*\lambda^+/Ne^2v_F$. Here the scattering time is related to the scattering length $\lambda = v_F\tau$, where $v_F = (2E_F/m)^{1/2}$ is the velocity of electrons at the Fermi energy. This indicates that the mean free path or scattering length of the majority carrier exceeds that of the minority carrier.

A different symbol, τ_{sf}, is used for the mean time between collisions that flip the direction of the spin, thus, for example, turning a minority carrier into a majority carrier. The spin-flip scattering time is related to the spin-flip scattering length $\lambda_{sf} = v_F\tau_{sf}$. It is generally the case that the resistivity mean free path $\lambda = v_F\tau << \lambda_{sf}$, so that a carrier will undergo many resistive scattering events before its spin direction is changed.

In the operation of the magnetic field sensors in the spin valve category, it is necessary that the spin projection of the typical carrier be preserved as it transits the device. So the condition for the GMR devices is $L < \lambda_{sf}$, where L is a typical device dimension, which qualifies these devices as examples of nanotechnology. As mentioned in Section 1.5, the width of the sensing sandwich may be 80 nm.

Finally a useful formula, due to N. F. Mott, for the resistivity in the situation that has been described is based on the idea that the currents and conductivities due to majority and minority carriers are additive:

$$\sigma = \sigma^+ + \sigma^-. \tag{8.7a}$$

So

$$1/\rho = (1/\rho^+) + (1/\rho^-) \tag{8.7b}$$

and

$$\rho = \rho^+\rho^-/(\rho^+ + \rho^-). \tag{8.7c}$$

8.6
The GMR Spin Valve, a Nanophysical Magnetoresistance Sensor

The operating element of the spin valve is a sandwich, comprising a hard magnet, typically Co, a conductor like Cu, and a soft magnet like NiFe. (See Figure 1.5, where the planar sandwich is perpendicular to the magnetic track, and carries current depicted as I_{in} and I_{out} in Figure 1.5.) The voltage across this (current-in-plane,

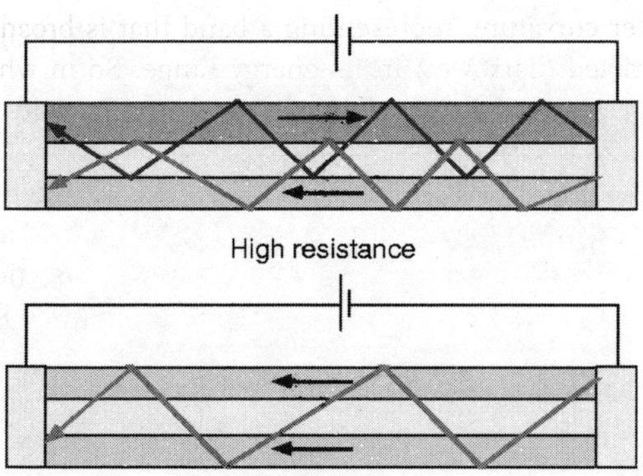

Figure 8.3 Sketch [6] of the CIP spin valve sensor (see also Figure 1.5) illustrating the shorter mean free path for carriers in the antiparallel electrode magnetization case, corresponding to higher spin valve resistance. (The horizontal arrows suggest the magnetization direction in the electrode, and the reversal of this direction in the upper electrode indicates that the upper (soft) magnetic layer has had its direction altered by encounter with the fringe magnetic field between two magnetic domains on the sample.) The device as shown is to be imagined as vertically positioned above a track of magnetic bits in a hard disk, with magnetization directions lying normal to the thickness *t* of the sensor, essentially the sandwich thickness (see Figure 1.5). Minimizing this thickness *t* is needed to efficiently capture the fringe magnetic field (up or down with respect to the printed page) and also to make the mean free path for spin-flip (see text) larger than the device dimension. (Reprinted with permission from Science, copyright AAAS.)

"CIP") sandwich is read to measure the (fringing) magnetic fields that point up or down at the joins between magnetic domains written into the disk. An enlarged sketch of the "CIP" spin valve sensor is given in Figure 8.3.

A typical trajectory for a carrier in the sandwich includes entry at one end, exit at length L, and several excursions of the carrier into hard and soft layers adjoining the central current-carrying Cu layer. The resistance of this device will depend upon whether the magnetizations of the outer magnetic layers are parallel or antiparallel. To estimate of the magnetoresistance MR = $(R_{AP} - R_P)/R_P$ of this device, one adds the resistivities that the carrier will experience in traversing the device, neglecting the resistivity in the copper layer and assuming equal excursions into the magnetic layers, where the resistivities are assumed to depend only on the relative spin orientations.

If magnetizations are parallel, there are two conductivities to be added:

$$\sigma_P = [1/(2\rho^+) + 1/(2\rho^-)], \tag{8.8a}$$

so

$$\rho_P = 2/(1/\rho^+ + 1/\rho^-). \tag{8.8b}$$

In the case of antiparallel magnetizations, then minority and majority carriers each have resistivity $\rho = \rho^+ + \rho^-$. After adding the two currents, the result is

$$\rho_{AP} = (\rho^{+} + \rho^{-})/2. \tag{8.9}$$

Based on this information, we find the approximate relation

$$\begin{aligned} MR &= (R_{AP} - R_P)/R_P \\ &= \{[(\rho^{+} + \rho^{-})/2] - [2/(1/\rho^{+} + 1/\rho^{-})]\}/[2/(1/\rho^{+} + 1/\rho^{-})] \\ &= (1-\alpha)^2/4\alpha \end{aligned} \tag{8.10}$$

where $\alpha = \rho^{-}/\rho^{+} > 1$.

8.7
The Tunnel Valve, a Better (TMR) Nanophysical Magnetic Field Sensor

The "*tunnel* valve", again a magnetoresistive circuit element (designated a "tunnel magnetoresistance" (TMR)), is a metal–insulator–metal tunnel junction (as described in Section 4.6.4) which comprises a soft magnet like Fe–Ni, an insulating tunnel barrier like Al_2O_3 (sapphire), and a hard magnet like Co. The orientation is described as CPP (current perpendicular to plane). In each ferromagnetic electrode, the energies of spin-up and spin-down electrons are relatively shifted by the "exchange splitting" $2J_E$ that gives ferromagnetism. This means that there will usually be different densities (often much different, see Figure 8.2) of spin-up vs. spin-down electrons at the Fermi energy.

The tunnel current from one electrode to the other preserves the spin direction of the electron, and the rule is that "spin-up" states in the opposite electrode can accept only "spin-up" electrons from the other side [7] as sketched in Figure 8.4.

The basic rule [7] that applies is that the tunnel current density is proportional to the product of the density of initial states times the density of final states at the

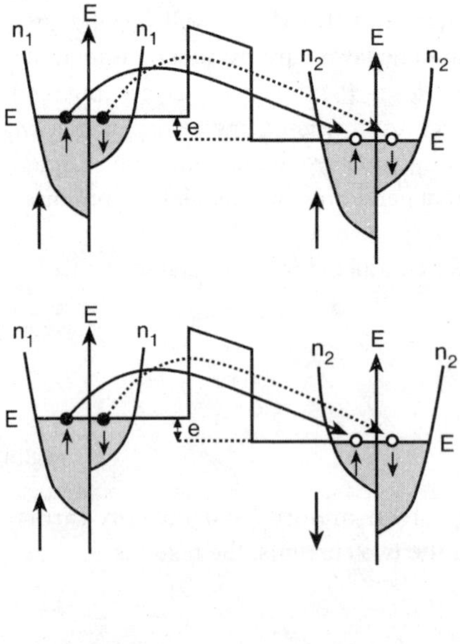

Figure 8.4 Illustration of the Julliere model [7] of the tunneling magnetoresistance. The currents of each spin projection are proportional to products of their initial and final densities of states, and spin direction is preserved in the tunneling process.

same energy. So the basic idea is that, by flipping the magnetization of the soft magnetic layer, its density of states for spin-up vs. spin-down electrons from the hard magnet Co layer are interchanged. So there may be a large change in the tunneling current at fixed voltage bias, in other words, a large tunneling magnetoresistance (TMR). These effects require exceedingly careful tunnel junction preparation, and are not easily observed. Strong effects were first reported [8] by Moodera et al. After much effort, larger effects have been found, e.g. see [9], and the TMR devices are now central to reading the information in hard drives.

In an applied magnetic field, the soft ferromagnet will easily flip the direction of its majority "spin-up" carriers (band), while the "hard" magnet will be stable against variation of ambient magnetic field, and will not change its majority spin direction. Since the "spin-up" and "spin-down" densities of states in each electrode may be significantly different, and the tunnel current is proportional to the availability of initial and final states, the total tunnel current at fixed bias (inverse to the tunnel resistance) can be significantly altered by a small ambient magnetic field. (A small magnetic field can redirect the ferromagnetic domain directions in a soft magnet, such as Permalloy, a Ni–Cu alloy. This is how the large observed permeabilities (4.94), on the order of 10^3 to 10^4, occur.) TMR effects of the order of 50% are observed at 300 K in recent magnetic tunnel junctions, which are incorporated into advanced magnetic disk drives as sensors, known as "tunnel valves".

In more detail (see Figure. 8.2) the tunnel current is proportional to the density of initial states at the Fermi energy (described above as n^+ for majority carriers and n^- for minority carriers) times the density of empty states of the matching spin polarization in the other electrode. Within a given electrode, the polarization $P = (n^+ - n^-)/(n^+ + n^-)$ (see equations (4.91)–(4.94)) is a useful description, and implies carrier densities

$$n^+ = N(P+1)/2 \text{ and } n^- = N(P-1)/2. \tag{8.11}$$

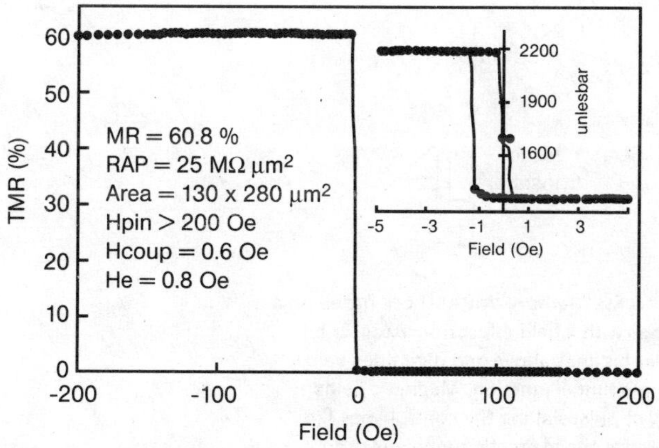

Figure 8.5 Curves [9] demonstrating 60.8% tunnel magnetoresistance appearing over a magnetic field range of about one oersted (0.1 mT) at room temperature in a tunnel junction device consisting of soft ferromagnet CoFeB, Al_2O_3 barrier, and FeCo hard magnet. The inset shows the resistance on an expanded magnetic field scale, clarifying the expected hysteresis. (Reprinted from IEEE Transactions on Magnetics with permission from IEEE.)

Using these definitions, the current density J can be seen to be the sum of tunneling contributions from spin-up and spin-down electrons, leading, after some algebra, to the formula [8]

$$\mathrm{TMR} = 2P_1P_2/(1 - P_1P_2). \tag{8.12}$$

Data like Figure 8.5 imply spin polarizations on the order of 0.5, which are reasonable. The TMR devices, based on the quantum-mechanical electron tunneling effect, with engineered tunnel barriers only a few nanometers in thickness, are presently used in the most advanced magnetic disk drives.

8.8 Magnetic Random Access Memory (MRAM)

8.8.1 Magnetic Tunnel Junction MRAM Arrays

Magnetic random access memory, sketched in Figure 8.6, based on tunnel junctions such as shown in Figure 8.5, offer advantages of nonvolatility, lower power consumption, and unlimited cycling lifetime [10]. At present cycling times on the order of 25 ns are reported for 4 Mb arrays, which are nearing the market stage. A prob-

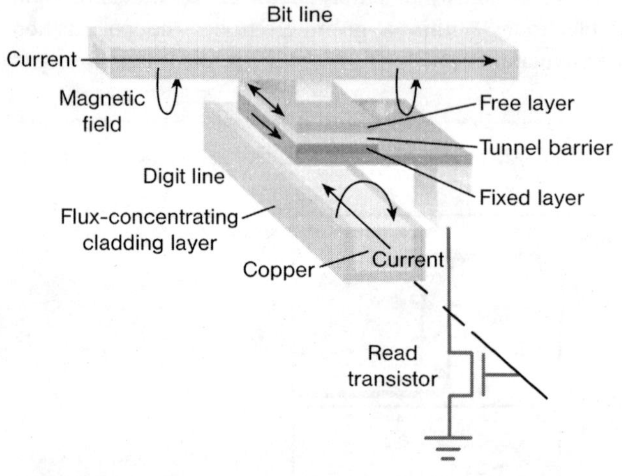

Figure 8.6 Magnetic Random Access Memory (MRAM) cell based on a magnetic tunnel junction in series with a field effect transistor for bit read selection [10]. Perpendicular bit lines above and digit lines below the magnetic tunnel select a single tunnel junction. Magnetic fields produced by simultaneous current pulses along the control lines flip the magnetization of the free layer of the magnetic tunnel junction. This memory is "nonvolatile" such that the information is retained indefinitely in the event of power loss. Such an element appears at each node of a cross-bar array, as shown in Figure 8.7. (Reprinted with permission from Science, copyright AAAS.)

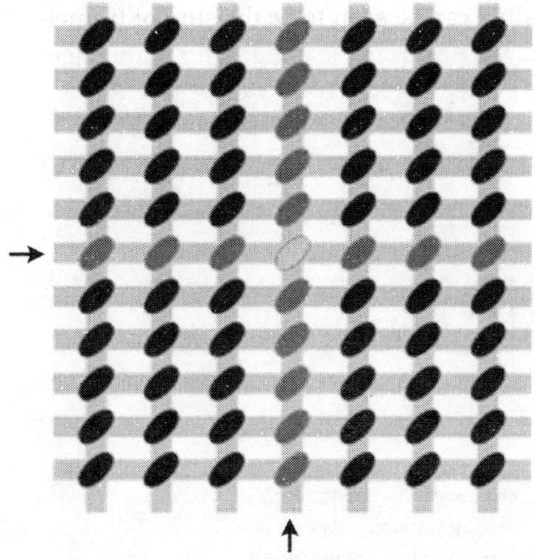

Figure 8.7 Top view [10] of an MRAM array, highlighting the fully selected cross-bar bit (arrows, and center position) and the ½ selected bits along each current-carrying write line. In toggle-MRAM, all bits are oriented at 45 degrees with respect to the write lines. Simultaneous current pulses in the two control lines are needed to flip the magnetization of the free layer at the intersection junction, and close control is needed to prevent errors if a flip occurs with only one pulse. It is stated that these nonvolatile random access memory arrays up to 4 Mb size have been produced. (Reprinted with permission from Science, copyright AAAS.)

lem in the technology is the control of parameters such that two simultaneous current pulses reliably flip the magnetization of the free magnetic layer (one electrode of the tunnel junction) while, equally reliably, a single pulse in one or the other control lines does not flip the magnetization. A concern also is the size of the output signal, even with 50% values of TMR, as mentioned in connection with Figure 8.5. It appears [11] that MgO-based tunnel junctions at room temperature can provide TMR of 230%, and that these, if successfully incorporated into the large arrays, may provide an important advance toward large-scale application.

8.8.2
Hybrid Ferromagnet–Semiconductor Nonvolatile Hall Effect Gate Devices

The hybrid devices sketched in Figures 8.8 and 8.9 use [12] a small (electrically isolated) sheet F of soft ferromagnet, whose magnetization is reversible by current pulses in an auxiliary control line (not shown), to create a voltage by means of the Hall effect (Section 5.4.1). The fringe magnetic field at the end of the small sheet of soft ferromagnet (F) has a sufficiently large normal component, about 1600 G (160 mT), perpendicular to the direction of the current flow, to produce a Hall voltage [12] on the order of 80 mV.

This device is a nonvolatile random access memory, and a prototype cell with micron dimensions is demonstrated with binary output states of 0 and 80 mV at room temperature. Device fabrication requires only two lithographic levels, one for the Hall cross and one for an electrically isolated, microstructured, bistable, ferromagnetic film F. The device is characterized [12] as having nanosecond read, write, and access times. Pulsed operation is expected to lead to larger output voltage. The device can perform functions of the traditional DRAM cell but is nonvolatile and does not consume energy for periodic refreshing of information, as is required in the DRAM case.

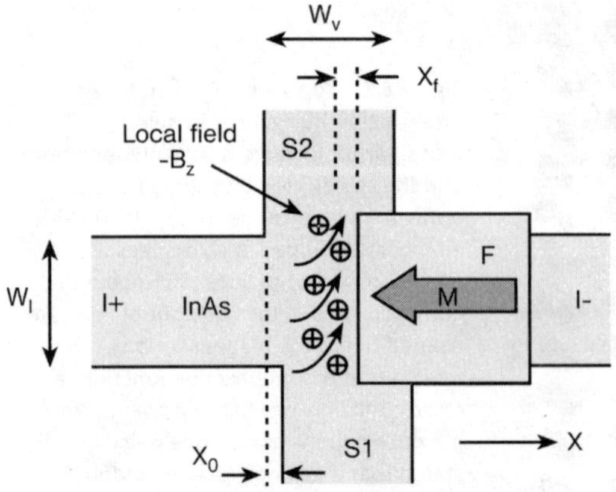

Figure 8.8 Top view [12] of hybrid ferromagnet–semiconductor switching device, based on the Hall effect (see Section 5.4.1). Magnetization M can be written into the small sheet of soft ferromagnet by means of a current pulse in an auxiliary current line, not shown. The circled plus symbols at the left edge of the ferromagnetic layer F indicate magnetic flux (local field $-B_z$) pointing down into the semiconductor. This magnetic field, up to 0.16 T, diverts current carriers in the InAs bar as shown, to produce a Hall voltage, which represents the output of the device. This device is nonvolatile and will operate if the size scale is reduced. (Reprinted from IEEE Transactions on Magnetics with permission from IEEE.)

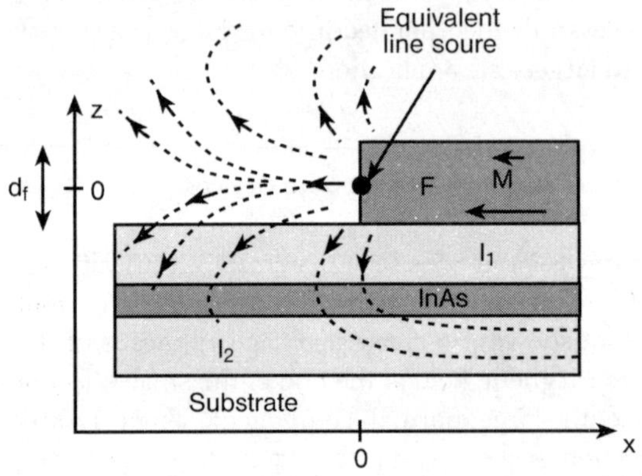

Figure 8.9 Side view [12] of hybrid ferromagnet–semiconductor switching device, showing detail of fringe magnetic field at the "active edge" of the ferromagnetic layer F which points down through the semiconductor layer to produce the Hall voltage. The insulating layer I_1 comprises 45 nm SiO / 3 nm InAs / 3 nm AlSb / 25 nm AlSb. Insulating layer I_2 is a 2 μm thick buffer layer of AlSb. (Reprinted from IEEE Transactions on Magnetics with permission from IEEE.)

8.9
Spin Injection: the Johnson–Silsbee Effect

Johnson and Silsbee [13] demonstrated experimentally that a non-equilibrium density of spin-polarized electrons can be transferred from a ferromagnet into a non-magnetic metal, like aluminum, under proper conditions. An important corollary, demonstrated experimentally in their pioneering work, is that the same type of ferromagnet–insulator–normal metal junction responds to the presence of non-equilib-

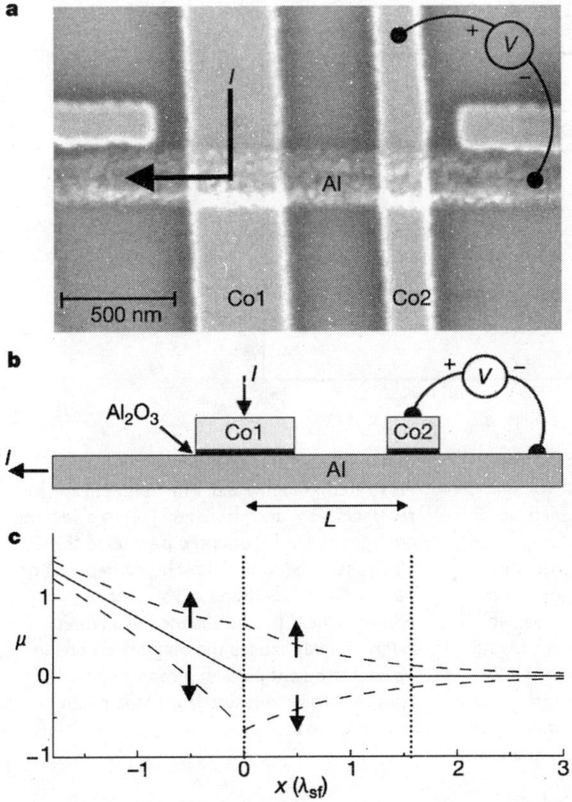

Figure 8.10 Geometry of the spin injection device [14]. (a) SEM image of device with spacing L of the two Co ferromagnetic crossing electrodes equal to 650 nm. Current flows from electrode Co1 through tunnel junction into the Al strip, and then flows only to the left end of the Al strip. There is no bias along the right side ($x > 0$) of the Al strip between the two Co electrodes. A voltmeter V monitors the potential difference between the Al strip and electrode Co2, as indicated. A voltage is seen to appear when non-equilibrium spin population diffuses along the (unbiased) Al strip the distance L from electrode Co1 to electrode Co2. (b) Cross section of the device, showing flow of current from ferromagnetic Co1 into Al and out the left end of the Al strip. (c) Plot of the electrochemical potential (voltage) in the device. The solid line is voltage in absence of spin injection. Dashed lines show spin-up electrons at higher potential than spin-down electrons, which reflects the energy shift of spin-up and spin-down bands in Co, suggested by Figure 8.2 (upper right panel). The horizontal scale, x coordinate, is measured in units of the spin diffusion length, which extends to the second tunnel junction between Al and Co2, where the non-equilibrium spin population produces an open circuit voltage V. (Reprinted from Nature with permission from Macmillan Publishers, Ltd.)

rium spin polarization in the non-magnetic metal by registering an open circuit voltage.

A nanometer-scale confirmation of the Johnson–Silsbee work has been reported [14] in the spin injection device shown in Figure 8.10.

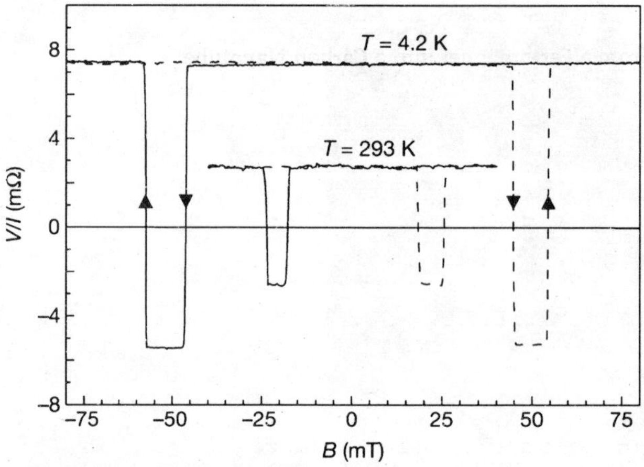

Figure 8.11 Observation [14] of the Johnson–Silsbee [13] effect is an open circuit voltage change that appears when the magnetization directions in the two Co electrodes are reversed. This means that the non-equilibrium spins that diffuse from Co1 to Co2 are oppositely directed to those in electrode Co2, which they sense through the tunnel barrier. Negative voltage sections of the measured V/I at 4.2 K and 293 K appear under external magnetic fields that make the magnetization directions in Co1 and Co2 antiparallel, reversing the roles of the spin-up and spin-down bands, as illustrated in Figure 8.3. The magnetization reversals were accomplished [14] by sweeping a magnetic field B oriented parallel to the Co electrode directions. The scheme depends on having Co1 wider than Co2, so that its magnetic field (in the absence of a larger external field) causes the magnetizations in 1 and 2 to be antiparallel. (Reprinted from Nature with permission from Macmillan Publishers, Ltd.)

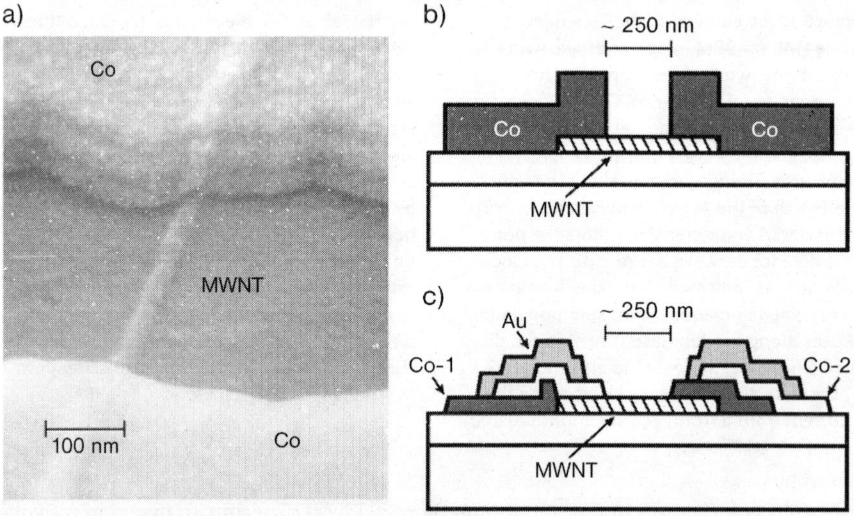

Figure 8.12 Co electrodes are shown [15] contacting two ends of a carbon multiwall nanotube, with a contact spacing of about 250 nm. The diameter of the MWNT was about 30 nm, ranging from 10 to 40 nm in other cases. The Co electrodes are polycrystalline Co deposited by thermal evaporation. The substrate is a semi-insulating Si wafer covered by a 200 nm thick layer of SiO_2. The magnetic domain size in the Co is reported as 50 nm. (Reprinted from Nature with permission from Macmillan Publishers, Ltd.)

8.9.1
Apparent Spin Injection from a Ferromagnet into a Carbon Nanotube

Coherent transport of electron spin transport in a carbon nanotube, suggesting a spin-flip scattering length of at least 130 nm, has been reported [15]. Figure 8.12 indicates two different experimental arrangements allowing electrons to flow from Co into a multiwall nanotube (MWNT).

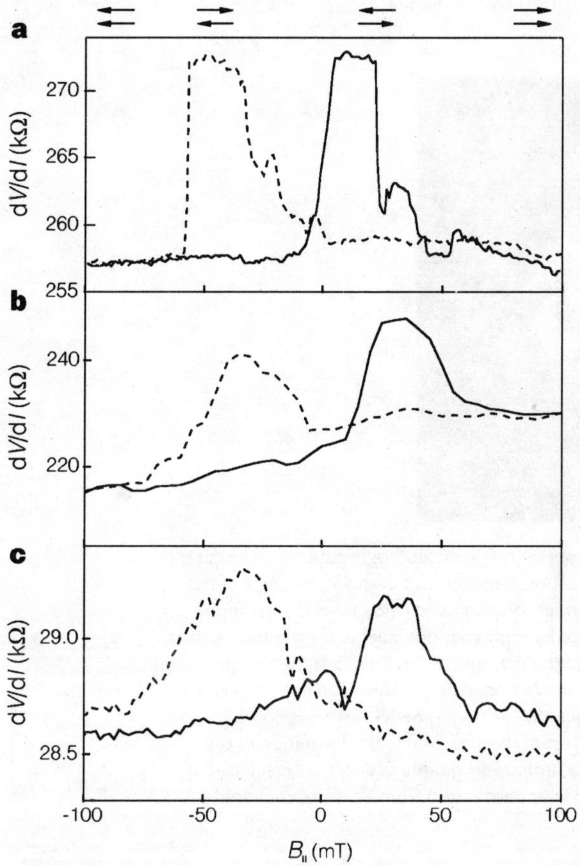

Figure 8.13 Measurements of the differential resistance of the Co/nanotube/Co device, consistent with the ideas of spin-coherent transport and with a coercive field (see Figure 5.18) in the Co on the scale of 20 mT. The high resistance regions are interpreted [15] as arising when the magnetizations of the two Co contacts are antiparallel. It is believed that the coercive fields of 50 nm domains in polycrystalline evaporated Co films are of this order. The magnitude of the resistance change is interpreted within the TMR formula given in connection with Figure 8.10, as consistent with spin polarizations in the Co on the order of 34%, which would give a resistance change of 21%. The observed change of resistance, 9%, is consistent with the above if the spin scattering length is shorter than the free length of the nanotube, giving the estimate for the spin scattering length of 130 nm. (Reprinted from Nature with permission from Macmillan Publishers, Ltd.)

8.10
Magnetic Logic Devices: a Majority Universal Logic Gate

It appears that a breakthrough has recently been reported [16,17] in the area of magnetic logic devices, of the class referred to as quantum cellular automata (QCA), based on interacting nanomagnets. The present devices are extremely promising, as they operate at room temperature and are straightforward in concept and in fabrication. The basic idea is clear from Figure 8.14, keeping in mind the nature of the dipole magnetic field, as exhibited by a classical bar magnet.

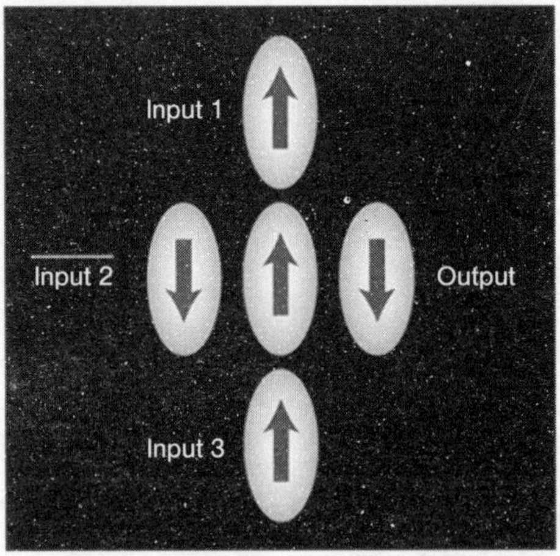

Figure 8.14 Schematic [16] of universal magnetic dot logic gate [17]. The central nanomagnet chooses its magnetization direction (up or down, because of its elongated shape) based on the *sum* of the magnetic fields from similar input nanomagnets 1, 2, and 3. The *output* simply is antiparallel to the central magnet, and thus represents the *negative* of the majority input. This is simple magnetostatics, playing with bar magnets, whose dipole magnetic fields lead to antiparallel near-neighbors in the *x*-direction, and parallel (ferromagnetic) near-neighbors in the *y*-direction. The only quantum aspect of the situation is the formation of the ferromagnetic domains, which is a profoundly quantum effect, as discussed in relation to Figure 5.20. (Reprinted with permission from Science, copyright AAAS.)

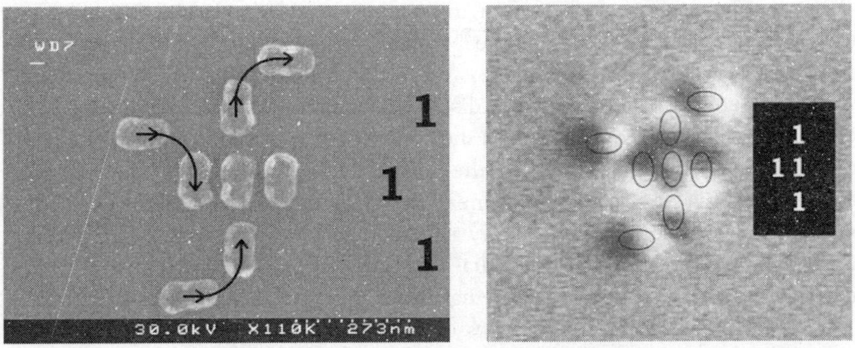

Figure 8.15 (Upper panel) TEM images [17] of planar nanomagnets of nominal dimensions 135 nm × 70 nm with 30 nm thickness of Permalloy film. Horizontal spacing between (vertically elongated) nanomagnets is 25 nm. The horizontally oriented magnets on the left are drivers, and those on the right read the output. An externally applied clocked *B* field arranges the driver magnet directions of magnetization. The idea is that the driver magnets are big enough to be dictated by the applied fields, while the smaller magnets respond only to their near-neighbors. Arrows suggest local magnetization directions and how these orientations were driven by the inputs. (Lower panel) Magnetic force microscope (MFM) image indicating the magnetization directions in the same structure as indicated in the upper panel. The location of the nanomagnets is drawn superimposed on the MFM data. (A larger set [17] of MFM images demonstrates the full majority logic gate functionality of the device.) It is stated that the magnetization switching time in these nanometer scale magnetic dots is about 100 ps. It is stated that this majority gate logic scheme would have an inherent operating speed of 100 MHz, with energy dissipation (heating) of about 1 eV per switching event. On this basis, an array of 1000 gates would have a total dissipation of 0.1 W, assuming each device switches once per cycle.

Table 8.1 Summary of logic states in the majority gate [17] for all input combinations (truth table). The logic state of the central nanomagnet is determined by the logical majority vote of its three input neighbors, of which the ferromagnetically coupled neighbors vote directly and the antiferromagnetically coupled neighbor votes inversely to its magnetic state. The logic state of the central magnet is written inverted to the output magnet by antiferromagnetic coupling. If programmed by the first-input bit value, a majority gate can function as a two-input NAND gate (upper four rows of the table) or as a two-input NOR gate (lower four rows of the table)

Logic state of input magnet	Logic state of central magnet	Logic state of output magnet
000	0	0
001	0	0
010	0	0
011	1	1
100	0	0
101	1	1
110	1	1
111	1	1

What has been demonstrated is a three-input inverting MAJORITY function. This function can be applied easily to perform the more familiar NOR or NAND functions. The work is noteworthy [16] also in that an "adiabatic" clocking scheme, in which the energy barriers between discrete data states are gradually lowered and then raised again, is used. This allows the system to move gradually from one computational state to another without the wasted energy inherent in conventional architectures. This feature addresses one of the most important problems of conventional microelectronics, the control of waste heat. Another advantage of the present universal gate is the idea that the actual function of the gate can be changed after the hardware is built. In principle, the hardware could be reconfigured within a few nanoseconds, allowing the microprocessor to adapt its very architecture to the best form for the computation in hand at that instant. It is suggested that this type of gate could operate on a similar time scale as the MRAM elements described above, namely, that sub-nanosecond switching times may be possible. It is anticipated [16] that this magnetic gate would likely find application in a hybrid arrangement where it would be imbedded in a silicon technology chip to do some class of functions.

8.11 Superconductors and the Superconducting (Magnetic) Flux Quantum

Superconductivity: a Macroscopic Quantum State

Superconductivity is less familiar (the effect requires a low temperature) and, perhaps, is more difficult to understand than ferromagnetism, the useful cooperative electron effect discussed earlier.

In superconductors, each electron spin is paired with its opposite member across the Fermi surface. Each pair $(\mathbf{k}\uparrow,-\mathbf{k}\downarrow)$ has zero momentum $\mathbf{p} = \hbar\mathbf{K}$ and zero spin: this pair is a Bose particle (like a photon, e.g. in a laser). An indefinitely large number of such particles (pairs) can exist in precisely the same $\mathbf{K} = \mathbf{k}\uparrow + (-\mathbf{k}\downarrow) = 0$ quantum state. Under an applied electric field, the whole (paired) Fermi surface (representing N^* electrons) *slightly* displaces from $\mathbf{K} = 0$ to $\mathbf{K} = \Delta\mathbf{K}$, so that there is a net macroscopic momentum, $\mathbf{P} = 2N^*\hbar\Delta\mathbf{K}$, a net velocity $\mathbf{V} = \mathbf{P}/m$, and a net current density $\mathbf{J}$. In such a superconductor, each pair contributes equally to provide the resultant current density $\mathbf{J}$, with no scattering and no loss of energy. The pair wavefunction

$$\psi = \exp[\mathrm{i}(\mathbf{K} - e\mathbf{A}/\hbar)\cdot\mathbf{r}] = \exp(\mathrm{i}\Theta) \tag{8.13}$$

of each pair is identical. The pairs are Bose objects, all in precisely the same quantum state, namely $\mathbf{K} = 0$. Here the symbol $\mathbf{A}$ is introduced for the "vector potential", to describe a magnetic field as

$$\mathbf{B} = \mathrm{curl}(\mathbf{A}) = \nabla\times\mathbf{A} \quad \text{(where } \mathbf{A} \text{ is the vector potential).} \tag{8.13a}$$

The superconducting electron pairs are all in a coherent quantum state, and the wavefunction $\psi = \exp[i(\mathbf{K} - e\mathbf{A}/\hbar)\cdot\mathbf{r}] = \exp(i\Theta)$ of each pair is reliable (coherent) over long distances and long times. [J. Bardeen, who led a student group (Cooper and Schrieffer) in formulating the "BCS" theory, the universally accepted basic theory of superconductivity, is quoted as having said that the superconducting phase Θ is "coherent over miles of dirty lead wire".]

Magnetic flux is expelled from the metal below the superconducting transition temperature, in the case of a "type I" superconductor. This is the Meissner effect, and can be described as perfect diamagnetism. For this reason, the current carried in conventional "type I" superconductors is confined to a layer near the surface, the depth being called the "penetration depth", and superconductivity disappears above a relatively small magnetic field. This means that "type I" superconductors are unsuited to environments which have large magnetic fields.

Fortunately, "type II" superconductors, which include Nb, Nb_3Sn, NbTi, V_3Ga, and all of the "high-temperature superconductors", allow penetration of (large) magnetic fields in the form of isolated vortex lines, maintaining a high transition temperature, T_c, even at high field. These superconductors are used in high-field applications such as the 1 T superconducting magnets used in the medical imaging MRI apparatus.

The Superconducting Magnetic Flux Quantum

The magnetic flux quantum can be used in a superconducting computer to represent the basic bit of information. Magnetic flux is described by $\Phi_M = \int \mathbf{B}\cdot d\mathbf{S}$. It is familiar that an electric field $\mathbf{E} = -\nabla V$, where V is the electric potential and ∇ represents the gradient operator, whose x-component, for example, is $\partial/\partial x$. Less familiar, but also of physical importance, is the magnetic vector potential $\mathbf{A}$, such that $\mathbf{B} = \nabla \times \mathbf{A}$. In nanophysics, the wavefunction for an isolated traveling electron is $\psi(r,t) = \exp[i(kr - \omega t)]$. However, in the presence of a magnetic field B, represented by the vector potential $\mathbf{A}$, this expression has to be modified (experiments confirm this) to $\psi(\mathbf{r},t) = \exp[i(\mathbf{k} - e\mathbf{A}/\hbar)\cdot\mathbf{r} - i\omega t)]$.

Imagine a metal loop, enclosing a magnetic field B, whose flux through the loop is $\Phi_M = \int \mathbf{B}\cdot d\mathbf{S}$. The wavefunction $\psi(\mathbf{r},t) = \exp[i(\mathbf{k} - e\mathbf{A}/\hbar)\cdot\mathbf{r} - i\omega t)]$ describes electrons in the metal loop. If $\psi(\mathbf{r},t)$ is correct, it must be single-valued, so that it will give the same probability, $P(\mathbf{r}) = \psi^*(\mathbf{r},t)\psi(\mathbf{r},t)$, for a given position, $\mathbf{r}$, independent of how many times around the loop the electron has traversed. Thus, the phase of the wavefunction must be the same, to an integer n multiple of 2π, upon traversing the loop. So, $e\int A/\hbar \cdot dL = n2\pi$. But since $\int A\cdot dL = \Phi_M = \int B\cdot dS$, it follows that $e\Phi_M = nh$, or $\Phi_M = h/e = 2\times 2.067\times 10^{-15}\,\mathrm{T\,m^2}$. Actually this value is twice the usually quoted value, $h/2e$, for the superconducting pair, because the effect is easiest to observe in a superconductor, where the charge carrier is two electrons (a pair). Thus, the superconducting flux quantum is

$$\Phi_0 = h/2e = 2.067\times 10^{-15}\,\mathrm{T\,m^2}. \tag{8.14}$$

A topic of interest is the area required in practice to store a flux quantum, as compared to the area required for other possible basic units of memory, such as the electron charge, the electron spin, or the smallest possible ferromagnetic domain (as used in current hard disk memory). This analysis has assumed that the phase of the electron is undisturbed during the period of measurement. The phase of the wavefunction contains the term $i\omega t = iEt/\hbar$. The time τ_i over which the energy of an electron is undisturbed is called the "inelastic scattering time", and for a pair of electrons in a superconductor this time is very long and effectively infinite. In normal metals and semiconductors the inelastic scattering time is short. For this reason it has been difficult to observe quantum interference effects in normal metals and semiconductors, but many observations have nonetheless been made which verify the quantum predictions.

8.12 Josephson Effect and the Superconducting Quantum Interference Detector (SQUID)

The superconducting wavefunction has a reliable phase, Θ, which applies to a macroscopic number of electrons. This regularity is enforced by the pairing phenomenon, which leads to a macroscopic number of pairs of electrons in precisely the same quantum state, analogous to the macroscopic number of photons occupying a single electromagnetic mode in a laser. (The phenomenon is also similar to a single ferromagnetic domain, where all electron spins are parallel.)

The reliability of the superconducting phase $(\mathbf{K} - e\mathbf{A}/\hbar)\cdot\mathbf{r} = \Theta$ leads to easy observation of interference effects even in large-scale devices. It is also important to notice that a magnetic field $\mathbf{B} = \nabla\times\mathbf{A}$ will lead to systematic variation of $\mathbf{A}$, and consequently the phase, over a superconducting region, even if the magnetic field is constant over that region. We have already seen above that the phase $(\mathbf{K} - e\mathbf{A}/\hbar)\cdot\mathbf{r} = \Theta$ changes by 2π in a path which encloses one quantum of magnetic flux.

A *Josephson junction* is a tunnel junction between two superconductors, which has a sufficiently thin barrier that the superconducting pair wavefunctions on the opposite sides significantly overlap, creating an energy E_J of coupling between the two superconductors across the junction. If this energy is larger than k_BT it "locks" the phase difference across the junction. A *Josephson supercurrent* flows through the junction which, remarkably, depends on the difference $\Theta_1 - \Theta_2 = \varphi$ of the superconducting phases on the opposite sides:

$$I_J = I_{J0}\sin\varphi. \tag{8.15}$$

It turns out that the Josephson coupling energy is essentially the same as the maximum Josephson supercurrent:

$$E_J = -\hbar I_{J0}/2e. \tag{8.16}$$

If there is a magnetic field parallel to the junction plane (B threads through the insulating barrier), then the phase difference φ, and thus the current density, becomes a function of position along the junction. Thus, the sign of the junction current will reverse every time φ increases by π! This has the effect of making the Josephson supercurrent $I_J = I_{J0} \sin \varphi$ oscillate with applied magnetic field B, with its first zero when the junction contains precisely one flux quantum, followed by added zeroes as additional flux quanta exist in the junction. This is the basis for a detector of magnetic field. In a loop containing two junctions, called a Superconducting Quantum Interference Device (SQUID), the area of B field intercepted increases from the cross sectional area of the tunnel barrier, to the large area enclosed by a circuit containing two Josephson junctions. This makes an extremely sensitive detector of magnetic field.

Rather similar coupling effects between two superconductors are also seen if, instead of a tunnel junction, the connection between the superconducting electrodes is a narrow wire, an example of a "weak link". A recent nanoscale example [18] of a SQUID, based on nanoscale weak links, is illustrated in Figure 8.17. (This device has several important differences in its behavior from the conventional SQUID, because its elements are so small as to allow some penetration of magnetic field, contrary to the usual Meissner effect, in which a magnetic field does not enter the superconductor, but is deflected around it.)

The fabrication is based on stretching DNA across a trench. The DNA is dissolved in water, dropped onto the sample at the trench, and the sample is dried in vacuum. Most of the DNA molecules reside completely on one side or the other, but some molecules dry with a small segment stretching across the trench. The entire chip is then coated with a few nanometers of superconducting MoGe using a sputtering method. The observed straight line segments crossing the trench are a consequence

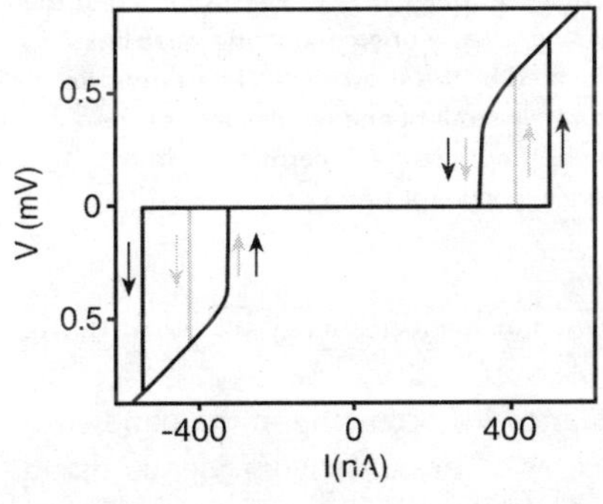

Figure 8.16 V–I measurements [18] on a DNA-templated quantum interference device at extremely low temperature, 0.285 K (such that the whole structure can become superconducting at $B = 0$), show hysteretic switching between fully superconducting $V = 0$ states, and two resistive states which are generated by application of slightly different perpendicular magnetic fields (about 0.25 and 0.5 mT). To generate these curves, the bias current was swept from large negative values to large positive values and return. (Reprinted with permission from Science, copyright AAAS.)

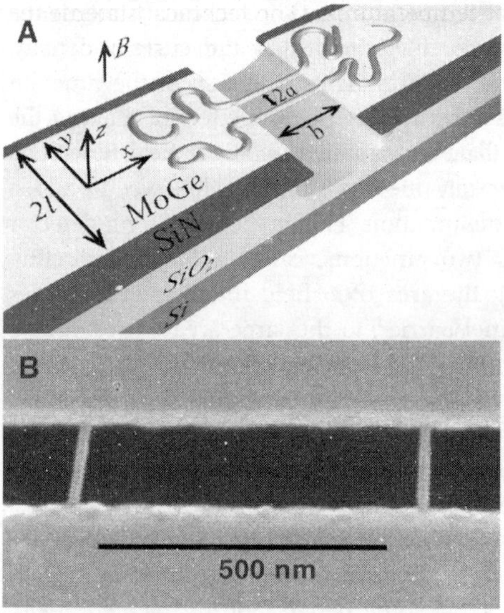

Figure 8.17 (A) Pictorial representation [18] of DNA-templated two-nanowire SQUID-like device. Two strands of DNA are stretched across a trench etched into SiN/SiO_2 on a Si chip. As suggested by the pictorial representation, long DNA strands are simply deposited (in a drop of water) on the SiN layer, above the trench; in the process of drying they self-assemble to pull the crossing sections tight across the trench, normal to the trench orientation. Subsequently, the molecules and the banks are coated with superconducting $Mo_{21}Ge_{79}$. Resulting are two weak links (MoGe-coated DNA strands) spaced by dimension $2a$ about 595 nm across a gap of width b about 120 nm, creating a (SQUID interferometer) area $2ab$ which is normal to an applied B field, in the z-direction. The included magnetic flux $\Phi_M = \int \mathbf{B}\cdot d\mathbf{S}$ is nominally $2abB$, but is found in the experiment to be much larger, because [18] the B field penetrates significantly into the "few nm thickness" MoGe contacting electrode "leads", whose width is $2l$. The field penetrates the leads to a distance on the order of 9–15 μm. Current flows to the device through these MoGe contacts of nm thickness and width $2l$. (B) Scanning electron microscope image of the device shows that the crossing wires (superconducting weak links), as located by the stretched DNA strands, lie accurately straight and parallel across the trench. (Reprinted with permission from Science, copyright AAAS.)

of the fact that the DNA can maximize its surface adhesion energy by minimizing the length crossing the trench [18].

The operation of this device is based on resistance V/I occurring in the nanowires as a consequence of "thermal phase-slip processes". The underlying idea is that a phase slip creates a voltage, and the one-dimensional nature of the nanowires weakens the pair binding energy, which is needed to enforce phase coherence, which gives zero resistance in a bulk superconductor and also across a Josephson tunnel junction for currents below the critical current I_{J0}. From the point of view of superconductivity, the detailed behavior of this DNA-templated nanoscale device requires a subtle and original analysis, provided by the authors [18].

The authors [18] point out that these devices may have an advantage (in addition to being nanometer scale), since, owing to the extreme narrowness of the wires, they operate over a relatively wide range of temperatures. (The technical statement is that the resistive transition in nano-superconducting wires becomes significantly broadened in temperature.) The authors [18] have demonstrated an unexpectedly simple way of generating wires of widths in the 5–15 nm range, which are accurately aligned, using DNA as a template. Since the thickness of the wires is easy to control, in the range of a few nanometers, by the sputter deposition process, the cross sectional area of these wires is unprecedentedly small, in the range of 15–45 nm^2. Supposing a thickness of 3 nm (~30 atoms) and width of 5 nm (~50 atoms), the cross section is made up by about 1500 atoms.

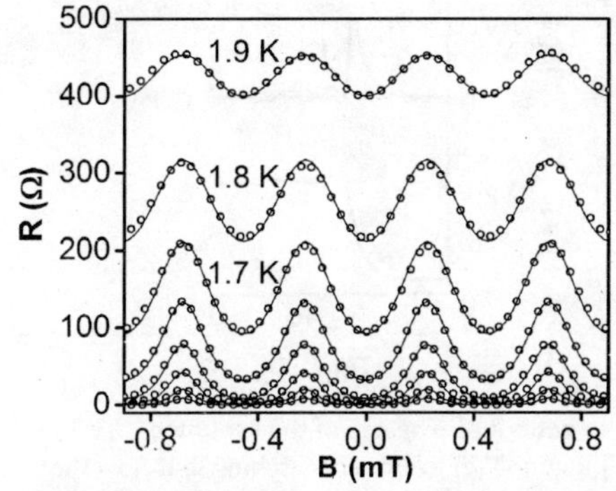

Figure 8.18 Resistance versus magnetic field (perpendicular to device) at temperatures 1.2–1.9 K. Each oscillation represents a change in flux through the device of one flux quantum $\Phi_0 = h/2e$. The solid lines are generated [18] by a model and assume a period B for the oscillations of 456 μT. This period is far short of the conventional value $B = \Phi_0/(2ab) \approx 25$ mT, which is explained [18] in terms of penetration of magnetic flux into a much larger area, due to the extremely thin nature of the contacting MoGe leads. The lack of zero resistance in this nanoscale superconducting device is attributed to "phase slips" in the narrow wires, an effect well known in superconducting nanowires. (Reprinted with permission from Science, copyright AAAS.)

8.13 Superconducting (RSFQ) Logic/Memory Computer Elements

The basic switching process in the (unshunted) Josephson junction circuit [19] occurs as a jump from the superconducting ($V = 0$) state (which was discussed above) into the voltage state, as illustrated in Figure 8.19(b). Switching out of the supercurrent state occurs when the applied current (in Figure 8.19(a) this is shown as $I_{in} + I_b$) exceeds the critical Josephson junction current I_c. As long as the Josephson junction (cross marked J in Figure 8.19(a)) is superconducting, all current flows through it and no current flows through the load resistor, R_L. The output voltage is thus *zero*. After the junction switches to its voltage state, at $V \approx 2\Delta(T)/e$, this voltage appears at the output of the device, representing a logical *one*.

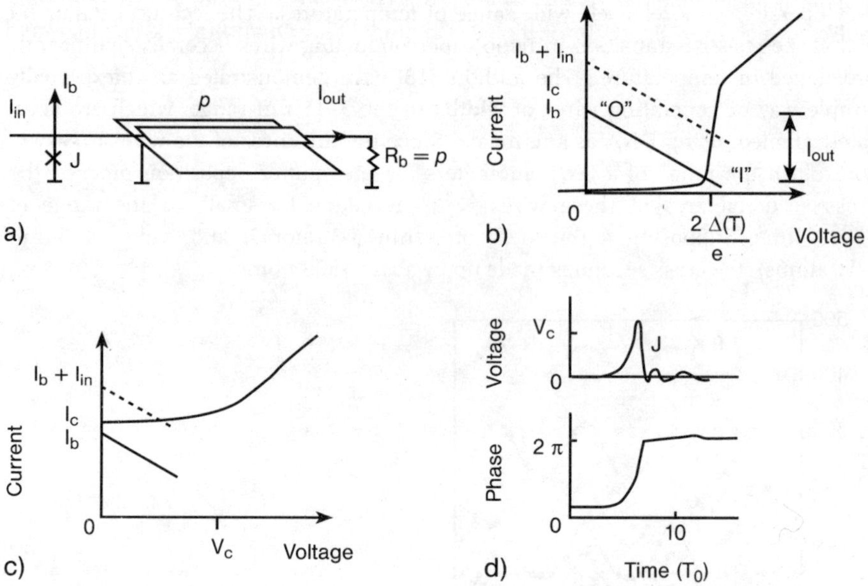

Figure 8.19 The simplest stage [19] of the Josephson junction logics. (a) Equivalent circuit and scheme of its operation. The input current is the signal plus the bias, and when the sum exceeds the critical current for the junction, it switches (passes one flux quantum) and the current (abruptly) is sent to the load resistor. How this happens is different when using (b) an underdamped (unshunted) and (c), (d) an overdamped (shunted) junction. The situation in (b) was found eventually to be detrimental, in that it was "latched", and hard to get out of. The use of the "shunted junction" (c) avoids the latching, but now the only message of the junction switching is the voltage pulse, sent down the transmission line, as seen in (d). This is the present state of the art, keeping track of the SFQ pulses. The curves shown of voltage and phase vs. time are the result of sophisticated numerical calculations [19], which are believed to be realistic. (Reprinted from IEEE Transactions on Applied Superconductivity with permission from IEEE.)

This form of Josephson junction superconducting logic was extensively investigated at the IBM Yorktown Heights laboratory during the early 1980s. Recently, this approach has evolved into a related but subtler scheme, based on *resistively shunted Josephson junctions*, leading to the $I(V)$ shown in Figure 8.19(c). The *shunting resistor* parallel to the Josephson junction has a small value (perhaps 1–8 Ω) so that the current I remains nearly constant after switching, but a transient voltage pulse (illustrated in Figure 8.19(d)) is carried along a transmission line (Figure 8.19(a)) to provide a *voltage pulse* at the output.

The Single Flux Quantum Voltage Pulse

The voltage pulse from the transit of a single flux quantum through the Josephson junction is

$$\int V(t)\mathrm{d}t = \Phi_0 = h/2e = 2.07\,\mathrm{mV\,ps} \quad \text{(RSFQ pulse)}. \tag{8.17}$$

This is just Faraday's law, $V = -d\Phi_0/dt$. These are the "single flux quantum" pulses, which form the basis for the "Rapid Single Flux Quantum" (RSFQ) logic and memory schemes. This result for the RSFQ pulse can be understood also using the Josephson junction relation

$$d\varphi/dt = (2e/\hbar)\,V(t) \quad \text{(AC Josephson effect)}, \tag{8.18}$$

where φ is the difference of the superconducting phase across the junction. The SFQ voltage pulse corresponds to an advance of this phase difference by 2π. In terms of the angle φ, the bias current, near to the critical value, as shown in Figure 8.20, corresponds to an angle a bit less than $\pi/2$ (from the relation $I_J = I_{J0} \sin \varphi$, where $\sin \varphi$ is nearly unity).

There is a well-known analogy between the current-biased Josephson junction and a pendulum whose deflection angle is φ. The arrangement in Figure 8.19(a) corresponds to a pendulum biased to an angle φ close to $\pi/2$, nearly horizontal, with the idea that the incoming pulse will kick the pendulum "over the top" to advance by 2π. Since the (equivalent) kicked pendulum is *overdamped*, it will make only one turn, leading to the single SFQ pulse.

According to calculations by the authors [19] the time scale for the SFQ pulses varies with the linear scale L of the junctions, with a time unit of 4 ps for $L = 5$ μm down to 0.5 ps for $L = 0.7$ μm. The numerical calculations (leading to waveforms as shown in Figure 8.19(d)) are advanced, and based on the assumption that the characteristic impedance of the transmission line (Figure 8.19(a)) matches the load resistance R_L.

The essential test of the scheme is whether the output SFQ pulse, when presented to another biased junction, can change that junction into the voltage state. The answer is "Yes", in that RSFQ circuits have demonstrated counting at 750 GHz, as described in Section 7.5. In the time since that result, which occurred in 1998, this technology has been quiescent, primarily as a consequence of the constraint of refrigeration to low temperature.

Analog-to-Digital Conversion (ADC) Using RSFQ Logic

One strong advantage of the RSFQ scheme is its fast operation, with 750 GHz counting as demonstrated in Figure 7.4. A more recent diagram for an analog-to-digital converter, based on the RSFQ scheme [20], is shown in Figure 8.20. It is stated that complete analog-to-digital systems with digital sampling rates of ~20 GHz have been demonstrated. The most immediate applications of such systems seem to be in acquisition of radar signals. What is involved is directly digitizing the incoming radio-frequency voltage. It is anticipated that advances in closed cycle cooling systems will lead to wider application of superconducting devices. It does not seem likely that the class of "high-temperature superconductors" will have a large impact in this area, because of the difficulty in fabricating reliable devices. The ADC converters are large systems, with up to 10 000 junctions. The conventional and reliable junction technology uses the $Nb/Al_2O_3/Nb$ "trilayer method" [21], with Josephson critical current densities on the order of a few kA/cm^2 in junction sizes on the order of several micrometers. The "trilayer method" involves depositing Al onto Nb, then oxidizing the Al (followed by evaporation or sputtering of the Nb counter-electrode), since Nb itself does not produce a suitable tunnel barrier [21]. Since damped junctions are needed, as mentioned above, the junctions require resistive shunts which are typically made using a metallic layer such as Mo or TiPd. The circuits also require inductors, which are usually Nb striplines, separated from a ground plane by an SiO_2 insulator.

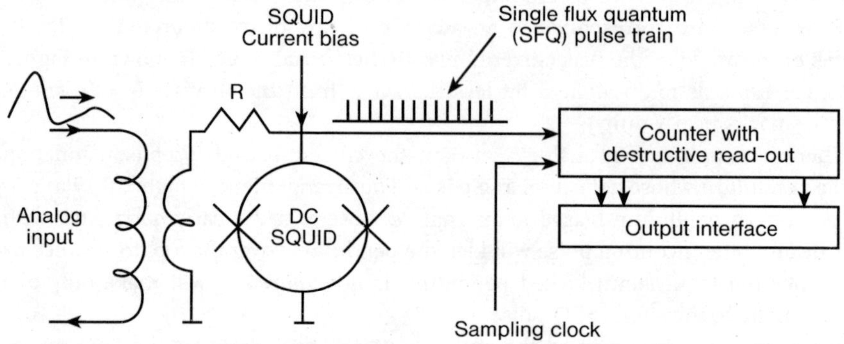

Figure 8.20 This is a schematic [20] of a sensitive *V/F* ADC based on the *V*-flux transfer characteristics of the SQUID and followed by an RSFQ binary counter. The train of SFQ pulses coming from the DC SQUID device is counted by an RSFQ counter of the type described in Section 7.5. (Reprinted from Proceedings of IEEE with permission from IEEE.)

References

[1] C. H. Bennett and D. P. DiVincenzo, Nature **404**, 247 (2000).
[2] "Strange Behavior", Economist, September 3, 2005, p. 72.
[3] G. P. Berman, B. W. Brown, M. E. Hawley, and V. I. Tsifrinovich, Phys. Rev. Lett. **87**, 097902 (2001).
[4] Y. Manassen, I. Mukhopadhyay, and N. R. Rao, Phys. Rev. **B61**, 16223 (2000).
[5] C. L. Hollenberg, A. S. Dzurak, C. Wellard, A. R. Hamilton, D. J. Reilly, G. J. Milburn, and R. G. Clark, Phys. Rev. **B69**, 113301 (2004).
[6] G. A. Prinz, Science **282**, 1660 (1998).
[7] M. Julliere, Phys. Lett. **54A**, 225 (1975).
[8] J. S. Moodera, L. R. Kinder, T. M. Wong, and R. Meservey, Phys. Rev. Lett. **74**, 3273 (1995).
[9] D. Wang, C. Nordman, J. M. Daughton, Z. Qian, and J. Fink, IEEE Trans. Magnetics **40**, 2269 (2004).
[10] J. Akerman, Science **308**, 508 (2005).
[11] S. S. P. Parkin, Nature Materials **3**, 862 (2004).
[12] M. Johnson, B. R. Bennett, M. J. Yang, M. M. Miller, and B.V. Shanabrook, IEEE Trans. Magnetics **34**, 1054 (1998).
[13] M. Johnson and R. H. Silsbee, Phys. Rev. Lett. **55**, 1790 (1985).
[14] F. J. Jedema, H. B. Heersche, A. T Filip, J. J. A. Baselmans, and B. J. van Wees, Nature **416**, 713 (2002).
[15] K. Tsukagoshi, B. W. Alphenaar, and H. Ago, Nature **401**, 872 (1999).
[16] R. P. Cowburn, Science **311**, 183 (2006).
[17] A. Imre, G. Csaba, L. Ji, A. Orlov, G. H. Bernstein, and W. Porod, Science **311**, 205 (2006).
[18] D. S. Hopkins, D. Pekker, P. M. Goldbart, and A. Bezryadin, Science **308**, 1762 (2005).
[19] K. K. Likharev and V. K. Semenov, IEEE Trans. Appl. Supercond. **1**, 3 (1991).
[20] O. A. Mukhanov, D. Gupta, A. M. Kadin, and V. K. Semenov, Proc. IEEE **92**, 1564 (2004).
[21] E. L. Wolf, J. Zasadzinski, J. W. Osmun, and G. B. Arnold, Solid State Commun. **31**(55), 321 (1979).

9
Silicon Nanoelectronics and Beyond

The hugely successful silicon electronics is moving forward with smaller and smaller size scales, as suggested (if not mandated) by Moore's Law (see Figure 1.2).

In confirmation, Intel [1] has recently announced a *test* chip with features of 45 nm, which will allow one billion (10^9) transistors per chip. This test device, following *present production* of devices with 65 nm features, is manufactured using traditional photolithographic technology, with glass and chromium metal masks (see Section 7.2.1). The smaller, 45 nm (about 450 atom diameter), feature resolution was achieved using light of 193 nm wavelength, but with "compensations in the design of the mask to compensate for the blurriness that can occur when the *wavelength of the light is larger than the mask features* it passes through" [1].

So, wave optics, supplanting geometrical optics, has been brought into play (according to [1]), in the advance to smaller scales. The normal delay between a test chip and a production chip is 18 months. So it is predicted that in July, 2007, Intel will sell chips containing a billion transistors, and this will not require a new type of fabrication facility, according to the press release [1]. This is good for all concerned, who already enjoy an amazing availability of computing power, at modest cost.

This progress is leading to devices whose sizes become comparable to the quantum-mechanical wavelength $\lambda = h/p$ of the electron. In this regime the wave behavior of electrons plays a role in device performance. The wave-like behavior of electrons is related to the notion of "ballistic particles", so that the particle is not scattered during its transit of the device. (This differs entirely from "diffusive behavior", in which many scattering events establish a local kind of equilibrium, the foundation of conventional device electronics.) In the ballistic regime, interference effects become important. In the following, the consequences of the wave behavior of electrons are first discussed. In later sections, aspects of wave behavior in existing devices are interposed with examples of new nanoscale elements which may be incorporated in hybrid technologies, mostly based on the existing silicon technology.

Nanophysics and Nanotechnology: An Introduction to Modern Concepts in Nanoscience. Second Edition.
Edward L. Wolf

ISBN: 3-527-40651-4

9.1
Electron Interference Devices with Coherent Electrons

A useful analogy exists between electromagnetic wave propagation in a waveguide or optical fiber and the transit of ballistic electrons through constrictions in (usually two-dimensional) channels for electrons. The constrictions are imposed by the potential energy of the situation. The experimental setup that is most commonly used to demonstrate electron wave effects is based on the "two-dimensional electron gas" (2DEG). A 2DEG is produced, for example, in a bulk GaAs sample, if a pure GaAs layer is epitaxially contained between two AlGaAs layers (barrier layers, of larger bandgap than GaAs), which are additionally doped with donor impurities. In this case, the electrons (from the donor impurities in the closely adjoining barrier layers) fall into the 2-d GaAs channel, causing it to become highly populated with electrons, metallic (degenerate) (see Section 5.7.1), with a high electrical conductivity which persists to extremely low temperature.

The electron mean free path and mobility in such "metallic" layers are large for three reasons: the whole structure is a single crystal, with constant lattice parameter as the layers are added; the effective mass for electrons in GaAs is small, leading to a high mobility; and third, the positively ionized donor sites, from which the electrons have been released to make the "metal" are spatially separated, lying in the AlGaAs barrier layers. This is the context in which most experiments on ballistic electron behavior are performed. The experiments require low temperatures, so the devices involved are not practical.

The appearance of these interesting and potentially useful effects in practical devices is connected with the size scale. With smaller devices (with similar lack of scattering events) the effects will be important at higher temperature. Below we will see typical wave behavior evident at slightly higher temperature, in a much smaller, defect-free system, the carbon nanotube.

A recent experimental demonstration of the electron wave effects in an elegant mesoscale device is shown [2] in Figure 9.1. This device is fabricated on n-type GaAs by molecular beam epitaxy. It consists of a 2DEG, a metallic conducting layer at a depth 57 nm below surface. This electron layer has a Fermi energy 16 meV, Fermi wavelength 37 nm, and mean free path 11 μm. The density of electrons is 4.5×10^{11} cm^{-2} with mobility $\mu \approx 10^6$ cm^2/V s. (The device is considered to be "mesoscale" in nature, and exhibits strong wave effects because the mean free path is so large.) This nearly perfect metallic sheet of electrons is contacted with ohmic source and drain contacts (typically operated at a small voltage difference, 0.2 mV). The electron current I and conductance dI/dV from source to drain are both measured. The current is controlled by two "point contact" gate electrodes, a constriction, operated at gate voltage V_g. The point contact gates (and the voltage chosen) provide a variable constriction between the two metallic regions surrounding the separate ohmic contacts. At negative V_g about −1.3 V (see Figure 9.1, panel B), the conductance becomes zero. A characteristic feature of the conductance of such a constriction, strikingly clear in this curve, is quantization in units of the conductance quantum,

$$G_0 = e^2/h = (25.8\,\text{k}\Omega)^{-1}. \tag{9.1}$$

The first plateau in the quantized conductance, at value $2G_0$, has onset at $V_g = -1.2$ V. It can be seen from Figure 9.1, panel B, that the second and third conductance plateaus have values $4G_0$ and $6G_0$ and have onsets at $V_g = -1.1$ V and -0.95 V, respectively. The quantum number here corresponds to distinct modes of electron propagation through the effective constriction, which becomes wider as the applied voltage is less negative.

The authors [2] have used an elegant scheme to map the electron flow pattern, using a negatively charged (scanned) AFM tip, which locally depresses the concentration of electrons in the underlying electron gas. The conductance dI/dV is measured and imaged as the position of the negative AFM tip is scanned. In regions of high current density the negatively charged tip has the biggest effect in depressing the conductance dI/dV. In this way, the authors [2] show the electron flow patterns of individual modes in the constriction. These modes are analogous to the different light intensity patterns in the optical fiber modes shown in Figure 4.2.

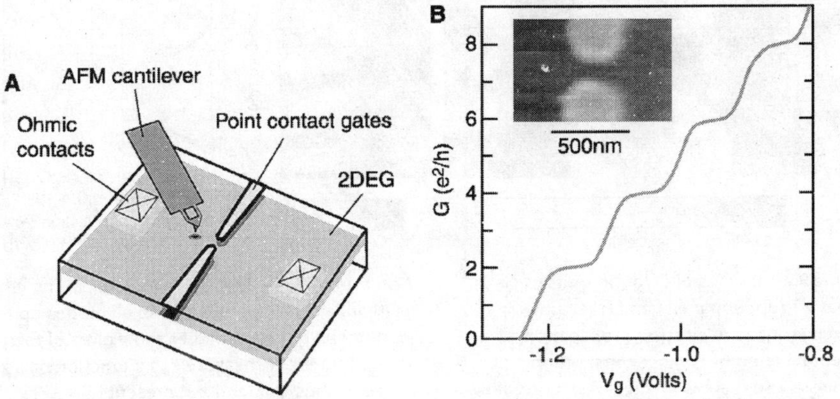

Figure 9.1 (A) Schematic of device and instrumentation, using a charged AFM tip (see text) to image the quantum point contact (QPC) conductance dI/dV, at small fixed source–drain voltage. (B) Quantum steps are observed in the device conductance dI/dV at 1.7 K (with no tip present) as a function of the voltage V_g applied to the point contact gates, a geometrical constriction. At −1.3 V the flow is shut off. The first mode of electron flow, corresponding to $dI/dV = 2e^2/h$, has onset at $V_g = -1.2$ V. The inset shows the geometry of the point contact gates. As they are made less negative, the effective width of the channel between them is increased, allowing additional modes of electron flow through the constriction [2]. (Reprinted with permission from Science, copyright AAAS.)

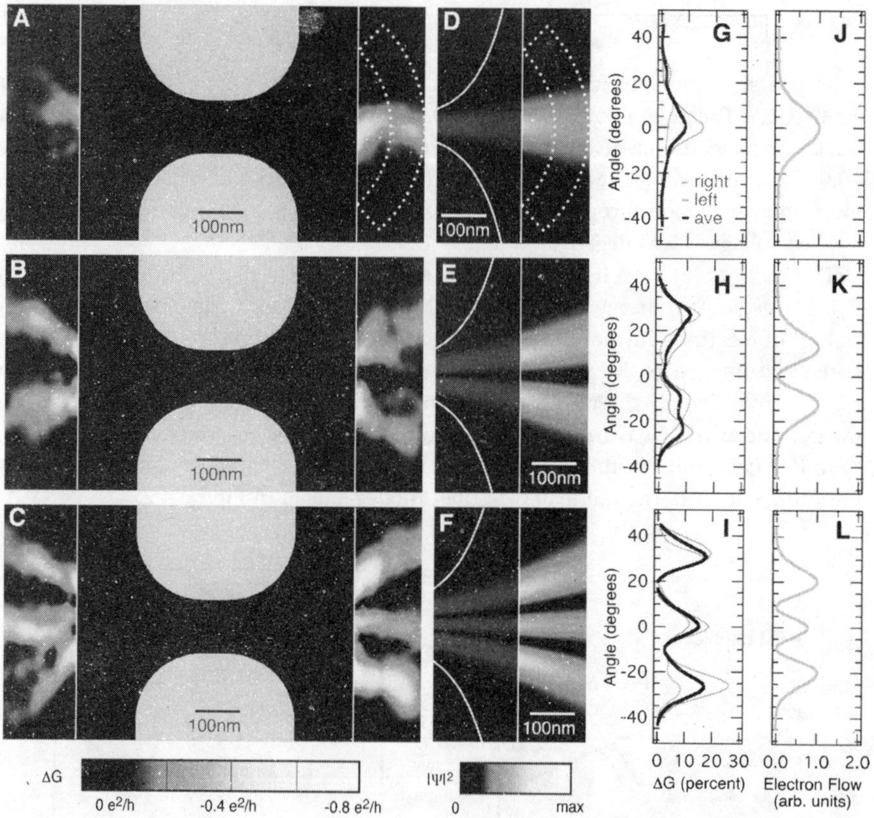

Figure 9.2 Images of dI/dV in units of G_0 of the $n = 1$ (upper panels) and $n = 2$ (lower panels) modes of electron propagation through the point contact constriction [2]. Panels A and B show images of electron flow on the left and right sides of the constriction. [In panel B, subtraction of a single-mode image has been made, to isolate the contribution from the $n = 2$ mode. In general, in the nth plateau, electrons propagate in all modes up to number n.] Panels G and H show plots of the current intensity measured as a function of angle. Panels D, E and J, K present corresponding models. (Reprinted with permission from Science, copyright AAAS.)

9.1.1 Ballistic Electron Transport in Stubbed Quantum Waveguides: Experiment and Theory

A more device-oriented line of research into wave behavior is exemplified [3] by measurements and theoretical study of a 2DEG geometry with a more complicated constriction, containing an internal cavity or "stub".

The authors have modeled the conductance through the stub device and find that the basic steps in units of the conductance quantum are predicted, as observed, but also that on each of the "plateau regions" there are a set of sharp minima in conductance. The sharp minima are interpreted as interferences of the electron wave from left to right with a set of internal (longitudinal) trapped modes, inside the cavity,

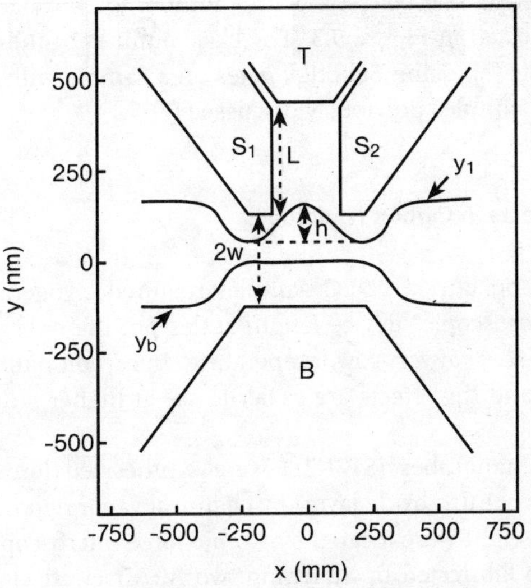

Figure 9.3 Stubbed quantum waveguide device [3]. Model of four-gate electron stubbed tuner is shown. Upper and lower portions show outlines of the four electrodes labeled S_1, T, S_2 and B. The curves labeled y_t and y_b are theoretical estimates of the upper and lower boundaries of the electron gas under typical bias conditions. The source and drain, as in the device of Figure 9.1, again located to the left and to the right, are biased with a small voltage, and the conductance dI/dV is measured and is also calculated. In this device the electron gas is at a depth of 80 nm, Fermi energy of 8 meV, and very high electron mobility and long mean free path. (Reprinted from Physical Review B with permission from APS.)

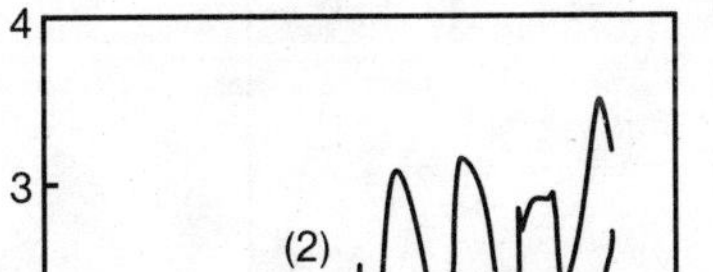

Figure 9.4 Calculated conductance dI/dV curves as function of gate voltage [3]. Curves (1) and (2) differ in the details of the biasing of the four electrodes of the gate. These curves clearly show the two plateaus for $n = 1$ and $n = 2$ as were observed in Figure 9.1. Here the finely-defined minima interrupting the basic plateaus in dI/dV are believed to arise from interference of the main propagating electron modes from left to right, with more closely spaced electron states resulting in confinement within the four-gate structure. Such fine structures are observed (not shown) at very low temperatures (0.09 K), but disappear from experimental measurements at higher (but still cryogenic) temperatures (2.5 K). (Reprinted from Physical Review B with permission from APS.)

related to the dimension h. These modes are believed by the authors to be modes along the current direction x, as indicated in Figure 9.3. The basic outlines of these features were observed in experiments [3], using Schottky gates on a sample with a buried 2DEG, basically similar to the samples previously discussed [2].

9.1.2
Well-defined Quantum Interference Effects in Carbon Nanotubes

The previously described works [2,3] on buried 2DEG samples required cryogenic temperature. The relatively large "mesoscopic" device length scales put the energy scale of the interferences below $k_B T$, except at very low temperature. In carbon nanotubes the length scale is nanoscopic and the effects are easier to see at higher temperature.

In this work, isolated single-wall nanotubes (SWNTs) were synthesized (on a degenerately doped silicon wafer with a 1 μm oxide layer) by chemical vapor deposition. Individual SWNTs with ~1 nm height were located by atomic force microscopy (AFM), and SWNT FET devices were fabricated by attaching two Au/Cr electrodes by electron-beam lithography. The electrical properties of the FET devices were characterized as a function of bias voltage V and gate voltage V_g. Oscillations as shown in Figure 9.5 were observed below 10 K. Of more than 100 such FET devices that were made and measured, more than 20 had room-temperature resistance below 15 kΩ, some as low as 7 kΩ, close to a theoretical limit (6.5 kΩ) for such a device

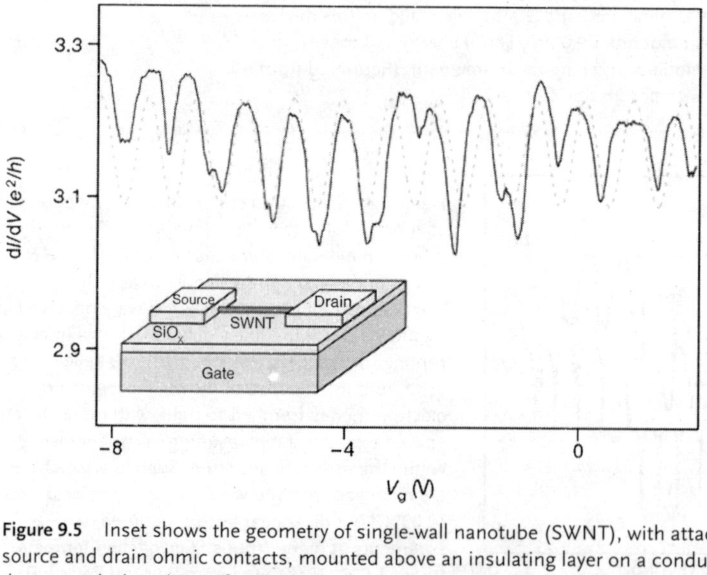

Figure 9.5 Inset shows the geometry of single-wall nanotube (SWNT), with attached source and drain ohmic contacts, mounted above an insulating layer on a conducting degenerately doped Si wafer which acts as gate electrode. The measured ($V = 0$) conductance dI/dV is shown (solid curve) vs. applied gate voltage V_g plotted in the range −8 V to about 1 V. The dotted curve shows a sinusoidal function with the same average period. The length of the nanotube is about 200 nm [4]. (Reprinted from Nature with permission from Macmillan Publishers, Ltd.)

with perfect ohmic contacts. The conductance quantum is (25.8 kΩ)$^{-1}$, as seen in Figure 9.1. The spacing of the observed oscillations in conductance was found to be inversely proportional to the tube length, L.

These devices show no evidence of Coulomb blockade: the electrons freely pass through these devices, with some reflections at each end. The interferences are more clearly shown in gray-scale images of conductance vs. V (source–drain voltage, always small) and V_g. The interferences are interpreted as between two propagating modes through the device, which interfere only with the ends of the device, and not with each other in the SWNT itself.

The authors [4] describe the effects as "Fabry–Perot" interference (the optical analog, see Sections 5.3.2 and 5.7.2) in a nanotube electron waveguide, which is a coherent molecular electron device, with (in the best cases) nearly perfect ohmic contacts. The resonant cavity is formed by electron scattering at the two (ohmic) nanotube-electrode interfaces.

An unexplained feature is the slight irregularity of the oscillations as a function of gate voltage. The authors tentatively assign this to effects of disorder on the propagating waves. Disorder may arise in the substrate or in the details of the nanotube–electrode interfaces. The authors suggest that the small irregularities do not arise from defects in the nanotubes themselves, as such defects would, in the opinion of the authors [4], have more drastic effects than those seen. The authors do not com-

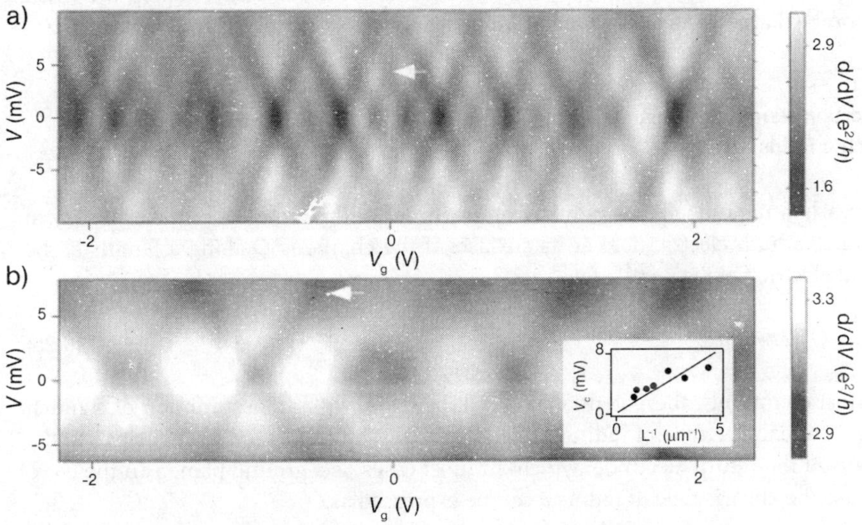

Figure 9.6 Conductance (dI/dV) interference images, with gray-scale calibration shown on the right [4]: (a) from a 530 nm device and (b) from a 220 nm device. Inset shows that the spacing of the oscillations with gate voltage is inversely proportional to device length, L. The oscillating features are more accurately described as dips in conductance, with spacings, V_c, 3.5 meV for the 530 nm device, and 6.5 meV for the 220 nm device. These V_c values are deduced from the locations (see arrows) along the V axis where adjacent positively and negatively sloping lines intersect. In the inset, the solid line has a slope equal to $h\nu_F$, where $\nu_F = 8.1 \times 10^5$ m/s is the Fermi velocity in the nanotube. (Reprinted from Nature with permission from Macmillan Publishers, Ltd.)

ment on the temperature dependence, nor on the absence of clear oscillations above $T = 10$ K. This restriction may well be an inherent feature of 200–500 nm length nanotube waveguide devices, in that the selected nanotubes and the contacts are nearly perfect. The aspect ratio (length divided by height) of the 220 nm device is nominally 220, still very large, so that shorter devices might show oscillations at higher temperature. On a molecular scale, these devices are still large.

9.2 Carbon Nanotube Sensors and Dense Nonvolatile Random Access Memories

The carbon nanotube is a nanoscale element with several useful properties, from an electrical point of view, but it is not envisioned that such elements, which can form individual devices, can be integrated to make a "carbon" computer. The electrical and also the mechanical properties are outstanding, which can be traced to the perfect periodicity, lack of defects inherent in a single molecule, and the consequent lack of scattering for carriers.

More likely are applications in the direction of extending Moore's Law in silicon: hybrid configurations of nanotube arrays attached to an underlying silicon configuration, perhaps a CMOS array.

Stand-alone applications, however, do seem likely in the role of the carbon nanotube as a sensor. Some such sensors, like the one described below, exploit the small nanotube diameter, to allow generation of a large electric field easily at low voltage.

9.2.1 A Carbon Nanotube Sensor of Polar Molecules, Making Use of the Inherently Large Electric Fields

The carbon nanotube has a radius a on the order of 0.5 nm, which makes it easy to produce a high electric field at its surface. If the charge is Q along a length L, the radial electric field is

$$E = Q/(2\pi\varepsilon_o L)r. \tag{9.2a}$$

Most commonly, the potential $V(a)$ will be fixed, with an assumption of a much larger coaxial electrode of radius b held at $V = 0$. (The results are not sensitive to the details of the return electrode, which, in most cases, is a ground plane.) In the coaxial case the electric field at radius a can be expressed as

$$E = V_0/[\ln(b/a)a], \tag{9.2b}$$

where V_0 is the potential at the nanotube surface, at radius a.

The important point is that the tube radius, a, is in the denominator of this formula for the electric field, which makes the field inherently large. For $b/a = 500$, $V = 1$ V and $a = 0.5$ nm, this is 0.322 V/nm. This field is not sufficiently large to

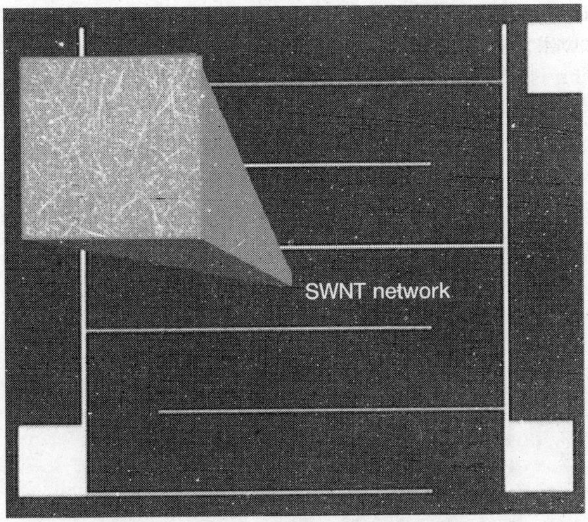

Figure 9.7 Optical micrograph [5] of a SWNT "chemi-capacitor". The region between the widely spaced electrodes is covered with an optically transparent but electrically continuous network of SWNTs (shown in the inset atomic force microscope image). The capacitance was measured between this top surface (the conducting nanotube network) and the underlying conducting Si substrate, using an AC method. The capacitance is sensitive to the effective nanotube diameter, which slightly increased by (E-field-induced) adsorbates. (The widely spaced electrodes were inter-digitated to allow simultaneous measurement of the network resistance, but this was not employed in the present work, and the two large-scale electrodes were electrically shorted). (Reprinted with permission from Science, copyright AAAS.)

ionize an atom, but it is large enough to provide binding to a polar molecule (see Section 5.3.1). This effect has been used [5] to provide a simple electrical sensor for polar molecules. In this case, a random but electrically continuous network of nanotube wires is sufficient. Such a network can be fabricated using the chemical vapor deposition (CVD) process. So this is a potentially immediate application.

Figure 9.7 gives a pictorial view of the device from the top, showing widely spaced wires to make contact with the nanotube network, and, in the inset, giving an impression of the random network of wires, on a much smaller length scale. The whole nanotube network is spaced from the conducting counter-electrode by a silicon oxide layer. The capacitance of the array to the ground plane is measured, and is sensitive to adsorption of polar molecules onto the nanotubes.

Following the discussion in Section 5.3.1, the energy U of an electric dipole moment $p = Ql$ in a field E is $U = -pE$, which is reached when the dipole is parallel to the field. Many molecules of interest have a fixed dipole moment, conventionally quoted in debye units (1 D = 3.3×10^{-30} C m). As examples, acetone and dinitrotoluene (DNT) have fixed moments 2.88 D and 4.39 D, respectively. In a field of 0.322 V/nm, DNT will experience a binding energy of 29.2 meV, which exceeds the thermal energy $k_B T$ at 300 K. *In this case the nanotube will be coated with oriented dipoles of DNT.* There are at least two means [5] by which this coating of the SWNTs will increase the observed capacitance between the continuous random network and the ground plane electrode, which is of course an experimental fact. (The effects are reversible, the coating evaporates when the voltage is removed.)

The sensor device shown in Figure 9.7 thus picks polar molecules out of its ambient atmosphere. The readout of the device is via the capacitance between the network of nanotubes and a ground plane (in place of a coaxial electrode of radius b). The capacitance

$$C = (2\pi\varepsilon_o L)/\ln(b/a) \tag{9.2c}$$

also depends on the radius a. The adsorption of polar molecules is indicated by an increase in capacitance of the random nanotube network vs. the ground plane of typically $\Delta C/C \times 100 = 0.01$ to 0.1%, changes which are readily measurable.

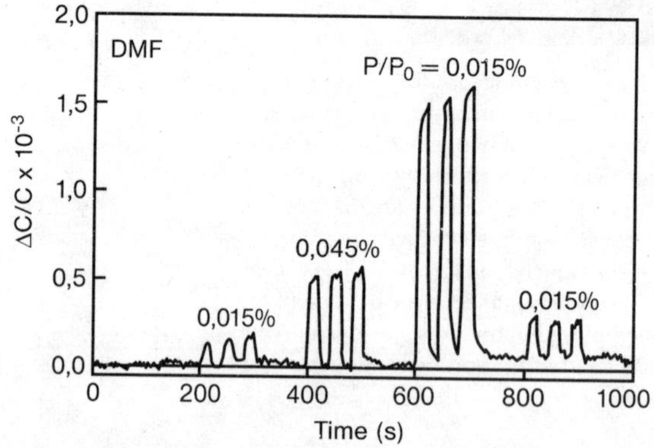

Figure 9.8 Capacitance changes on a 0.1% scale are clearly registered with the adsorption of a typical polar molecule, DMF (dimethylformamide) administered in 20 s doses at the indicated concentrations [5]. The DMF molecule has a relatively large fixed dipole moment, 3.62 D. (Reprinted with permission from Science, copyright AAAS.)

The mechanism of increase of capacitance is not entirely obvious. One simple picture is to imagine increasing the radius of the nanotube, a, by the length, l, of the dipole. Alternatively, the local dielectric constant may be increased [5], which will also increase the capacitance.

9.2.2
Carbon Nanotube Cross-bar Arrays for Ultra-dense Ultra-fast Nonvolatile Random Access Memory

An ambitious proposed use of carbon nanotubes as RAM has been described [6]. An $m \times n$ cross-bar array of nanotubes is proposed, providing nm intersections (junctions). The two parallel sets of nanotubes are displaced vertically, so the spacing be-

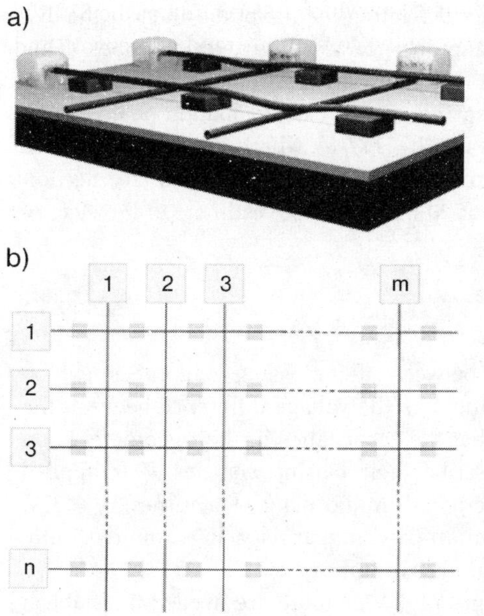

Figure 9.9 Schematic of carbon nanotube cross-bar array as RAM [6]. Orthogonal control lines terminate at the edges of the array. The *m* ground-level nanotubes are arranged accurately forming a grid lying directly on the (oxidized) ground plane surface. The *n* elevated nanotubes are supported, mid-way between nanotube intersections, by an accurately located square array of *mn* support pillars. The tube diameters may be around 1 nm, and the pillar spacing may be as small as 20 nm. Once an upper tube approaches closely enough to the crossing nanotube immediately beneath, it can lock into a near-neighbor (small tunneling resistance) configuration by short-range van der Waals attraction. A subsequent electrical voltage pulse (simultaneous along the *m* and *n* control lines) can restore the upper tube to its relaxed, linear and high-resistance configuration. (Reprinted with permission from Science, copyright AAAS.)

tween the (upper and lower) nanotube surfaces (a design parameter), ranges from 0.6 nm to 3 nm.

The limiting performance of the *nonvolatile* RAM based on this nanotube technology proposal is stated as nearly 10^{12} devices/cm^2, capable of operation at 100 MHz. These parameters far exceed the parameters of existing technology.

Each junction serves as a latching (tunnel junction) switch of extremely small size and fast action. The orthogonal sets of nanotubes are vertically displaced by an initial separation (spacing between the upper and lower nanotube surfaces) ranging from 0.6 nm to about 3 nm. In the ground state of the junction ("OFF" condition), each tube is straight, and the initial separation sets a high tunneling resistance between the two tubes. The energy of the crossing two-nanotube system is given as

$$E = E_{\text{vdW}} + E_{\text{elastic}} + E_{\text{electrostatic}}. \tag{9.3}$$

The first term is attractive, as discussed in connection with Figure 5.3. The crossing-cylinder case is not included specifically in Figure 5.3, but two surfaces of area *S*, spaced by a distance *s* in Figure 5.3, are assigned an energy

$$E_{\text{vdW}} = -S/(12\pi s^2), \tag{9.3a}$$

where *A*, the Hamaker constant, is typically 10^{-19} J.

The elastic energy can be crudely estimated (for larger separations) as

$$E_{\text{elastic}} = \frac{kx^2}{2}, \tag{9.3b}$$

where k is the effective spring constant for deflecting the nanotube and x is the displacement of the center of the upper nanotube (toward the fixed nanotube). (This neglects a tube–tube repulsion energy at extremely small spacing, evident from the curves in Figure 9.10.) The value of the spring constant k for equation (9.3b) can be estimated from the formula (see Section 2.1) $k = 3YI/L^3$, where Y is Young's modulus (reported as 1 TPa), I is the 2nd moment of the cross-section area (see Section 2.1), and L is the length of the tube, about 20 nm. A simple estimate of the electrostatic energy is

$$E_{\text{electrostatic}} = \frac{CV^2}{2}, \tag{9.3c}$$

where C is the effective capacitance between the crossing nanotubes (which increases as their spacing is reduced), and V is the voltage difference between the crossing nanotubes. The authors state that the capacitance in their geometry is on the order of 10^{-7} pF. The authors describe their biasing voltages V as applied between the separate tubes vs. the underlying ground plane. For values $V_1 = 3$ V, $V_2 = -3$ V, the authors find a stable minimum (latching) at a spacing around 0.3 nm, while at $V_1 = 20$ V, $V_2 = 20$ V the minimum has disappeared.

The authors state that attractive voltages of 4 V or more are needed to establish latching, and that repulsive voltages of 20 V or more are needed for release.

The net binding energy for 4 V attractive bias is about 4.5×10^{-18} J = 28.1 eV. It appears that the calculations are based on (10,10) single-wall nanotubes. The diameter for the (n,m) nanotube is

$$d_{\text{tube}} = \pi^{-1} a(nn + mm + nm)^{1/2},$$

where $a = 0.249$ nm, and n and m are integers defining the axis about which the graphene sheet is rolled to form the tube. This gives the diameter of the (10,10) tube as 1.37 nm.

If a voltage pulse is applied to the upper tube, owing to its capacitance in crossing the lower tube, there will be an attractive force, which will quickly draw the two tubes together. At a smaller separation, perhaps 0.3 nm, van der Waals attractive forces (see Section 5.3.1) between the two tubes will lock the two tubes together in the "ON" condition. These actions have been confirmed [6] in observations of a single junction. In the latched "ON" state, the tunneling resistance of the junction is substantially reduced, perhaps from 1 GΩ to 0.1 GΩ. The latched distortion is stable, until a second pulse is received to drive the two tubes apart.

Studying this carefully considered Figure 9.10, it is plausible that a single switch might be 20 nm in linear dimension. In this case, the number of devices per cm is 0.5×10^6, so that the number on a chip (1 cm^2) might be 2.5×10^{11}, an exceedingly high density. The switching speed of these devices is potentially high: the authors suggest 100 GHz.

Simplified analysis of these electromechanical switches is easily approached from classical physics. The interested reader is directed toward the Exercises.

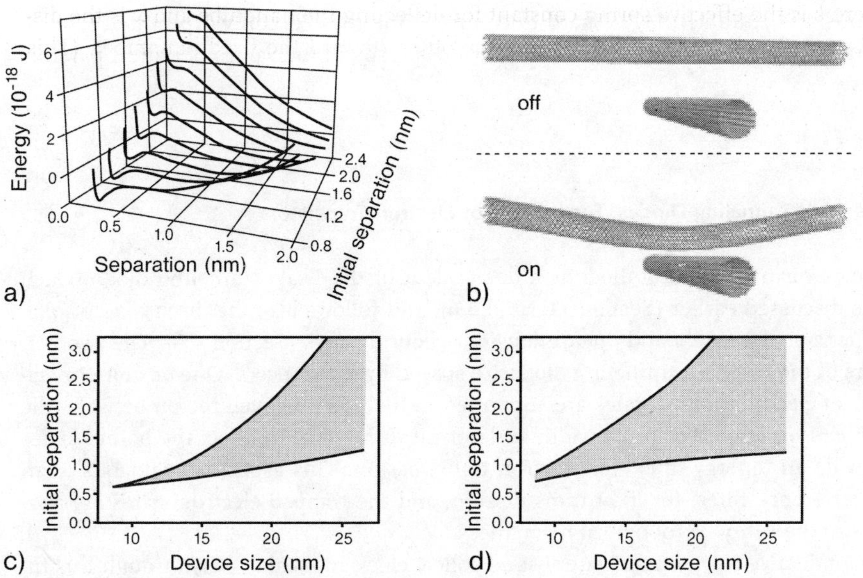

Figure 9.10 (A) The first curve (the front curve in the plot), for initial separation 0.8 nm, suggests a locking energy of about 10 eV, at separation of about 0.3 nm. (B) The sort of distortion involved in the locking via the van der Waals interaction is indicated in the lower portion of this panel. The latching occurs because the van der Waals attraction exceeds the energy cost resulting from distorting the upper nanotube, when the tubes are close, but still not really in physical contact. (C) This plot addresses the sensitivity to precise initial location to the size of the device in question. This panel may suggest that 15 nm device size is needed to allow even a 0.5 nm range in initial separations, which may be difficult to achieve. (D) This plot, similar to (C), again addresses issues of device size and initial tube spacing [6]. (Reprinted with permission from Science, copyright AAAS.)

The *fabrication* of such a revolutionary nonvolatile RAM is a formidable challenge. The assumption is that the long individual nanotubes are mechanically bonded to the underlying substrate by elastic supports whose precise spacing is 20 nm. Producing the array of supports, which might be done by a nano-printing process, more conventionally would require electron-beam writing, a slow serial process. Long single-wall nanotubes are available from chemical vapor deposition (CVD) processes, as confirmed by the preceding discussion of the chemical sensor device, but placing such nanotubes on a prepared lattice is more difficult. There is evidence that an array of pillars, produced by lithography (but on a scale too large for the presently envisioned device array) can be capped by catalysts, and that suspended nanotubes, as required in the present proposal, can be grown [7] between the catalyst sites.

Still, the questions facing such a revolutionary proposal are many. How would it be integrated into the presumed underlying CMOS structure? Some might say that a "mechanical" device is unreliable and subject to failure. (It appears that the nanotube is extremely robust, but its point of anchor and its individual description for radius and chirality may be hard to control.) One may certainly expect that fabrication of such a tiny array would lead to wide spreads in the values of critical parameters.

9.3
Resonant Tunneling Diodes, Tunneling Hot Electron Transistors

The resonant tunneling diode is a descendant of the Esaki tunneling diode which was discussed earlier (Sections 1.3 and 5.6), and follows also the discussion of one-dimensional barriers and one-dimensional bound states (Section 4.6). The two barriers of the resonant tunneling diode are spaced by a distance L. One or more (metastable) electron bound states are supported in the potential well region between the barriers, which are of finite width and permit tunneling. In effect the bound states provide an "energy filter" for electron transmission. This aspect is analogous to an "interference filter" for light transmission, and the trapped electron states are analogs of the Fabry–Perot optical resonances.

Conductive n-type layers supply and collect electrons which pass through this filter. The primary new feature of the current–voltage response is one or more nonlinear "current peak" features which arise when the bias voltage roughly matches the energy of the bound state (or states) of the electron in the region between the two barriers. So $I(V)$ will contain one or more peaks [8], each peak followed by a negative resistance region. In a resonant tunneling transistor, the double barrier filter is used to inject carriers into a region which corresponds to the base of a conventional npn bipolar transistor. The fabrication of such devices is done in molecular beam epitaxy, so that succeeding layers having different bandgap energies are precisely grown with no change in the crystal lattice. This makes nearly ideal devices whose characteristics can be usefully modeled in a simple one-dimensional potential $V(x)$. The most common example of such growth may be a GaAs layer sandwiched between two thin GaAlAs barrier layers.

An example of this approach [8] is illustrated by a three-barrier, double potential well, resonant tunneling diode (TBRTD) device, which exhibits a multivalued $I(V)$ characteristic (Figure 9.11). At zero gate voltages, the $I(V)$ has two peaks. The application of gate voltages to G_1 and/or G_2 leads to three states of the device.

The vertical field-effect resonant tunneling transistor, a triple-barrier, double-well resonant tunneling diode (TBRTD) is demonstrated. The device with different gate voltage choices is shown [8] to yield both binary and ternary logic and memory functions.

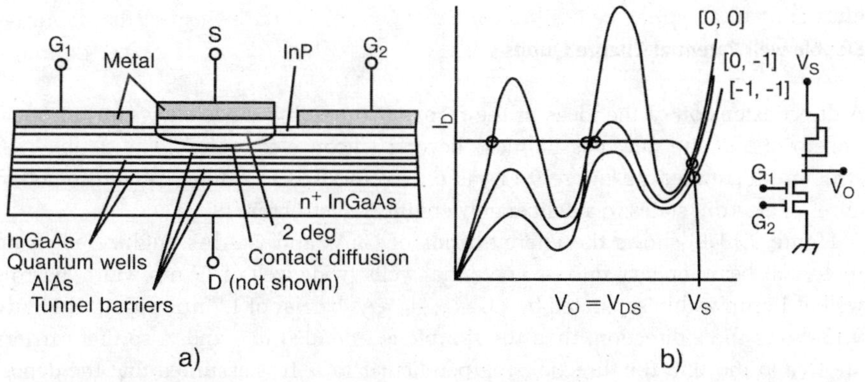

Figure 9.11 (a) Schematic cross-section of double-gated, double-well resonant tunneling TBRTD transistor [8]. Source–drain current flows through two quantum wells formed in InGaAs with AlAs barriers. The effective barrier height is about 0.5 eV. Gates G_1 and G_2 allow changes in the device characteristics. (b) Source–drain *I–V* characteristics with three different choices of gate potentials: zero voltage on both gates (0,0) gives low voltage intercept at target current; bias (0,–l) gives a second, higher, source–drain voltage intercept at the target current; while bias (–1,–1) gives a third, and even higher, voltage intercept at the target current. Three different states of the device are selectable by the gate potential choices. (Reprinted from IEEE Transactions on Electron Devices with permission from IEEE.)

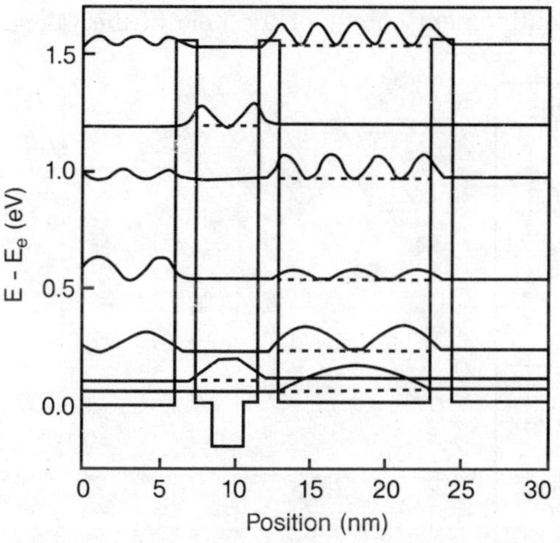

Figure 9.12 Computed [8] energy band diagrams and wavefunctions of the AlAs/InGaAs/InAs triple-barrier resonant tunneling diode (TBRTD). The resonant wavefunctions have been scaled, such that each has the same peak value. For the n = 2, 3, 4 and 5 states in the wide well, the amplitudes of the wavefunctions in the quantum well are comparable to the respective amplitudes in the emitter region; hence interferences between the incoming and reflected waves are apparent. The remaining wavefunctions, however, are sufficiently amplified in the respective well regions that reflected waves in the emitter region are not discernible. (Reprinted from IEEE Transactions on Electron Devices with permission from IEEE.)

9.4 Double-well Potential Charge Qubits

A direct extension of the ideas of the resonant tunneling diode allows an introduction to one of the new possibilities "beyond silicon electronics". The double-well structure illustrated in Figure 9.13 was one of the first to exhibit "quantum coherence" of electron states in a deliberately engineered structure [9].

Figure 9.13 [9] shows the energy bands of GaAs and GaAlAs engineered using molecular beam epitaxy into two potential wells (wide well of 17 nm width, narrow well of 12 nm width, separated by a $Ga_{0.65}Al_{0.35}As$ barrier of 1.7 nm width). If Figure 9.13 shows the x-direction, then the sample is extended in y and z, so that carriers are free to move in the directions perpendicular to x. It is assumed that the deposited material includes no shallow donor or acceptor impurities, so the only carriers in the device initially are the relatively small numbers of electrons and holes due to thermal electron–hole ionization. So the initial structure is nearly free of mobile charge (see Section 5.7.1).

The slight tilt of the energy bands in Figure 9.13 (upper left) illustrates application of an electric field to relatively shift the trap energies in the wide and narrow wells, to achieve resonance, in which case the linear combination states will be time-independent. The arrow shows a photon of light creating an electron in the left trap (wide well): the photon energy is about 1.53 eV. This photon energy represents the bandgap energy of GaAs plus the confinement energy of the hole in the valence

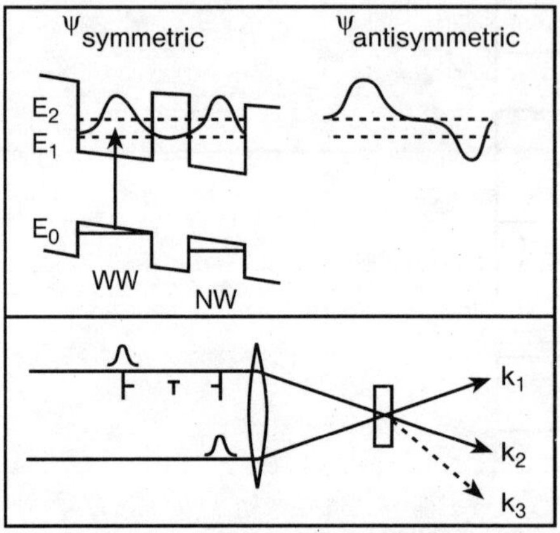

Figure 9.13 (Top) Schematic [9] of the double-well structure at "resonance" (an electric field has been used to line up the bound-state energies in the two wells), showing the symmetric (left) and antisymmetric (right) delocalized eigenstates, at energies E_1 and E_2, respectively. The vertical arrow indicates the excitation conditions for creation of a wave packet in the wide well (WW). The laser energy equals the bandgap energy plus the localization energies of the hole in the valence band and the electron in the conduction band (Section 4.6.1). (Bottom) Geometry of the experiment. The variable time difference T (between the pump and probe light pulses) is adjusted by slightly different path lengths L for the light, using the conversion $L/T = c = 0.2998$ mm/ps. The second (probe) pulse is used to detect the motion of charge into the right well after the excitation. (Reprinted from Physical Review Letters with permission from APS.)

band plus the confinement energy of the electron in the conduction band. Focus on the electron, which is created at time $t = 0$ in the bound state of energy nearly E_1 (as shown in Figure 9.13). But, actually the energy of the electron at $t = 0$ is not precisely E_1! The electron at $t = 0$ is confined to the left wide well, while energy E_1 is indicated as the energy of the symmetric combination of electron states in the right and left wells. Consider the situation of the initially injected electron, with reference to Figure 9.14.

It has a chance to escape into the right well by tunneling through the barrier. Eventually, one expects that the electron will arrive in the right well, leaving the left well empty. This oscillation (which will continue periodically, until there is an energy loss from the electron) between left and right wells is a fundamental feature of quantum mechanics. These experimenters were the first to verify this effect in an engineered semiconductor structure. The period of the oscillation [9] was 1.3 ps, which agrees with the energy splitting $\Delta E = (E_2 - E_1)$ of about 3 meV. The agreement is based on the theory prediction of $f = h/\Delta E$, where f is the frequency of oscillation of the charge between the two states. Both quantities were measured [9] in this pioneering work.

The physics of this situation is similar to the physics of the hydrogen molecular ion, and to the physics of the "charge qubit", which is one of the approaches to the "quantum computer". The rules are:

$$\Psi_S = 2^{-1/2}(|L\rangle + |R\rangle), \quad \Psi_A = 2^{-1/2}(|L\rangle - |R\rangle), \quad f = h/\Delta E. \tag{9.4a}$$

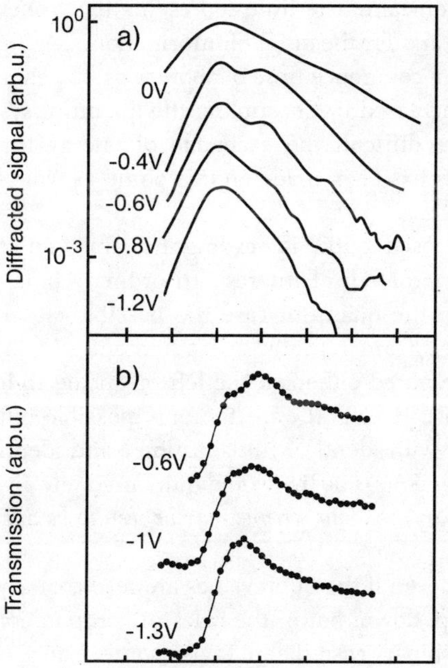

Figure 9.14 (a) Diffracted signal and (b) change in transmitted probe intensity vs. delay time for different bias fields (energy shifts between wells). These measurements are indicative of shift of the charge from the left well to the right well. It is seen that for shifts –0.4 V and –0.6 V a second peak in diffracted signal occurs at about 0.7 ps delay. This is interpreted as evidence of the oscillation of the coupled wavefunction from its initial position in the left well, toward the right well [9]. (Reprinted from Physical Review Letters with permission from APS.)

It follows from this that

$$|\mathbf{L}\rangle = 2^{-1/2}(\Psi_S + \Psi_A), \quad |\mathbf{R}\rangle = 2^{-1/2}(\Psi_S - \Psi_A). \tag{9.4b}$$

The state $|\mathbf{L}\rangle$ is often used as the logical 0, the state $|\mathbf{R}\rangle$ as the logical 1, because these two states are the easiest to read out by a capacitive sensor of the presence of an electron charge, or no charge. In this discussion it is assumed that higher energy states of the electron in the double well can be ignored.

The symmetric (S) and antisymmetric (A) combinations are the two basic solutions to the coupled-well nanophysics problem, i.e., to Schrodinger's equation (4.37). The reason physically (see wavefunction sketches in upper panel of Figure 9.13) that the symmetric combination is the lower energy state (and the *only* observed state in the hydrogen molecular ion) is that in the symmetric case the electron charge has nonzero probability midway between the two wells, which allows it to sample the negative potential energy of both wells at once. The antisymmetric combination has a node at the center point, and misses this binding energy.

If Ψ_S and Ψ_A are valid solutions to the differential equation (Schrodinger's equation), then it follows that any linear combination $\Psi = a\Psi_S + b\Psi_A$, where $a^2 + b^2 = 1$, is also an equivalently valid solution. Such a general linear combination contains information, in the many choices of the coefficients a and b, equivalent to the number of points on the unit circle! The qubit can also be expressed by an angle, via $e^{i\Theta} = \cos\Theta + i\sin\Theta$ for $0 < \Theta < 2\pi$. This is called the "quantum bit" of information, the qubit. The information in a qubit lasts as long as the energy of the electron in the linear combination state is precisely constant. The time represents the "coherence time" for the linear combination state and for the qubit of information.

The experiment in Figure 9.13 exhibited a coherence time of about 1 ps, too short for computation! The coherence time is increased by decoupling the quantum system from its environment, making more difficult the exchange of energy that destroys the encoded information. Progress has been made on this point, as will be seen below.

Another important aspect of this proposed qubit is reading the final qubit, presumably containing the answer to the problem of interest. In ordinary binary computation, the 0 or 1 is plain to see. In the quantum case the possibly closely spaced values of a and b must be inferred from repeated measurements. In the charge qubit case, the electron will be measured either on the left or on the right (most likely by a single electron transistor). $|\mathbf{L}\rangle$ or $|\mathbf{R}\rangle$ are the only possible outcomes, but a million measurements (each with identical initializations, and identical computation steps) giving 640,000 $|\mathbf{R}\rangle$ and 360,000 $|\mathbf{L}\rangle$ would quite precisely give a value $\theta = \tan^{-1}(b/a) = 53.13°$. *This (necessary) readout scheme may be seen to be difficult, slow, and inexact.*

On the other hand, it may transpire that even if the qubit values are read coarsely, for example resolving only three states (up, down, both), the rules of computation may still be significantly favorable over the binary case. There is also some chance of sampling these outcomes at GHz frequencies. The expectation is that for a quantum computer a very significantly reduced number of gates would be required than in strictly binary computers. (As mentioned above [1], Intel has announced a demonstration chip containing 10^9 transistors, which is getting expensive, especially in the fabrication facilities that will soon be needed.)

In spite of the several, and possibly pivotal, difficulties (cogently summarized by Laughlin [10]), "quantum computing" is one of the presently most active research areas. There are classes of computing problems, of great practical importance, that are better solved by a proposed class of "quantum computers". Some solutions have already been demonstrated, and it appears, even with small numbers of qubits, that the difference in the algorithms makes the approach useful. The potential applications include schemes to decode encryption used by banks and governments to keep vital information secure. From an operational view, the large research budgets in this area are evidence of potential importance.

9.4.1
Silicon-based Quantum Computer Qubits

A significant fraction of the ongoing quantum computation work is on silicon-based systems. The phosphorus (P) donor in Si is a scaled version of a hydrogen atom, as we discussed earlier (Section 5.7); see also an eloquent discussion in [10]. If two P donors (see Section 5.7) are arranged near each other (spaced by about 20 nm, each located in pure Si perhaps 15 nm below a SiO_2 insulator, and each under a control nano-electrode), and one electron is removed, then the resulting single electron system has the essential features of the charge qubit as discussed above. This D–D^+ system is analogous to the H_2^+ hydrogen molecular ion, known to chemists as an example of the "one-electron bond". The D_2^+ donor case differs, in that the antisymmetric states (which do not exist for H_2^+ in vacuum) are bound for D_2^+ in silicon, thus available for use in the computational schemes.

The proposed donor qubit is small in all three dimensions (compared to the situations in Figures 5.17 and 9.13, where the transverse dimensions are large). Although only planar arrays of these 3-d donor atoms and molecules are being considered, a huge density of quantum qubits is possible.

The density of planar packing is probably limited by the size of electrodes which are needed to perform operations on the qubit from the outside world. In more detail, electrodes will be placed above the L and R sites, which can be used to ionize the electron from the R site, to localize the single remaining electron into the L site, which is conveniently associated with logic 0.

The control electrode in the center can be used to adjust the transparency of the tunneling barrier between the two sites. To "pre-initialize" the qubit, the left electrode may pull the electron to the left while the center electrode is adjusted to prohibit tunneling out of |**L**> into |**R**>), which enforces the localization of the electron in |**L**>. The qubit device arrangement will also have two "single electron transistor" (SET) readout devices, one sensitive to charge in the |**L**> location and one sensitive to charge in the |**R**> location. The "pre-initialization" operation will also be the time to calibrate the SET sensors, noting the readings that correspond to one electron on the left and zero on the right. A second "initialization" step may be to provide a positive pulse of specified voltage and time duration to take the electron from the L state to the equal admixture state. In analogy to the corresponding pulses in a magnetic spin-½ qubit, this pulse may be referred to as a "$\pi/2$ pulse", which would mean taking the spin from vertical spin-up to horizontal, halfway toward spin-down. This sets up rotation (precession) of the magnetization vector around the magnetic field direction at the Larmor frequency, generating a radio-frequency signal which is detected, for example, in the MRI imaging apparatus. In the charge qubit, free oscillations at frequency $f = h/\Delta E$ will occur after releasing the electron (increasing the barrier transparency) following the localization into |L>. These oscillations will continue for the coherence time. After the "$\pi/2$ pulse" in the double quantum well qubit, which takes |L> precisely into Ψ_S, the configuration will be stable until an energy loss, decoherence, occurs. Logic operations may be carried out after the several qubits involved are "initialized" in this symmetric fashion. Before a more detailed discussion of how charge qubits might be organized in a computer, we need to understand more about the readout, which most likely involves the single electron transistor (SET) or even the "radio-frequency single electron transistor" (RFSET). Both devices can detect charges only a small fraction of the single electron charge, $e = 1.6 \times 10^{-19}$ C, and in the RFSET the sampling rate can be fast, up to 10 MHz.

9.5 Single Electron Transistors

The "single electron transistor" (SET) is configured with a source and drain like a field-effect transistor (FET), except that the active region, whose potential is controlled by the gate electrode, is so small, and its capacitance is so small, that its capacitive charging energy enforces a single electron rule of occupancy. In the simplest terms, $e^2/2C >> k_BT$, and the role of the gate voltage V_g is to adjust the residual charge on the "island" to the magic value $e/2$. At this residual charge level the addition of one single electron ($-e$) takes the total charge to $-e/2$, with precisely the same Coulomb energy $k_Ce^2/2C$. This transfer requires no energy, and this condition of half-integer residual charge, selectable by the gate voltage, defines the "ON" state of the device.

The transfer process for charge, from source, to island, to drain, is in each case tunneling through a tunnel barrier whose resistance must be at least 25.8 kΩ, the von Klitzing resistance. This condition is needed to make the charge on the island a well-defined quantity, adjustable only in units of e by the random tunneling events. A great deal is known, both theoretically and experimentally, about SET devices. The basic difficulty is to achieve the operating condition $e^2/2C >> k_BT$. For elements generated on silicon by electron-beam lithography, the capacitances are still so large that the operating temperature must be below 4 K, liquid helium temperature. The other possibility is to make the island much smaller.

An example of a sub-electron-beam lithographic scale for the "island" has been reported [11] using a single C_{60} "Buckyball" molecule, located between source and drain electrodes. Adopting a classical metal sphere approximation for the C_{60}, the capacitance of an isolated Buckyball, $C = \varepsilon_0 a$ (where a is the radius, 0.6 nm), is 5.31×10^{-21} F. Following this, a classical estimate of the "single electron charging energy" ("Coulomb blockade energy") for an isolated C_{60} molecule is 15.1 eV. This leads to a naive estimate of the on/off resistance ratio of the device as $\exp[-e^2/(2Ck_BT)]$, about exp(–603) at 300 K. This hypothetical device is completely Coulomb blockaded at charge e. In actual practice, however, the appropriate capacitance is not that of the isolated sphere, but is the sum of the capacitances from the island to the source, drain and gate. This will be a much larger capacitance, depending on the details of the design, so that the charging energy is much smaller.

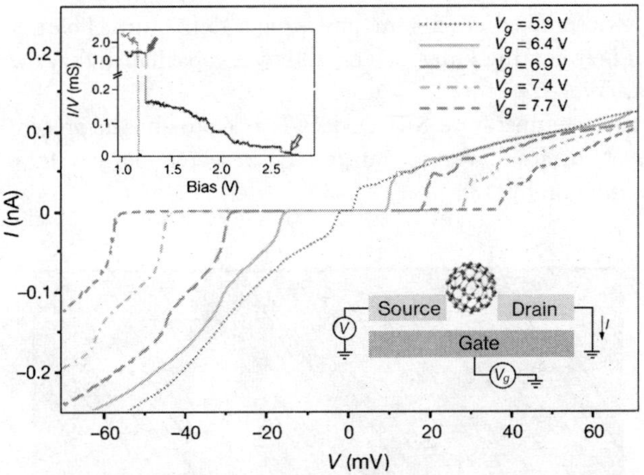

Figure 9.15 Current–voltage (*I–V*) curves [11] obtained from a single-C_{60} transistor at 1.5 K. Five *I–V* curves taken at different gate voltages V_g are shown. Single-C_{60} transistors were prepared by first depositing a dilute toluene solution of C_{60} onto a pair of connected gold electrodes. A gap of about 1 nm was then created in the gold electrode using electromigration-induced breaking of the electrodes. (Upper inset) A large voltage *V* was applied between the electrodes while the conductance *I*/*V* through the connected electrode was monitored (solid, irregularly decreasing curve). After the initial rapid decrease (solid arrow), the conductance stayed above ~0.05 mS up to ~2.0 V. This behavior was observed in most single-C_{60} transistors, but it was not observed when no C_{60} solution was added (dotted, sharply descending curve in upper inset). The bias voltage was increased until the conductance *I*/*V* fell low enough to ensure that the current through the junction was in the *tunneling regime* (open arrow, in upper inset). The low bias *I–V* measurements shown in the main panel were taken after the junction formation process was completed. (Lower panel) An idealized diagram of a single-C_{60} transistor formed by this method, which shows the biasing and measuring schemes. The diagram suggests that the capacitances are well in excess of the estimate of the capacitance for an isolated spherical C_{60} molecule, allowing the charging energy to be significantly decreased. The *I–V* curves indicate that the *on condition* (half-integer charge on the island) is achieved in the particular device near a gate voltage of 5.9 V. The simplest possible theory would predict for the maximum conductance *I*/*V* in the on condition: conductance $= I/V = \frac{1}{2}e^2/h \sim 0.5/(15\ \text{k}\Omega)$. (This would correspond to two 15 kΩ resistors in series.) The data indicate a conductance of about $(1.66\ \text{G}\Omega)^{-1}$, smaller by a factor of nearly 10^5 than the idealized model at its high-conductance limit. The discrepancy can be accounted for simply by assuming that the molecule–electrode tunnel barriers have much higher resistance than 15 kΩ. As has been noted, the resistance of a tunnel barrier is strongly dependent upon the thickness and the barrier height. (Reprinted from Nature with permission from Macmillan Publishers, Ltd.)

The fact that the device is only reported as operating at 1.5 K means that the Coulomb energy is small, evidently measured in units of k_BT at 1.5 K. This means that the capacitances are larger than that of an isolated C_{60}, probably reflecting capacitances to the nearby electrodes.

Still, the conductance of this device may be too small to be useful, perhaps because the van der Waals interaction in holding the molecule against each electrode is weak and makes a large tunnel resistance. This device, in spite of the tiny size scale of the island, for one reason or another does not operate above 1.5 K.

Operation of SET devices has been achieved at room temperature under extraordinary conditions. In one well-documented (but impractical) example, a single carbon nanotube [12] can be "twice kinked", each kink providing a 15 kΩ tunnel barrier and the conducting region between the kinks having so low a capacitance that the SET operating condition is achieved.

Reliable production of room-temperature SET devices is not possible in present silicon technology. However, useful low-temperature SET research devices have been fabricated by e-beam lithography [13].

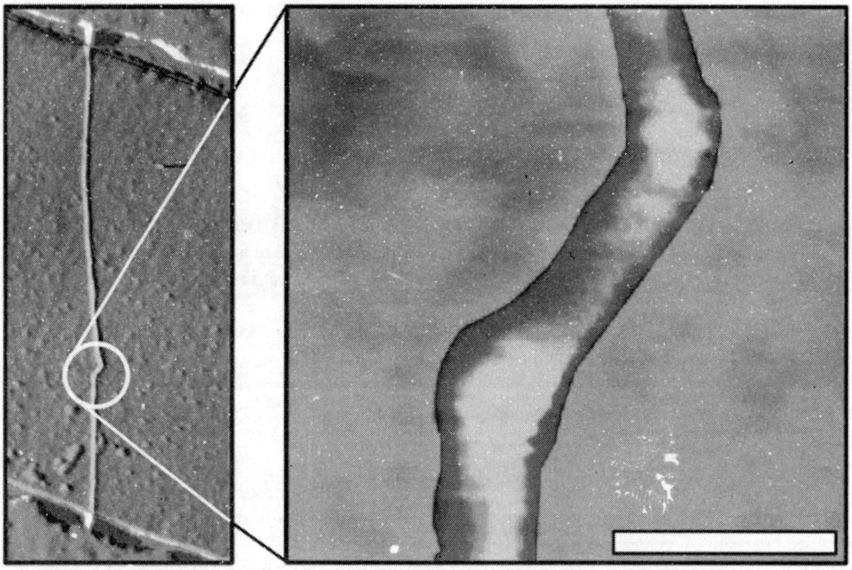

Figure 9.16 Fabrication [12] of a room-temperature single electron transistor within an individual metallic carbon nanotube by manipulation with an AFM. (Left) AFM image of [12] "twice kinked" nanotube. The nanotube connecting the upper and lower gold electrodes lies on a Si/SiO_2 substrate/gate, and initially has a gate-voltage-independent resistance of 50 kΩ. The AFM tip is manipulated to produce two kinks in the tube, as seen more clearly in the right panel. After the two kinks are produced, the room-temperature resistance of the tube is increased ten-fold to ~0.5 MΩ. The white sca'e bar is 20 nm.

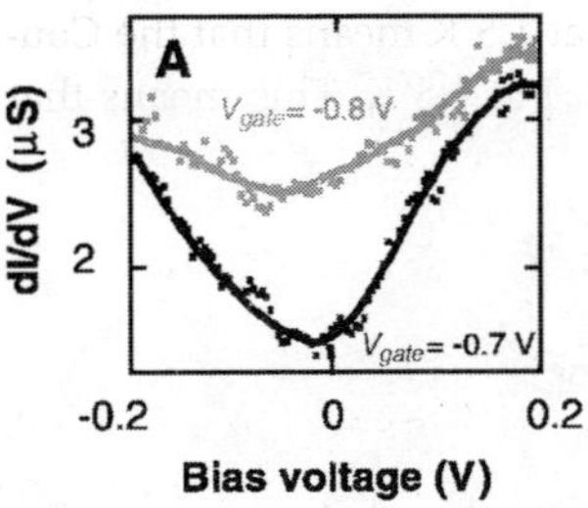

Figure 9.17 Measurements at 300 K of the dI/dV of the "twice kinked" nanotube, which acts as a single electron transistor (SET) [12]. These data indicate that the SET gap closes at about –0.8 V on the gate, corresponding to half-integer electron occupancy of the isolated section of the nanotube, between the two kinks (strong tunnel barriers). The capacitance (with respect to a ground plane), and the capacitive one-electron charging energy of a 20 nm length of nanotube of diameter 1.4 nm, can be estimated using elementary physics (see text and Figure 9.7), and are expected to give a charging energy $e^2/2C$ larger than $k_B T$ at 300 K (which is about 0.025 eV). This work demonstrates SET action at room temperature, but does not suggest a clear road to practical applications. (Reprinted with permission from Science, copyright AAAS.)

9.5.1
The Radio-frequency Single Electron Transistor (RFSET), a Useful Proven Research Tool

The SET is needed to read the final state of a charge qubit. For this purpose, capacitive coupling, which can be reliably estimated from device geometry and the known dielectric constants, is provided between the island of the SET and one of the sites of the charge qubit. The RFSET [14] provides a scheme for quickly reading out the charge on the island, using methods proven in other areas of physics. The scheme is similar to the conventional detection of electron spin resonance (ESR), using the change of reflection of (high-frequency) microwaves from a high-Q resonant circuit (cavity) when the cavity experiences internal losses (damping).

9.5.2
Readout of the Charge Qubit, with Sub-electron Charge Resolution

The SET is used to accomplish readout of the final configuration of the qubit in the double-well *charge qubit* cases. The gate electrode is capacitively coupled to one of the |**L**> or |**R**> sites of the charge qubit structure. The sampling rate of the SET is increased by making the source–island–drain portion of the SET a resistive (shunting) element in a separate tuned high-frequency sensing circuit. The Q of the tuned high-frequency (GHz) sensing circuit can be monitored on a fast time scale, limited only by its high resonant frequency, and this configuration, called RFSET, has been used to measure an electric current, flowing through the SET circuit by counting the number of times the island in the SET is occupied by an electron. Such a scheme might allow the many repeated samples of the outcome of the quantum computing calculation that was mentioned above.

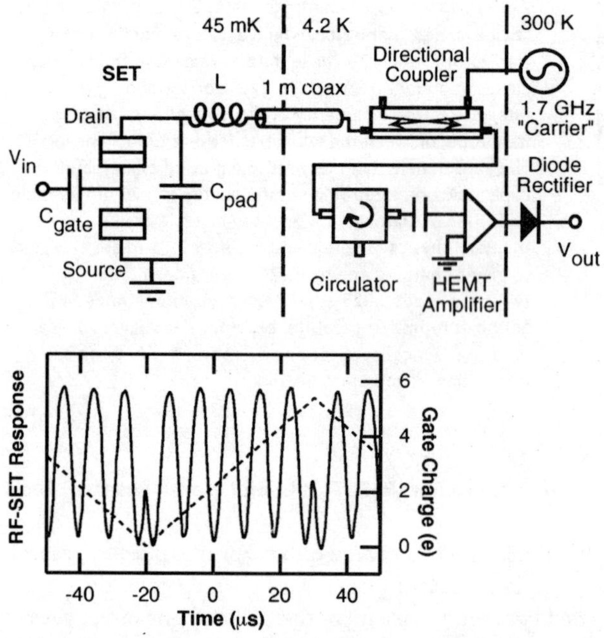

Figure 9.18 (Upper panel) Schematic of apparatus used for measurement of the high-frequency conductance of the SET, which allowed very fast and sensitive operation of the RFSET [14]. The idea is to use radio-frequency (RF) to monitor the *drain–source impedance*, which quickly takes different values as the charge on the island (between the source and the drain) goes from half-integer (low impedance) to integer values (high impedance, due to Coulomb blockade). This quickly varying *drain–source impedance (which contains information about the island charge)*, shunted by capacitor C_{pad}, is made part of a series-resonant high-Q circuit (including inductor L) driven by the 1.7 GHz "carrier" fed down in a coaxial line. The shunting of the resonating capacitance C_{pad} by the "ON" SET conductance substantially changes the portion of the 1.7 GHz signal reflected back toward the source. The reflected signal is recorded, as a measure of the electron occupation of the SET island. In the upper panel, one sees that the reflected 1.7 GHz signal is detected via the directional coupler, circulator, capacitor, high electron mobility transistor (HEMT) and diode detector to get a DC signal. This technique allows high-frequency readout of the SET island occupation. This method is standard for detecting electron spin resonance from a high-Q microwave cavity whose internal loss is modulated when the resonance conditions are satisfied. The impedance of the SET source–drain connection inherently varies with an extremely short time constant, and this elegant readout method [14] does not degrade the time response. (Lower panel) Dashed line represents an applied 30 kHz triangle wave about 5.5 electrons on island, peak-to-peak [14]. It is seen that the RFSET response (solid curve) easily and quickly records the island (gate) charge in units of e, with very small noise. The gain of this device is nearly constant up to 10 MHz. (Reprinted with permission from Science, copyright AAAS.)

9.5.3
A Comparison of SET and RTD (Resonant Tunneling Diode) Behaviors

It is useful to compare the operation of the SET and the resonant tunneling diode (RTD). In the SET the electron is momentarily localized on the island. The particle is definitely localized on the island, and the chance of going forward rather than back is enhanced somewhat by V_{SD}.

Localization on the island is enforced by tunnel barrier resistances larger than 25.8 kΩ. Having only one electron at a time is enforced by large values of the single-electron charging energy, which occur only in the limit of small island capacitance.

In the resonant tunneling diode (RTD), localization is not assumed to occur, but rather a continuous flow of probability (charge) without accumulation in the central region. This may be described as an electron wave behavior, where the wavefunctions are described by the potential $V(x)$ within the device, but no extra consideration is needed for Coulomb interaction between electrons within the device. The behavior might be suggested by Figure 9.1, with the additional idea that between *two such constrictions* there would be an enhanced occupation. The analogy might be cars moving on a freeway between the locations of two accidents, where only one lane is open: the car density will rise! Still, there is no going back, in such a case (in the SET case, the chances of going back and going forward are nearly equal).

One important distinguishing factor is the large cross-sectional area in the typical well region of the RTD resonant tunneling diode, which makes the capacitance large, and the charging energy negligible. The only small dimensions in the double quantum well described above are the barrier widths and the well widths; the y and z dimensions are much larger. So to put one more electron into the well does not involve an appreciable charging energy. This distinction in modes of operation carries over into other types of nanoscale electronic devices we may consider. Carbon nanotubes have been demonstrated as conducting channels in FET transistors. Here the nanotube plays the role of a high electron mobility conductor, and the gate voltage changes the number of carriers in the channel, as in the usual FET. The overall length of the nanotube in such an FET is large enough that the charging energy effect will be small. If we return to Section 9.2 (equations (9.2a,b,c)) we see that the capacitance of a nanotube of length L is $C = 2L/[k_C \ln(b/a)]$, where k_C is the Coulomb constant. If $L = 1$ μm and $b/a = 500$, then $C = 36$ aF, which, although small, is not small enough to give an important charging energy at room temperature. However, if the nanotube is kinked (see Figure 9.16), tunnel barriers are introduced, enclosing a shorter length L, more like 20 nm, for which the much larger charging energy will make it act like the island of an SET.

There are further classes of devices in which single molecules are placed essentially between a source and a drain, with the possibility of a gate electrode nearby. Of course, molecules come in a variety of sizes and properties. It appears that, depending on the molecule and how it is connected to the source and drain, these systems may behave as wave devices or as SET devices. A single benzene ring has been considered as the active element of an FET, with special attention paid to the fact that delocalized π-orbitals around the ring act like electron conductors. If the delocalized π-orbitals are in some way tightly coupled to the source and drain, then wave motion through the whole device may be expected, and this is the case that has been theoretically assessed [15,16]. On the other hand, if molecules have a conducting central ring structure (island) which is isolated by linear networks of CH–CH bonds, more likely to act as tunnel barriers, connecting to the source and drain, then the device will probably act more like an SET. In other types of molecular devices a bistable molecule (having different isomeric states) connecting source and drain may act more like a latch, which may be activated by a voltage pulse. The conformation (local structure) of the molecule may be changed, substantially altering its conductance between the source and drain. In various cases the latch can be opened, perhaps again by a voltage pulse, or by changing its charge state.

9.6
Experimental Approaches to the Double-well Charge Qubit

The charge qubit, within the framework of silicon technology, has been recently approached by two different schemes. The first that we will mention [7] is based upon a commercially available silicon-on-insulator substrate which provides a buried conductive layer of 35 nm thickness (5.7×10^{19} P donors/cm^3) at a depth of 10 nm, above a 50 nm thick SiO_2 insulator, provided over the whole wafer. The upper protective 10 nm layer can be removed using electron-beam lithography and

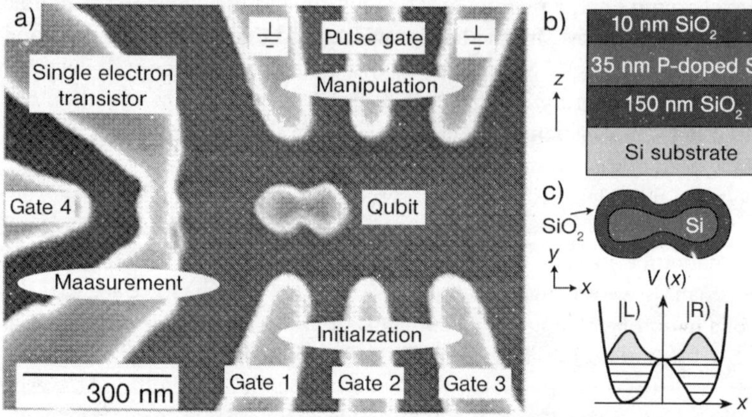

Figure 9.19 The charge qubit operation of an isolated double quantum dot [17]. (a) Scanning electron micrograph of the sample. The device was fabricated by high-resolution electron-beam lithography and reactive ion etching to "trench-isolate" the elements, which all lie in the conductive plane of the sample. Thermal oxidation was carried out to reduce the dimensions of the silicon regions and to passivate the surface-charge trap states. (b) Material profile, showing the 35 nm thick active region between the insulating layers. The dopant (phosphorus) concentration was 5.7×10^{19} cm^{-3}. (c) Schematic view of the isolated double quantum dot after oxidation, with the active region surrounded by the oxide layer. Localized charge configurations |**L**> and |**R**> of the electron at the highest occupation level are illustrated schematically. (Reprinted with permission from Science, copyright AAAS.)

etching to expose the conductive layer. Secondly, an e-beam lithography and reactive ion beam etching process can be used to "trench-isolate" selected areas in the conductive region itself; these include the electrodes and the active double-well area.

In Figure 9.19 this scheme is indicated, where the reactive ion beam etching has defined double quantum wells separated by a constriction of about 20 nm width. In this case the total number of electrons in each "quantum well" may approach 2000 (in other cases the number per dot has been estimated as 25–30). However, the rules of filling available states (two electrons per size-quantized state in the well, as in our discussion of the quantum dot, see equation (4.67)) acts to provide a few electrons at the highest filled level in each dot. This scheme can be adjusted to provide a single electron sharing two adjacent wells connected by a tunnel barrier, making use of electrodes which allow shifting the potential of one well with respect to the other, and for adjusting the coupling strength between the two wells. In this scheme all of the control electrodes and the SET readout device are deposited on adjacent areas of the conductive layer, so that it is never the case that an active area is buried beneath an electrode.

Looking at Figure 9.19(a) the basic features [17] are the double dot, consisting of the |**L**> and |**R**> wells coupled by the constriction, and the single electron transistor (SET) located to the left of the quantum wells. Note first gate 4, to the left of the SET, which can be used to adjust the average charge on the SET, to bring it to a suitable operating point for the detection of a single electron charge on the L well. Six capacitively coupled electrodes (three above and three below in Figure 9.19(a)) are available to initialize and manipulate the single charge on the double quantum well.

The scheme for qubit initialization, manipulation and measurement is summarized in Figure 9.20. The coherent oscillation of the charge between wells is shown in Figure 9.21 [17].

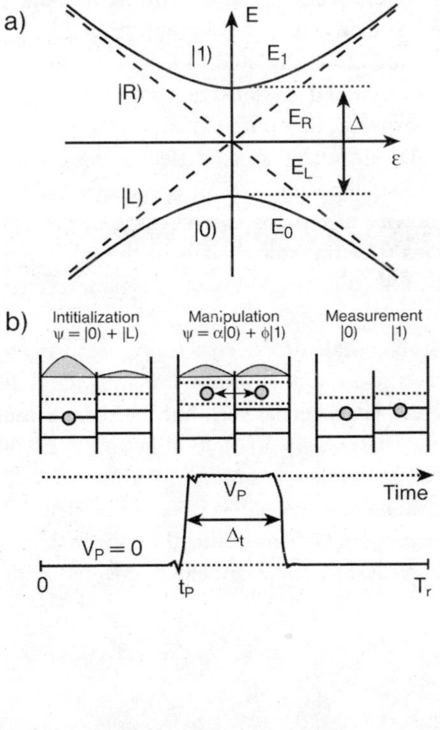

Figure 9.20 (Upper panel) The heavy curves labeled E_1 and E_0 represent the energies, respectively of the antisymmetric and symmetric combinations of L and R states, plotted against ε, which is the relative energy shift of the wells, bringing their highest filled levels to equality "resonance" at $\varepsilon = 0$. (Lower panel) Schematic of the wavefunction (pre-initialization has put the electron in L; the voltage pulse V_p applied for a time Δt brings the electron into a state of superposition between L and R; and finally at a later time T_r, the sequence is repeated) [17]. (Reprinted from Physical Review Letters with permission from APS.)

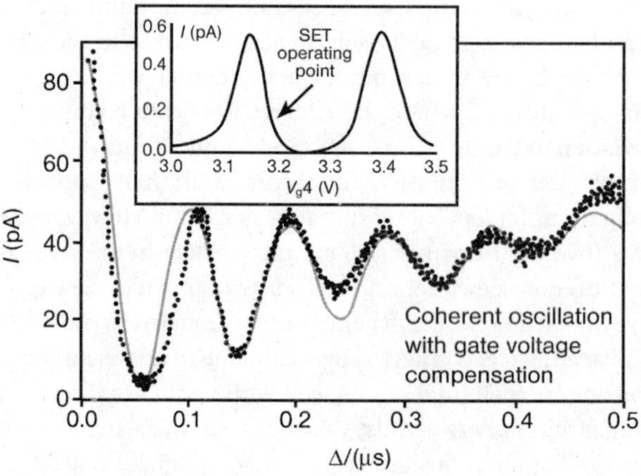

Figure 9.21 SET readout of coherent electron oscillation in the coupled double-well system [17]. Inset shows the biasing point of the SET measuring device used to obtain the curve in the main panel. The fit to the damped oscillation indicates a characteristic decay time of about 200 ns. The authors [17] point to the isolation of the active elements in the device, with only capacitive coupling to the control system, as important in achieving a rather long decay time near 0.2 μs (at temperatures in the vicinity of 20 mK). (Reprinted from Physical Review Letters with permission from APS.)

These data are fit to conclude that the energy spacing is 40.5 neV between bonding and antibonding states, with coherence time 220 ns. This time greatly exceeds the value, about 1 ps, inferred in the first device shown in Figure 9.14. The present device operates only at cryogenic temperature.

Recently [18] a double-well system with two electrons (an analog of the hydrogen molecule) has been demonstrated as a qubit, and an effective coherence time of 1 μs was achieved. This scheme uses states that are closely related to the bonding and antibonding states in di-hydrogen, which were discussed in connection with Figure 5.1.

In a second approach [19] to the single charge qubit, a pure silicon crystal is carefully *ion-implanted* with individual donor D^+ impurity ions, closely spaced with precision, to become double-well qubits. After deposition of an epitaxially grown covering layer of pure Si, a pattern of nanowire electrodes is necessary, in precise registry with the underlying qubit impurities. The individual wells in the following schemes are shallow donor impurities (D) in Si, such as P, As and Sb, as described in Section 5.7.

Elaborate plans [19] for building up such nanoscale device structures, requiring a scanning tunneling microscope to pattern the locations where the individual P donors will be incorporated, have been described. Alternatively, but with the same general framework, a thick resist may be patterned by STM to allow only a few exactly spaced locations where a beam of energetic P^+ ions impinging on the surface will be able to penetrate into the silicon. Methods have been described for electrically detecting the incorporation of individual energetic P^+ ions into the silicon layer. Again, this precise buried array of double wells has to be annealed, covered with more epitaxial silicon, and then a pattern of nanoscale electrodes deposited in close registry using electron-beam writing. The electrodes will be used to ionize the double-well P–P$^+$ pairs so each pair has only one electron. The readout of this scheme also involves single electron transistors. The approach here in terms of the quantum states, manipulation and readout is closely similar to the situation described in Figure 9.20, except that (i) the quantum wells are single donor impurity atoms, rather than a small cluster of atoms, and (ii) the controlling electrodes are deposited on the crystalline silicon surface directly above the buried donor impurities.

The authors [19] propose two different approaches to *coupling adjacent charge qubits*, to allow the interaction that is needed in the action of a quantum computer. Furthermore, they demonstrate a scheme for coupling a linear array of *N* such charge qubits. Their first two approaches to coupling adjacent qubits, spaced a distance *d*, are shown in Figure 9.22.

Schemes for coupling charge qubits [19] are shown in Figure 9.22. In two different configurations, the phase of one qubit is shown to affect the phase of a second qubit. In the left side the phase of the first qubit is shown as altering the barrier height of the second qubit. In the right side of Figure 9.22, the phase of the first qubit alters the energy of one of the sites in the second qubit. The authors [19] pres-

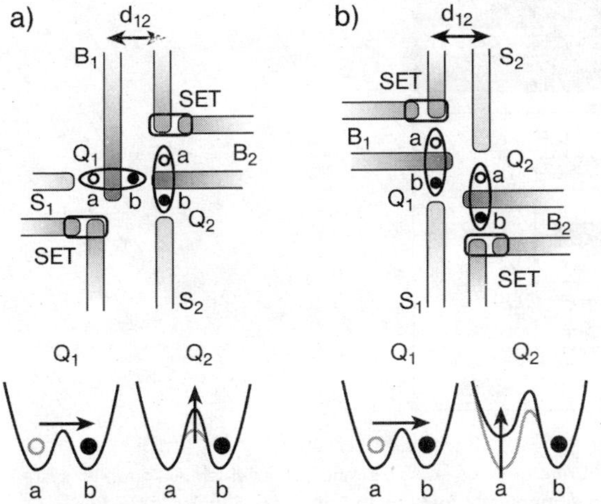

Figure 9.22 Charge qubit coupling schemes based on the Coulomb interaction [19]. (a) Orthogonal qubit arrangement of two single charge D_2^+ qubits, here labeled Q_1 and Q_2, each with equivalent wells "a" and "b". At the bottom of panel (a) the coupling is illustrated, such that driving Q_1 into its right well "b" raises the barrier for motion of Q_2, shown as locking the position of the electron in well "b". Note the locations of SET readout transistors at the "a" sites of each qubit. Note also control gates S_1 and S_2 addressing wells "b" of each qubit, and also B gates, which facilitate control of barrier heights. (b) Parallel qubit arrangement of two single charge D_2^+ qubits, here labeled Q_1 and Q_2, each with equivalent wells "a" and "b". In the lower portion of panel (b) the action of moving the charge from well "a" to well "b" in Q_1 is seen to raise the potential of charge Q_2 in its well "a", and thus to drive the charge to well "b". The qubit spacings d_{12} are assumed in the range 30–60 nm, and are estimated to lead to coupled qubit operation times of 0.1–1 ns. (Reprinted from Physical Review B with permission from APS.)

ent an extension of this plausible coupling scheme to a sequence of N gates. The authors discuss schemes for using the control gates to initialize the qubits and the readout. The coupling schemes adopted have been discussed earlier [20,21]. One might point out that the coupling of the qubits [19] has been proposed and analyzed, but not observed.

Both of these schemes for implementing double-well charge qubits in silicon technology require a cryogenic environment, because the binding energy of the donor states is quite small (see Section 5.7), and for a conventionally produced SET or RFSET readout to work.

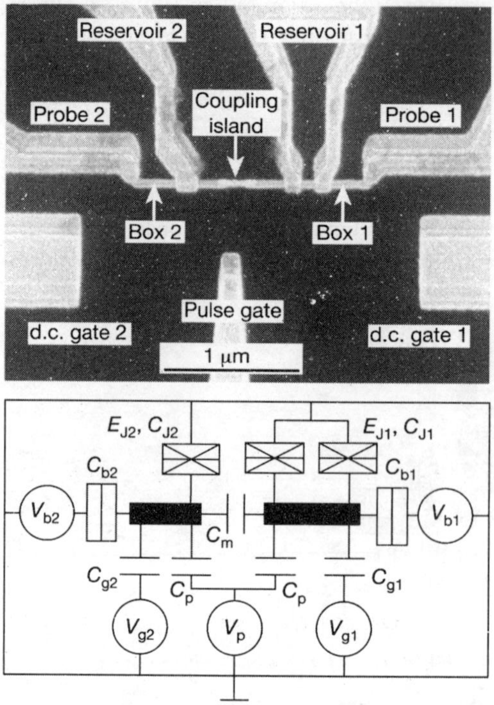

Figure 9.23 (Upper panel) Annotated microphotograph of the device [22] used to demonstrate two capacitively coupled charge qubits. Note the 1 μm marker. (Lower panel) Schematic diagram of the device. [In this diagram the symbol for the Josephson junction is a crossed rectangle (cross representing the Josephson supercurrent; see Section 8.13) with a line indicating parallel capacitance C, an inherent feature of any Josephson junction. Pure capacitances are uncrossed rectangles (junctions too wide to have a locked Josephson current.)] The *dark rectangles* in the lower panel correspond to "Cooper-pair boxes" (a quantum dot confining an electron pair; denoted as Box 1 and Box 2 in the microphotograph above) and contain the qubits. In each "Cooper-pair box", a single qubit (which is a coupled system of zero and one Cooper pairs in the box, analogous to the symmetric vs antisymmetric delocalized states of the charge qubit) can have its phase Θ (related to coefficients a and b, in the charge qubit case) tuned via the Josephson junction couplings (represented by E_{J1} and E_{J2}: see Sections 8.11–8.13). The coupling between the two qubits is via capacitance C_m (lower panel), realized in e-beam lithography as the "coupling Island" in the upper panel. The importance of this work is the demonstration of coupling (interaction) between two qubits, an essential feature needed for quantum computing. (Reprinted from Nature with permission from Macmillan Publishers, Ltd.)

9.6.1
Coupling of Two Charge Qubits in a Solid State (Superconducting) Context

Coupling of two charge qubits in a solid state context has been reported [22]. This two-qubit device uses *electron pairs* (see Sections 8.11–8.13) in superconducting Al, shown schematically in Figure 9.23.

Figure 9.24 shows measurements which clearly demonstrate coupling of the two charge qubits. In each case, a measured damped sinusoidal waveform is shown on

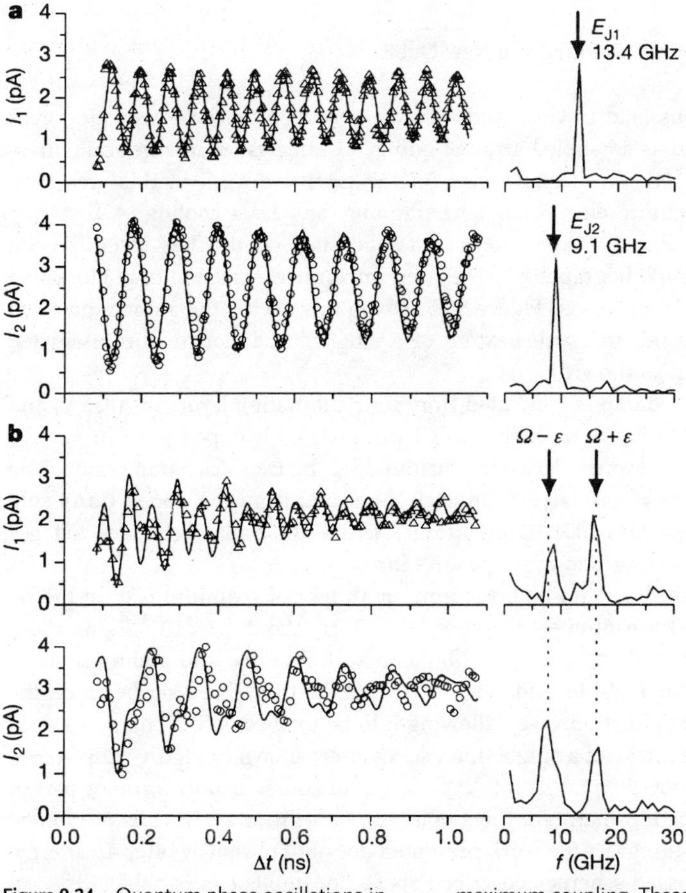

Figure 9.24 Quantum phase oscillations in qubits [22]. (a) Probe current oscillations in the first (top) and second (bottom) qubit when the system is configured for decoupling. In each case the right panel shows the corresponding spectra obtained by Fourier transform. In both cases the experimental data can be fitted to a damped cosine dependence (solid lines) with an exponential decay with 2.5 ns time constant. (b) Probe current oscillations in the qubits when configured (by electrode potentials) for maximum coupling. These spectra each contain two frequency components, revealed by the Fourier transforms on the right. The lower two traces (panel b) show the effect of coupling on these phase oscillations, including clearly resolved splittings in the spectra caused by the coupling. Such coupling is the means by which calculations are carried out within the quantum computer. (Reprinted from Nature with permission from Macmillan Publishers, Ltd.)

the left. The corresponding Fourier transform, revealing the underlying frequency components, and hence the possible qubit coupling, is shown on the right.

These results show the largest number of coherent oscillations, the highest Q, of the several qubit examples that have been shown. This feature makes the separation of the uncoupled and coupled frequencies quite convincing. Even though the device is superconducting, however, it seems that the coherence times are not larger than in other cases, and the measurements also require cryogenic temperature.

9.7
Ion Trap on a GaAs Chip, Pointing to a New Qubit

Linear arrays of ions held in close and accurate spacing by electromagnetic forces have been proposed as a coupled array of qubits, candidates for a large-scale quantum computer [23,24]. The technologies for the extensive work in this area have been ultra-high vacuum, electromagnetic trapping, and laser cooling. A first step toward realizing a linear array of ions in a solid state context has recently been reported [25], using a lithographically prepared micrometer-scale ion trap, fabricated on a monolithic chip, shown in Figure 9.25. Using this device, the authors go on to confine, laser-cool, and measure heating of a single $^{111}Cd^{+}$ ion in the integrated radio-frequency trap as shown.

As can be seen, the trap is fabricated from four alternating layers of AlGaAs and GaAs. The two GaAs layers of thickness 2.3 μm are highly doped ($\sim 3 \times 10^{18}$ cm^{-3}) and formed into cantilevered electrodes surrounding the free-space trap region. The electrodes are electrically isolated from each other and from the doped GaAs substrate by the interleaved AlGaAs layers of thickness 4 μm. Radio-frequency voltages are applied to the electrodes to trap a positive ion.

The device is placed in ultra-high vacuum, in a vapor of cadmium (Cd) in the region of the trap, with estimated pressure 10^{-11} Torr (about 1.3×10^{-9} Pa or about 10^{-14} bar). A laser method is used to photoionize Cd atoms, and additional laser beams are used to cool and to load single ions into the trap. The laser beam on the ion leads it to emit light (fluoresce), allowing it to be imaged with a charge-coupled device camera. The image of a single fluorescing ion is shown in Figure 9.26.

The semiconductor element in this experiment of course is only a minor part of the apparatus needed to maintain the single ion, or a linear array, as is of interest for quantum computing. So the work presented does not obviously offer an alternative to the silicon-based schemes described above, for qubits compatible with existing silicon technology. Refrigeration is relatively simple compared with the present requirements [25] of ultra-high vacuum, laser beams for ion cooling, and radio-frequency technology for ion trapping. Achieving the requisite level of vacuum in the interior of a chip is unlikely without the basic ultra-high-vacuum chamber. There are, of course, other applications for micrometer-size electromagnetic traps, including miniature mass spectrometer arrays, and compact atomic clocks. The small dimensions greatly reduce the radio-frequency power necessary to do the trapping of single ions.

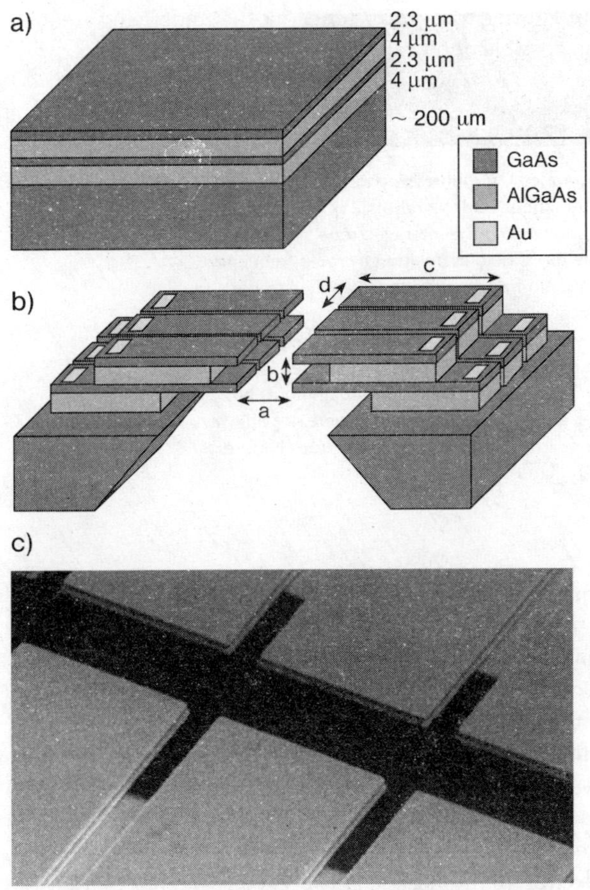

Figure 9.25 Three aspects of the fabrication of the micrometer-scale electrodynamic (radio-frequency) ion trap by molecular beam epitaxy on a single-crystal GaAs substrate [25]. (a) Layers of AlGaAs as grown epitaxially on the GaAs substrate. Gold (Au) is used as a contact metal. (b) Trap structure after steps of etching from below (to provide optical access), use of inductively coupled plasma etch to give access to the submerged GaAs layers, and deposition of gold/germanium bond pads. Further plasma etching isolates the cantilevered electrodes, and hydrofluoric acid etch undercuts the AlGaAs insulator material between the electrodes. (c) Scanning electron microscope image of a single ion-trap segment, showing the upper and lower GaAs layers separated by $h = 4$ μm. The spacing of the electrodes across the gap is 60 μm. (Reprinted from Nature Physics with permission from Macmillan Publishers, Ltd.)

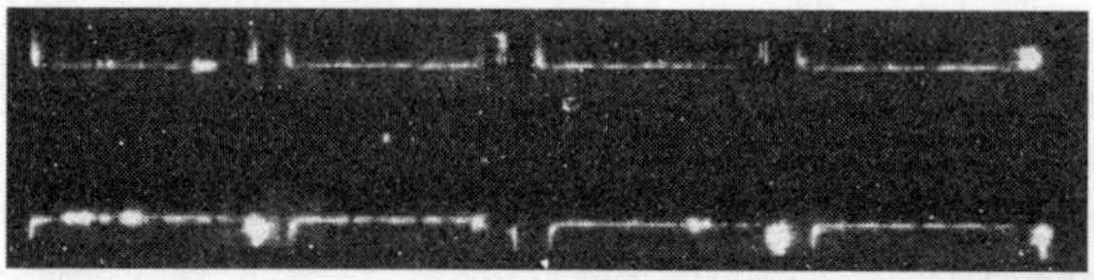

Figure 9.26 An image of a single trapped Cd^+ fluorescing ion along a view perpendicular to the chip plane, after about 1 s integration [25]. The ion is seen in the second of the four sections shown, and the inter-electrode spacing (vertical on the page) is 60 μm. The ion is held in position by radio-frequency electromagnetic fields arising from radio-frequency voltages applied to the nearby electrodes. The ion fluoresces with laser radiation nearly resonant with an S–P electronic transition in the ion, at 214.5 nm wavelength. The profile of the electrodes is seen as scattered radiation from an additional laser. With continuous laser cooling, the ion remains in position for up to an hour; but with no laser cooling, the lifetime is about 0.1 s. The authors were not able in this apparatus to load more than one ion at a time. (Reprinted from Nature Physics with permission from Macmillan Publishers, Ltd.)

9.8 Single Molecules as Active Elements in Electronic Circuits

The current through a single molecule need not be immeasurably small. In the early days of the scanning tunneling microscope (STM), the idea that 1 nA could pass through a *single atom* at the apex of the tip (which was required by observation of single atoms in STM topographs) initially seemed outlandish. In reality it is quite possible. After all, 1 nA is 6.25×10^9 electrons per second, corresponding to a residence time of 0.16 ns per electron tunneling event. *On an atomic scale this is not a short time.* In fact, the basic times for electrons in atoms are much smaller than 0.16 ns. For example, the orbital period of an electron in the 1s state of hydrogen is about 8.1×10^{-15} s.

In electron flow through any kind of atom or molecule, the actual electron transfers are by tunneling. The rate at which electron tunneling occurs has been discussed above, in connection with the time for an electron to tunnel out of a square well (as in the resonant tunneling diode, Section 4.6.4), in connection with the charge qubit devices (starting in Section 9.4), and also in connection with the time for hydrogen atom ionization by an electric field (Section 5.10). In all cases it is clear that the time for such processes is strongly (exponentially) dependent on the barrier width and height, both of which can be dependent on the applied electric field (or bias voltage).

However, there is no difficulty in finding times for tunneling much less than 1 ns. In connection with the hydrogen atom ionization (see Exercise 5.12) it was found that a field of 25 V/nm (perhaps 2.5 V across the atom itself) gave a lifetime ~2 fs (2×10^{-15} s), very short indeed, much shorter than would be needed in a molecular electronic device. So a characteristic of tunneling transmission is rapid variation of the rate (tunnel current) with the applied potential; but high rates of transmission are possible if one is careful not to destroy the device (destabilize the binding of the relevant atom).

Chemical bonds act as *tunnel barriers* (examples of tunnel barriers are shown in Figures 5.11 and 5.17), so that the electron wavefunction decays in an exponential fashion across the bond. For example, it has been discovered experimentally that the resistance of a linear chain of single-bonded molecules, alkanes, between two gold electrodes, increases exponentially with the number of bonds in the chain. In other chemical bonds (typically double or triple bonds), the wavefunction propagates across with little or no decay, as in a metal or across a Josephson junction. Electrons are essentially free to move around a benzene ring, for example.

Some chemical groups (see for example Figure 9.27 [26], where some rings enclose metal atoms, such as Co) act like small reservoirs or "islands" where electrons can localize. The potential at such "islands" can be changed by an external electric field from a "gate electrode". Single electron tunneling devices seem quite possible in this category, if the tunneling barriers represented by the covalent bonds have resistances in excess of 25.8 kΩ. If the tunneling resistance is much less, then the chance for resonant tunneling devices is present, including the chance to modify the transmission by a field electrode. *Within the large array of existing molecules, many opportunities exist for molecular devices performing the same functions as diodes, FETs and more complicated devices.* Reliable fabrication, self-assembly and integration into larger scales of such molecular devices are larger challenges.

The larger challenges to making molecular electronics are to locate the right kind of molecule, with the right kind of chemical bond (to act as an ohmic contact, for example) at the assigned location with the requisite *reliablility*.

A recent and cogent assessment of the success of the existing silicon technology [10] emphasizes the importance of reliability of the binary logic. The author argues that "emergence" in the silicon technology of simpler concepts ("quasiparticles" in Si, for example, "emerge" from the underlying quantum mechanics of many interacting electrons to provide an essentially Newtonian, classical picture of electrons and holes, with mass and charge and predictable motion) has been essential. This success of the deterministic, binary, silicon computation technology, in this view [10], is because it is an extension from the abacus and mechanical computation schemes, which are inherently error-free. This essential reliability is threatened in alternative approaches: for example, by the *inherent uncertainty* in any quantum approach; additionally, by the lack of large mean free paths, and of reproducibility, in polymeric replacements for silicon; and by the large defect rates expected in any large-scale assembly of molecular elements.

In contrast, the proponents of molecular computing approaches point to demonstrated methods to identify and bypass random failures in device structures. The famous example is the "Teramac", a fully deterministic silicon computer, which was deliberately loaded with defective elements. The important contribution was demonstration of software successful to identify and bypass the defects. The deliberately defective hardware was diagnosed and corrected in software.

Similar approaches are proposed to accommodate expected errors in molecular computer fabrication steps. Any molecular computer almost necessarily must involve self-assembly at some stage. (Proposals for tip-induced assembly of large systems have certainly been refuted (see Section 10.1.1).)

Self-assembly brings in errors, predicted from the statistical mechanics, relating to binding energies of particular molecules (in particular orientations) to particular sites, versus the thermal energy at the temperature in question. In these regards, concerning rates of error in any self-assembled structure, the basic probability factor is the Boltzmann factor, $\exp(-E/k_B T)$, where E is the binding energy, k_B is Boltzmann's constant, and T is the absolute temperature. While it is agreed that biology provides reliable self-assembly at 300 K, the process designer for molecular electronics will need to be careful.

There is optimism that defects in "wiring" that will likely come from molecular self-assembly can be compensated in software, but this has not been proven in any molecular computing method at present.

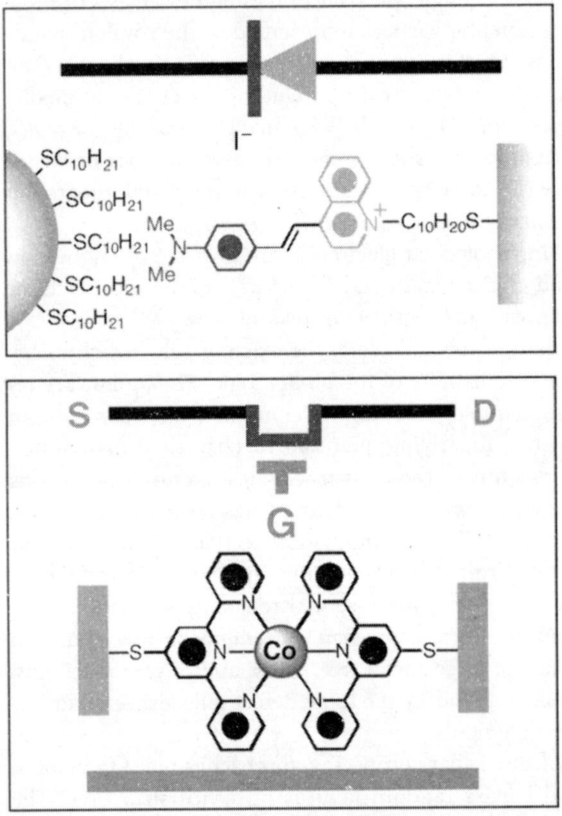

Figure 9.27 Molecular electronic devices [26]. (Top) A molecular rectifier constructed from donor (left ring), bridge (connects the two ring structures), and acceptor (right double ring). The active element is connected to gold electrodes by means of chain molecules with sulfur substitutions at the ends which bond to gold. (Lower) A single-molecule transistor constructed from a symmetrical Co complex. Here S, D and G label source, drain and gate electrodes, respectively. (Reprinted with permission from Science, copyright AAAS.)

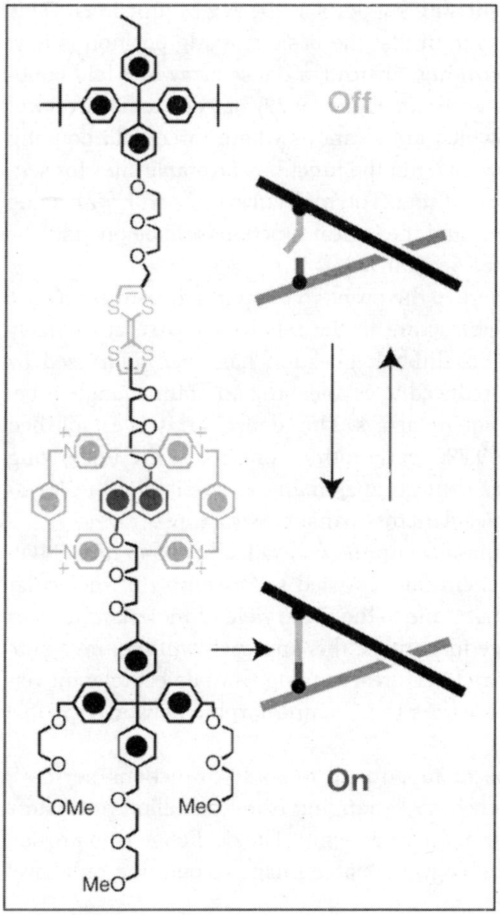

Figure 9.28 A two-terminal molecular switch tunnel junction [26] in its "OFF" and "ON" states. (Left) Structural formula of the "ON" state of a bistable [2]rotoxane. This molecule, at a specific voltage, switches from one stable structure (isomer) to another metastable isomer with a different conductivity, and remains (latched) in the latter state until either another voltage pulse is applied or a thermal fluctuation causes return to the original isomer. (Reprinted with permission from Science, copyright AAAS.)

9.9 Hybrid Nanoelectronics Combining Si CMOS and Molecular Electronics: CMOL

This approach to extending the existing silicon chip technology proposes an ultra-dense superstructure of molecular logic/storage devices to be "epitaxially" (deposited above AND connected to) a larger-scale (perhaps 45 nm nodes) conventional Si CMOS chip. The general schematic of the proposal is indicated in Figure 9.29. The "dense superstructure" is envisioned as a cross-bar array of gold nanowires, with vertical displacement provided by insulating supports at the $N \times M$ junctions. (This geometry differs from that in Figure 9.9, in that the basic crossing position is now blocked by the insulating support, providing, instead, a dense array of 3-fold bonding sites.) Thus the $N \times M$ cross-bar array (in Figure 9.29) of gold nanowires will provide $N \times M$ 3-fold junction sites, which are locations where three gold bonding surfaces are provided above and below, making the junctions favorable sites for self-assembling molecules floated in from solution. The molecular units will, with some error rate, select these connecting sites, and the logical functions will be provided by the choice of the molecular entities (see Section 9.8).

The word "epitaxy" is generalized here to the problem of optimally connecting the ultra-dense molecular cross-bar superstructure to the relatively coarse but reliably periodic underlying CMOS array. This difficult question has been addressed by Likharev [27]. The problem has been reduced to establishing an optimal angle α between the "dense" and "ultra-dense" square arrays. The "dense" array is established by "vertical plugs" shown in Figure 9.29, representing contacts to the underlying CMOS, and the optimization requires contacting as many as possible of the $N \times M$ crossing sites in the "ultra-dense" molecular cross-bar superstructure array.

This is an ambitious high-level proposal, counting on availability of 45 nm CMOS [1], and on advances in the area of chemically assisted self-assembly of molecular logic devices. The author [27] states that: "due to the finite yield of molecular devices and their sensitivity to random charge impurities, this approach will require a substantial revision of integrated circuit architectures, ranging from defect-tolerant versions of memory matrices and ... processors to ... neuromorphic networks capable of advanced image recognition ...".

The potential of the CMOL [27] is to support up to 3×10^{12} functions per cm^2, which may be compared to earlier benchmarks ranging from "a million transistors per chip", to the projection [1] of 10^9 transistors per chip. The challenge is to provide performance, in an enhanced Si technology, to enable image recognition on a level comparable to that of humans.

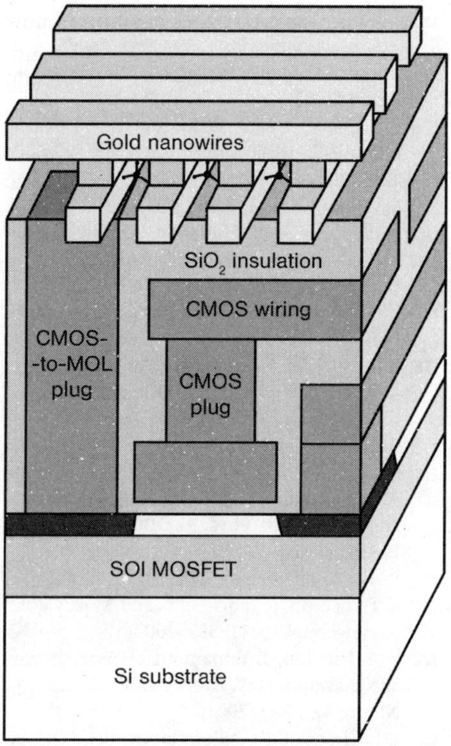

Figure 9.29 General concept of CMOL (CMOS / nanowire / molecular hybrid) circuit [27]. The lower level of the device is occupied by the CMOS subsystem, with transistor density suggested as 10^9 cm^{-2}, which could be achieved with the "45 nm node" technology [1]. Vertical plugs are installed to connect this level of the hybrid structure to the upper-level nanowires. The nanowiring is envisioned as extremely narrow, explicitly patterned, gold wires of widths around 3 nm half-pitch. The upper orthogonal set of nanowires is insulated and supported by an array of insulating posts. Finally, the molecular active elements self-assemble into the 3-fold interstices of the crossing nanowire array, as sketched. It is argued [27] that this scheme could support up to 3×10^{12} functions per cm^2. (Reprinted with permission from Elsevier.) [28]

References

[1] K. Greene, "The Impact of Emerging Technologies: Moore's Law Lives", TechnologyReview.com, January 26, 2006.

[2] M. A. Topinka, B. J. LeRoy, S. E. J. Shaw, E. J. Heller, R. M. Westervelt, K. D. Maranowski, and A. C. Gossard, Science **289**, 2323 (2000).

[3] P. Debray, G. E. Raichev, P. Vasilopoulos, M. Rahman, R. Perrin, and W. C. Mitchell, Physical Review **B61**, 10950 (2000).

[4] W. Liang, M. Bockrath, D. Bozovic, J. H. Hafner, M. Tinkham, and H. Park, Nature **411**, 665 (2001).

[5] E. S. Snow, F. K. Perkins, E. J. Houser, S. C. Badescu, and T. L. Reinecke, Science **307**, 1942 (2005).

[6] T. Reuckes, K. Kim, E. Joselevich, G. Y. Tseng, C.-L. Cheung, and C. M. Lieber, Science **289**, 94 (2000).

[7] A. M. Cassell, N. R. Franklin, T. W. Tombler, E. M. Chan, J. Han, and H. Dai, "Directed Growth of Free-Standing Single-Walled Carbon Nanotubes", J. Am. Chem. Soc. **121**, 7975 (1999) (see Figs 1 and 2).

[8] C. H. Mikkelson, A. C. Seabaugh, E. A. Beam III, J. H. Luscombe, and G. A. Frazier, IEEE Trans. Electron Devices **41**, 132 (1994).

[9] K. Leo, J. Shah, E. O. Gobel, T. C. Damen, S. Schmitt-Rink, W. Schafer, and K. Kohler, Phys. Rev. Lett. **66**, 201 (1991).

[10] R. B. Laughlin, "A Different Universe: Reinventing Physics From the Bottom Down" (Basic Books, New York, 2005), pp. 59–70, especially pp. 64–65.

[11] H. Park, J. Park, A. K. L. Lim, E. H. Anderson, A. P. Alivisatos, and P. L. McEuwen, Nature **407**, 57 (2000).

[12] H. W. C. Postma, T. Teepen, Z. Yao, M. Grifoni, and C. Dekker, Science **293**, 76 (2001).

[13] H. Namatsu, Y. Watanabe, K. Yamazaki, T. Yamaguchi, M. Nagase, Y. Ono, A. Fujiwara, and S. Horiguchi, J. Vac. Sci. Technol. **B21**, 1 (2003).

[14] R. J. Schoelkopf, P. Wahlgren, A. A. Kozhevnikov, P. Delsing, and D. E. Prober, Science **280**, 1238 (1998).

[15] M. DiVentra, S. T. Pantelides, and N. D. Lang, Appl. Phys. Lett. **76**, 3448 (2000).

[16] M. DiVentra, S. T. Pantelides, and N. D. Lang, Phys. Rev. Lett. **84**, 979 (2000).

[17] J. Gorman, D. G. Hasko, and D. A. Williams, Phys. Rev. Lett. **95**, 090502 (2005).

[18] J. R. Petta, A. C. Johnson, J. M. Taylor, E. A. Laird, A. Yacoby, M. D. Lukin, C. M. Marcus, M. P. Hanson, and A. C. Gossard, Science **309**, 2180 (2005).

[19] C. L. Hollenberg, A. S. Dzurak, C. Wellard, A. R. Hamilton, D. J. Reilly, G. J. Milburn, and R. G. Clark, Phys. Rev. **B69**, 113301 (2004).

[20] R. Landauer, Science **273**, 1914 (1996).

[21] L. Fedichkin, J. Yanchenko, and K. A. Valiev, Nanotechnology **11**, 387 (2000).

[22] Y. A. Pashkin, T. Yamamoto, O. Astafiev, Y. Nakamura, D. V. Averin, and J. S. Tsai, Nature **421**, 823 (2003).

[23] J. I. Cirac and P. Zoller, Nature **404**, 579 (2000).

[24] D. Kielpinski, C. Monroe, and D. J. Wineland, Nature **417**, 709 (2002).

[25] D. Stick, W. K. Hensinger, S. Olmschenk, M. J. Madsen, K. Schwab and C. Monroe, Nature Physics / Advance Online Publication / December 2005.

[26] A. H. Flood, J. F. Stoddart, D. W. Steuerman, and J. R. Heath, Science **306**, 2055 (2004).

[27] K. K. Likharev, "Electronics Below 10 nm", in J. Greer et al. (eds.), "Nano and Giga Challenges in Microelectronics" (Elsevier, Amsterdam, 2003), pp. 27–68.

[28] J. Greer et al. (eds.), Nano and Giga Challenges in Microelectronics, Elsevier, Amsterda, 2003.

10
Looking into the Future

It is clear that nanophysics will increasingly contribute to the development of technology as the decreasing size scales require quantum concepts. Those working in the various related areas of materials science, silicon technology, device design and fabrication will benefit by understanding the quantum phenomena which will inevitably show up on the smallest size scales.

An example from the foregoing is the quantum dot, where very useful behavior, a size-dependent shift in wavelengths absorbed and emitted, is adequately explained starting from the simplest quantum concept, the trapped particle in one dimension. Exploiting this uniquely nanophysical behavior has contributed to development in biological research and in laser development, as well as in electron device technology.

On a practical basis, it should be clear that a more detailed understanding of the basic phenomena of nanophysics, a topic not often emphasized in contemporary college science and engineering in the US, can indeed be relevant and profitable to a worker interested in nanotechnology. The conceptual thread of nanophysics underlies the properties of atoms, molecules, solids, electron devices, and biological science, including notably its instrumentation.

Nanophysics also bears on some conceptions, expectations, and misconceptions of nanotechnology. Nanoscience touches on some of the larger issues of human existence as well as on smart ways to make new machines and new industries. Nanotechnology has even been depicted as a threat to life. What about this?

10.1
Drexler's Mechanical (Molecular) Axle and Bearing

Tetrahedral bonding as occurs in diamond, silicon and germanium is the basis for Drexler's proposed family [1] of "diamondoid" nanostructures. These are large and complicated, covalently bonded, molecules. Start with a molecular axle and bearing (sleeve bearing) as shown in Figure 10.1. There is little doubt that this structure, and others like it, would be stable and robust, as calculated, *if it could be formed.*

The suggestion seems to be that such molecular structures could be designed *and eventually built* to form nano-replicas of common mechanical devices starting with

Nanophysics and Nanotechnology: An Introduction to Modern Concepts in Nanoscience. Second Edition.
Edward L. Wolf

ISBN: 3-527-40651-4

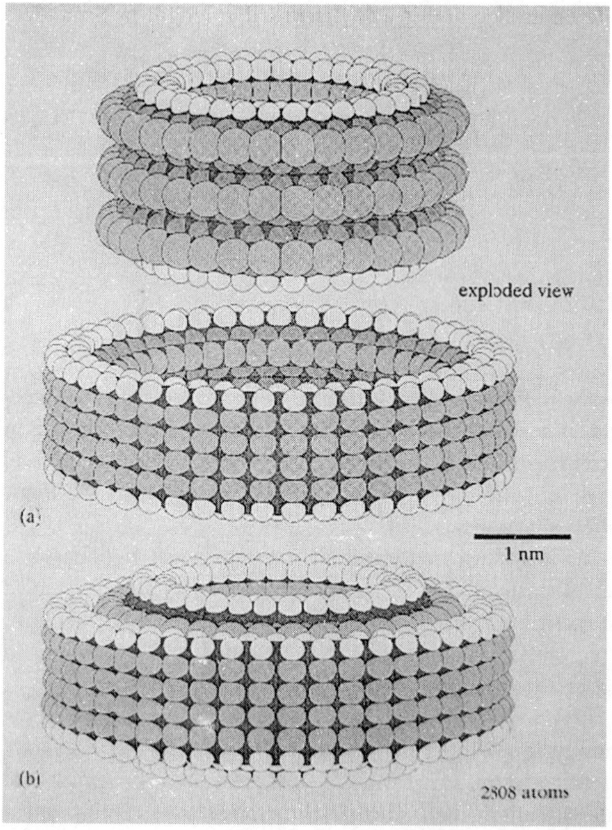

Figure 10.1 Proposed sleeve bearing composed of 2808 diamondoid atoms [1].

elementary bearings; planetary bearing, pumps, and building up to molecular scale robotic devices, machines, and factories.

10.1.1
Smalley's Refutation of Machine Assembly

As has been clearly pointed out by Smalley [2], such structures (e.g., see Figure 10.1) *cannot be fabricated using STM tips.* Richard E. Smalley, who won the Nobel Prize in chemistry for his work in discovering and fabricating C_{60} and carbon nanotubes, is an authority on nanotechnology. There is no answer to Smalley's basic point, that these large single molecules have interior locations that are completely inaccessible to such a relatively large object as an STM tip. Smalley also points to the inadequate functions of any such single tip, suggesting that several tips might have to work together to achieve the needed functions, making the tip size even larger and more of a problem. The total inadequacy of the rate of atom-by-atom assembly for making bulk matter has been considered earlier, and is also discussed at length by Smalley [2].

Additional arguments support Smalley's position, including the tendency, noted in connection with Figure 3.9, for an inserted atom to choose its own preferred bonding configuration, rather than the one required by the minimum energy for the overall structure.

Another way of stating this point is to note that the structure in Figure 10.1 has many strained bonds, to achieve its curvature. If the atoms were put into place in sequence, the deposited atom would choose an unstrained bond, tending toward a straight beam rather than a beam bent into a circle.

These structures cannot be assembled atom by atom [2]. On the other hand, structures such as this presumably stable molecule, might self-assemble, in the traditional mode of synthetic chemistry, in the right environment. A good environment for growing a structure must allow modifications of the structure to be tested for their relative stability. This means that the thermal energy must be sufficiently high to allow alternative structures to form, at least on a transient basis. A rough measure of that thermal energy is the melting temperature of the type of solid structure that is involved. The melting temperatures of strong solids are in the range 2000 K – 3000 K. To grow a diamondoid structure such as Figure 10.1 would require a temperature well above 1000 K, an environment completely devoid of oxygen, probably rich in hydrogen, and containing stoichiometric amounts of the elements for the particular structure. The structure in Figure 10.1 is stated to contain 2808 atoms.

Synthetic diamonds are commercially available, and are used in abrasives and other applications. Their formation requires high pressure and high temperature. Physicists and materials scientists have had a difficult time in controlling the growth of these small crystals, in any shape. Synthetic chemists have an honorable and productive history and a good intuition as to what kinds of molecules can be formed and how. However, synthetic chemists have not notably solved the problem of making diamondoid molecules or solids.

A step in this direction has recently been reported by Scott *et al.* [3], in an article entitled "A Rational Chemical Synthesis of C_{60}". The process involves 12 steps from commercially available starting materials by rational chemical methods. The final step is "flash vacuum pyrolysis at 1100 °C" (1373 K). Here again, the pyrolysis shows the need for a really energetic environment to close such a strongly curved structure. It appears that C_{60} is the only product in the reported synthesis. The authors comment that the approach they have used "should make possible ... preparation of other fullerenes as well, including those not accessible by graphite vaporization". The decomposition temperature of graphite is stated as 3300 K.

To make graphite from heavy hydrocarbons requires extended exposure to an extremely high temperature in a reducing atmosphere. To make a C_{60} molecule or carbon nanotube, conventionally requires enough energy to produce elemental carbon, as in a carbon arc, or ablated carbon plasma, corresponding to a temperature certainly in excess of 1000 K (the decomposition temperature of diamond is listed as 1800 K, and of SiC is 2570 K). The outstanding stability and relative simplicity C_{60}

explains its presence, along with C_{70} and other relatively stable species, as the atomic carbon condenses from a discharge.

In the condensation of a high-temperature, oxygen-free, atomic vapor containing the correct amounts of diamondoid elements, the single-molecule wheels and gears of Drexler (similar but more complicated than the sleeve bearing of Figure 10.1); or, perhaps, variants of such designs containing fewer strained bonds, will certainly have a chance of forming. That chance, however, may well be negligibly small.

Catalysts, if such can be found, would certainly improve the prospects of certain species. It is known that once the radius and twist of a nanotube is established, for example, on the surface of a tiny catalyst, perhaps a nano-sphere of carbide-forming metal, then the tube tends to grow rapidly along its length. Similar linear growth seems to occur for silicon, and other nanowires (see Figure 6.3 for an InP nanowire). A nanotube contains only one element, however, while the typical envisioned wheels and gears contain a range of covalently bonding elements.

In summary, the diamondoid structures cannot be produced on an atom-by-atom basis [2]. The extremely strong bonding and chemical variety in these structures makes their synthesis very difficult, except in cases of the fullerenes, which have only one element and very symmetric structures. Diamond-like carbon molecules of arbitrary shapes would also seem to be inaccessible to synthesis, as certainly are the diamondoid molecules of reference [1].

10.1.2
Van der Waals Forces for Frictionless Bearings?

Frictionless motion seems almost attainable, between flat or regularly curved surfaces which are nonreactive, attracted by van der Waals forces and repelled by atom–atom overlap forces. The best examples are graphite, in which the individual planes rather easily slide past each other, and in nested carbon nanotubes. In the case of nested nanotubes, translation and rotation have been carefully observed. These surfaces are non reactive (no dangling bonds) and the spacing is so small that no foreign atoms can exist in the gap. To the degree that the surfaces are commensurate, one would expect a locking tendency, but with weak attractive forces this may be negligible. In theoretical work [4] non-commensurate surfaces have been suggested to minimize any tendency to locking. These theoretical ideas can be tested against measurements on nested nanotubes.

10.2
The Concept of the Molecular Assembler is Flawed

The molecular assembler and the "self-replicating molecular assembler" are ideas that have been much discussed. From the point of view of nanophysics, such devices are not possible.

The molecular assembler is imagined to perform the functions of a hand or a robot arm on single atoms. The proposed device selects a particular kind of atom, grasps the atom, orients it suitably, inserts it into an atomic site in a molecule or solid being assembled atom-by-atom. It is necessary for the atom in question to be detached from the assembler arm and deposited in the proper site.

Molecular assemblers have been presented as able to assemble large machines out of diamondoid elements on reasonable time scales. Such projections must assume vast numbers of assemblers, considering the rate analysis discussed in Chapter 7. But, no matter, for no such tip-like device working on single atoms is possible. Any such tip is too large to allow access to atomic sites in a complicated structure. A small enough tip does not have facility for orienting an atom and for adjusting its property to grasp and then to release the atom. The rate at which any such device could operate is too slow to be of use in producing gram quantities of matter on an atom-by-atom basis. The device is said to work on any atom and to build any structure.

The argument advanced as an existence proof for molecular assemblers is nature itself, which we know has complicated and diverse molecular scale assembly capabilities. *Yet, there is nothing in nature that remotely resembles the imagined molecular assembler, in form, appearance, or general philosophy.* In nature, very specific enzymes apply to very specific molecules, to catalyze their formation or to cut them at particular locations. The idea is more like a lock and a key. Rather than working on atoms, the assembly processes in nature, work from a large inventory of molecules and polymers, which are formed (self-assembled) in traditional modes of chemistry and polymer chemistry. Rather than being general, and applying to a wide variety of structures (as the molecular assembler is imagined to do), the key, or decision-making, assembly processes in nature are extremely specific. There is an extensive hierarchy of such assembly processes, but they are independent and individually they are extremely specific, working on entities much larger than atoms.

Look at Figure 5.7, showing, in outline, the process of splitting a double helix DNA strand into two identical strands (the DNA replication fork). In this figure, all of the constituents: the bases A, C, G, and T; the deoxyribose cyclic sugar molecule, and the ionic triphosphate; are self-assembled molecules resulting from well-known biochemistry. *Even nature does not attempt to build anything atom by atom.* The decision-making "assembly process" in this figure is extremely specific, following from the lock-and-key fit of only two specific base-pairs across the helix from one strand to the other. Only the A will fit the T in bridging the strand, and only the C will fit the G.

To conclude, from the point of view of nanophysics, the molecular assembler is not possible. This of course means that self-replication of such imagined objects cannot occur. Nanotechnology (as differentiated from biotechnology) is therefore not to be feared from the point of view of possibly generating self-replicating organisms. These alarming ideas have come from a basically flawed concept and can be rejected as impossible.

While it is conceivable that entirely different chemical environments might allow the evolution of distinct forms of biology and life, no chemical environment will allow the imagined molecular assembler. This can be rejected simply on the properties of atoms. Atoms are the same throughout the universe.

What can occur, and has occurred to some extent already, is that great understanding and control of molecular biology allows modification of viruses, genes, and forms of life. Everyone agrees that these are topics in genetic engineering and biotechnology.

10.3 Could Molecular Machines Revolutionize Technology or even Self-replicate to Threaten Terrestrial Life?

In view of the discussion above, as long as "molecular machines" are understood to be the "assemblers" which apply to the diamondoid elements previously discussed, the answer to the question is "no".

Unfortunately, there is a history of misplaced concern that needs to be addressed. The field of nanotechnology has suffered from alarming warnings from its inception. The early popular book "Engines of Creation" [5] starting from an assumption of "replicating assemblers … able to make almost anything (including more of themselves) from common materials", goes on to warn that "dangerous replicators could be … too rapidly spreading to stop" and refers to the scenario of imagined replicators spreading like a cancer as the "gray goo problem" (p. 172). This attention-getting notion seems distinctly misleading from the present view of nanophysics, in which there is absolutely no chance of molecular assemblers ever existing apart from biology and plausible variations on biology.

This alarming possibility [5] was recently recalled [6] by Bill Joy in an essay entitled "Why the future doesn't need us". Joy mentions "genetics, nanotechnology and robotics" (GNR) as three dangerous areas of research that prudently should be banned.

Fukuyama, in his book "Our Posthuman Future" [7] (primarily a rational analysis of ethical issues of genetic engineering), casually links "nuclear weapons and nuclear energy … perceived as dangerous from the start …" to "nanotechnology – that is, molecular-scale self-replicating machines capable of reproducing out of control, and destroying their creators." Both categories, he says (p. 8) are "threats … easiest to deal with because they are so obvious." The answer for nuclear weapons and nanotechnology, according to Fukuyama [7], is strict control on an international basis. Fukuyama, dismissing nuclear power and nanotechnology as being taken care of by strict control, goes on to consider genetics and biotechnology in careful detail.

From the point of view of nanophysics, a *link of nuclear weapons and nanotechnology is ludicrous.* There is no particular danger in small-sized particles of matter. Fear of nanotechnology is not justified. Such fears appear to be based on the imagined, fallacious, but widespread notion of the "self-replicating" molecular assembler.

The imagined danger of the "molecular assemblers" of nanotechnology is a red herring, for no such assemblers will ever be possible. The assembly processes of nature, of course, do allow the spread of dangerous organisms, in the form of viruses and bacteria. "Genetic engineering" and "biotechnology" of viruses and bacteria, and even of higher life forms, do, certainly present threats as suggested by Bill Joy [6]. Genetic modifications and biotechnology are real and active areas of research and production that have major present and potential impacts, including the possibility of a "posthuman" future.

Nanotechnology, as distinct from genetics, biotechnology and robotics, is a natural interdisciplinary area of research and engineering based on nanophysics, chemistry, materials science, mechanical and electrical engineering and biology, and has no particular risk factors associated with it. Nanometer-sized particles of all sorts have been part of nature and the earliest forms of industry for all of recorded history. Nanometer-sized chemical products and medicines are routinely subject to screening for safety approval along with other chemicals and medical products.

10.4 What about Genetic Engineering and Robotics?

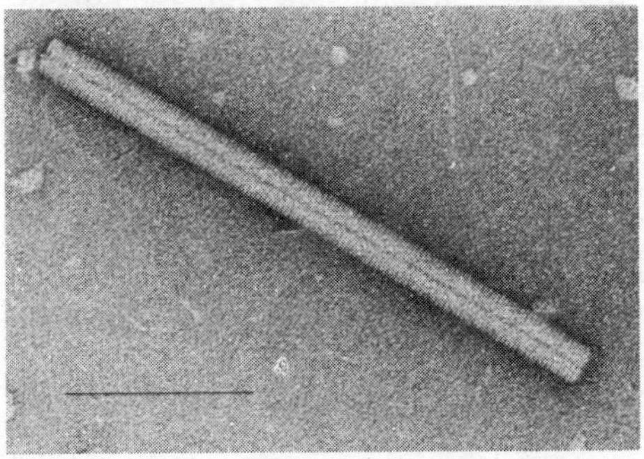

Figure 10.2 TEM image of the Tobacco Mosaic Virus (TMV) [8]. Note scale marker 100 nm. The TMV is 18 nm by 300 nm. This is the earliest identified virus, not a living object, but capable of parasitically entering living cells. Great facility now exists worldwide for modifying such viruses.

Figure 10.2 shows a TEM image [8] of the Tobacco Mosaic Virus, the first virus to be identified. A virus is not a living object. Rather it is a chemical structure which is capable of entering an appropriate living cell, parasitically altering the operation of that cell, and ultimately replicating within the host organism. Viruses are the causes of many diseases, including polio and anthrax.

The New York Times recently reported [9] the experience of John Tull, a native of Santa Fe, New Mexico, who was infected by bubonic plague, apparently by a flea on his own property in suburban New Mexico, in the year 2002. Mr. Tull became gravely ill while visiting New York City, but his life was saved, at great effort and sacrifice on his part, in New York City hospitals. The effects of the (initially very few) *Yersinia pestis* bacteria on Mr. Tull's system were so great that the doctors had to induce Mr. Tull into a coma for some weeks, and, finally, to amputate his feet, simply to keep him alive.

Great facility now exists, all over the world, in the production, alteration, and application of viruses. Modified viruses can be used to produce modified cells including modified versions of bacteria such as *E. coli* and *Yersinia pestis.*

The New York Times of August 19, 2003 reports [10] a medical procedure on a Parkinson's disease patient. The patient's skull was opened to allow insertion of *an artificial virus.* The virus was a naturally available virus, which had been modified (engineered) to carry a specific DNA sequence, which in turn was designed to counter the effects of Parkinson's disease. This experimental procedure had been approved by the Federal Drug Administration for 12 patients with advanced Parkinson's disease.

Research workers at State University of New York at Stony Brook [11] reported synthesis of the polio virus from starting chemicals available in the modern biological laboratory. The only starting information used was knowledge of the DNA sequences in the polio virus. The implication is that this virus could be built to altered specifications, perhaps including those for which the Salk vaccine would be useless.

A large industry exists today based on biological molecules with modified properties. Enzymes are central to the biological breaking down of large molecules into smaller molecules. Enzymes have been modified and successfully introduced into common household detergents with the purpose of breaking down dirt and difficult stains. Corn seed is widely available which has been genetically modified to provide, in effect, its own pesticide. Considerable furor greeted the potential introduction of seed corn genetically engineered in such a way that the resulting seed would be infertile. Marketing of this product, engineered with the so-called "terminator gene", eventually was abandoned.

The perceived [6] threat from robotics and artificial intelligence is related to the hugely increasing computing power, and the continuously advancing field of artificial intelligence.

A corollary to Moore's Law (see Figure 1.2) is the notion that at some not far-distant time the computing power of a computer may exceed that of a human being. This comparison does not make sense in many ways, because the type of computation in the human and in the silicon chip is so different. There is a push to find ways to make these two types of intelligence more compatible, to provide a "broadband connection" to the brain. Scenarios along this line are considered at length by Kurzweil [12]. The idea that, somehow, a huge computer system might become willful (The Age of Spiritual Machines), is very intriguing. Whether this is possible is not obvious, but some scientists expect it may eventually be. It is certainly assumed in recent Hollywood films such as The Matrix and Terminator.

On the one hand, any self-empowered computer system will presumably be constrained to live in its air-conditioned box, limiting the damage it might inflict. The option would seem to exist to pull the plug. On the other hand, a huge parallel-processing system controlling large sums of money in a bank, if it were to undergo a phase transition and declare independence, might, in its own territory, be able to do a lot of financial mischief.

A recent and thorough analysis and projections in the robotics/artificial intelligence area have been given by Rodney Brooks [13]. Brooks is well qualified for this analysis, as Director of the Artificial Intelligence Laboratory at MIT and also a chief officer in the iRobot Corporation.

A steady series of increasingly sophisticated robots have been designed specifically to interact with humans. Some have been laboratory devices (Kismet and Genghis). Others have been sold for the amusement of children (Furby, My Real Baby). And, primarily for the amusement of adults, there is AIBO, the Sony robot dog. iRobot Corporation sells a robotic home vacuum cleaner which is both fun and useful.

It is well known that the IBM chess playing computer Big Blue bested Gary Kasparov, the world's leading chess player. But, as noted by Brooks, Mr. Kasparov, in defeat, took satisfaction from the fact that Big Blue showed no evidence of elation in having won the match! So, clearly, Big Blue, lacking such an obvious emotion, is not human. Brooks [13] considers where this ongoing improvement may lead.

The sequence of robots mentioned does lead toward robots of human equivalent calculating power. Let's make the assumption that the AI/computer science community will eventually discover how to program such a supercomputing robot to make it willful and independent. (This idea is not new. Isaac Asimov's book, famously entitled "I, Robot", was written in 1950.)

In a Hollywood-like scenario, the intelligent robots become so much more advanced than the humans that they eventually decide to do away with the humans.

Brooks gives a set of hidden assumptions for such a (disastrous) outcome:

1. The machines can repair and reproduce themselves without human help.
2. It is possible to build machines that are intelligent but which do not have human emotions and, in particular, have no empathy for humans.
3. The machines we build will have a desire to survive and to control their environment to ensure their survival.
4. We, ultimately, will not be able to control our machines when they make a decision.

(End of quote from p. 200 of [13])

The status of these four assumptions is analyzed by Brooks, with a summary to the effect that one may expect many decades before any such disaster can occur.

It is also suggested, as by Fukuyama, that international rules of conduct may become appropriate (e.g., warlike aggressor robots are forbidden, not so different from the (presently void) ABM treaty).

To take a positive point of view, the benign areas of nanophysics and nanotechnology, are nonetheless likely to make contributions toward developments in the (possibly regulated, but doubtless active) areas of biotechnology and AI/robotics. In both medical and computer instrumentation, physics has been a strong source of innovation and insight. This is likely to accelerate, providing opportunities for nanotechnologists to be employed and to make contributions to advances.

10.5 Possible Social and Ethical Implications of Biotechnology and Synthetic Biology

An article entitled "Futures of Artificial Life" appeared in *Nature* in 2004 [14]. This article suggested that biologists, who have "for decades ... been transferring modified genes into microbes, nematodes, and mice", are increasingly secure in believing that nothing greatly harmful is likely to arise, even from expansion of the activity into what is now called *synthetic biology*. (This is the research area which led to genetically modified corn, a useful product, but arousing considerable resistance in some consumer communities.)

The emerging field of *synthetic biology* is described as an outgrowth of genetic engineering, an area heartily endorsed, for the amazing results and products that have been obtained, in the biochemical and chemical engineering communities. The results include (in addition to the shunned self-destructing corn seeds) detergents that enzymatically digest stains (left-over food), spectacularly improving results, with no particular complaints.

In recent examples of synthetic biology, bacteria and yeast have been manipulated genetically to build proteins that are impossible in Nature, differing in part by incorporation of synthetic amino acids. It is reported [15] that several groups are working toward assembling simple cells from basic components, i.e., to *create artificial living matter*. It is anticipated that a synthetic biologist may soon be able to produce an artificial living cell, along the same lines that a group earlier announced [11] that they had *synthesized the complete poliovirus* from "mail-order DNA" [15].

The potential for positive outcomes of synthetic biology is large. For example, work has progressed on genetically modifying the common bacterium *E. coli* to synthesize the powerful antimalarial drug artemisinin [15]. (This drug is conventionally available, in small supplies and at high cost, from the wormwood plant.) The synthetic biology approach [15] is to import 10 genes from other organisms, including wormwood and brewer's yeast, into *E. coli*, which is then cultured to produce the desired drug artemisinin cheaply and in volume. Other researchers have found ways of introducing more than 80 unconventional amino acids into proteins [15,16], suggesting un-natural forms of life.

Some responsible persons may see a disconnection between the activity of the synthetic biology community, going ahead with energy and optimism to improve on Nature, and the potential, in an age of security risks, of enabling biological poisons. (The capacity already exists, perhaps not widely recognized, to create new viruses, as demonstrated by the synthesis [11] of the poliovirus. It may seem, from one naive point of view, that what is now of concern is merely an extension, from one to two classes of infectants: maybe only an incremental change.) And it may well be that there are larger, unaddressed risks already, e.g., those that may be included in shipping containers arriving uninspected at urban ports.

Yet, the revolutionary ability *to modify life and to create new life* (even in activities conceived benevolently) has many consequences, which are hard to predict. The area including synthetic biology, to begin, should be explored more fully in public discourse [14]. A possible outcome of synthetic biology activity is that a condemning edict may appear (as has already occurred in regard to research based on particular kinds of cells). Those with fixed beliefs about life may well be alarmed. This activity may well be viewed as heretical in many quarters, representing, perhaps, half of humanity.

The scope of biotechnological industrial activity is broad. This is indicated by the number of patents, in the US alone, on aspects (new forms?) of life. A recent report [17] states that the number of US patents per year on nucleic acids (DNA and RNA) has been in the range 3000 to 4000 per year over the past decade, and that the largest holder of such patents, the University of California, holds 1018 patents. (On the whole, patents on "life" are an industrial phenomenon.) This suggests that *modifications of life*, if not starting new forms of life, are an ongoing industrial activity, whatever the consequences.

10.6 Is there a Posthuman Future as Envisioned by Fukuyama?

The "Posthuman Future" [7] is a phrase to suggest a time period when the human condition will substantially change. The scenario particularly troubling to Fukuyama has to do with genetic engineering, and the possibility that humans may be subject to breeding not unlike breeding of dogs and other animals. This prospect has great opportunity and great danger. Roughly speaking, there has been success in genetically engineering new crops, new mice, and cloning of many animals has become routine. Why not engineer new humans! If one wants an athlete, there are various animals whose genes might make an addition! There is a lively debate between proponents and opponents of human intervention in what is called the "human germline".

One starting point is in the area of in-vitro fertilization, which has been beneficial to a small fraction of advanced human populations in overcoming infertility. The next step in this process is to perform more advanced screening on the genes of the several embryos that are often produced, and to offer the option of informed selection among these to promote an offspring with desired characteristics. Such screening is already used to exclude embryos that have undesired properties. (Of course, ultrasound screening of the normal in-vivo embryos has been used all over the world and has led in many places to very unnatural ratios of male to female births.) The opportunity seems to exist for an accelerated type of breeding of humans, not different in principle from the breeding of dogs or other animals.

There is further discussion of artificial genes being designed as a package that might be offered to augment any embryo, in the hope of achieving longer life or higher intelligence or other properties. The new genetic makeup would then propagate, and be perpetuated in offspring of such "super-human" or "post-human" beings. Some see this genetic engineering possibility as a potential breakthrough, to raise the capabilities of the human race, an opportunity not to be missed, a bit like populating outer space. Others see this as a chance for grave injustices, not to mention terrible mistakes. It seems it cannot occur without seriously violating the perceived rights of privacy and continuity of many, religious groups for example, probably constituting a majority. The issues are such that they may be easily misrepresented and easily misunderstood, or not understood at all. These are socially explosive issues.

The opportunity to breed a superior managerial class and an inferior laboring class would seem possible for a totalitarian regime, not to forget the attempt of the Nazi regime in prewar Germany to eradicate the Jewish population. Fukuyama [7] gives a careful discussion of this situation, and concludes that international controls of such developments are important. He quotes (p. 9) notably from Thomas Jefferson: "The general spread of the light of science has already laid open to every view the palpable truth, that the mass of mankind has *not* been born with saddles on their backs, nor a favored few booted and spurred, ready to ride them legitimately, *by the grace of God*" (Italics added). Most people disapprove of human slavery, which is at issue.

The analysis of Fukuyama [7] is restrained and statesmanlike. In contrast, the analysis of Brooks, in the area of germline engineering (changes in the human genome), is written more from an operational point of view, and imparts feelings of inevitability and urgency.

Quoting from Brooks [13] (pp. 235–236):

"It is clear that Robotic technology will merge with Biotechnology in the first half of this century…We are on a path to changing our genome in profound ways. Not simple improvements toward ideal humans as are often feared. In reality, we will have the power to manipulate our own bodies in the way we currently manipulate the design of machines. We will have the keys to our own existence. *There is no need to worry about mere robots taking over from us. We will be taking over from ourselves with manipulatable body plans and capabilities easily able to match that of any robot.* **The distinction between us and robots is going to disappear**" (Italics added).

So, one well informed answer to the question, "Will there be a posthuman future", is YES! Yes, our "posthuman future" seems very likely, and may well call for regulation!

This book has been about the benign and multidisciplinary science of nanophysics and the corresponding nanotechnology. These areas underlie the development of all futuristic scenarios, but they are not directly at the forefront (as we have demonstrated) of genetic engineering, artificial intelligence, or robotics. There is thus a large and ethically unencumbered opportunity for nanophysicsts and nanotechnologists to contribute, if so inclined, to "our posthuman future".

References

[1] From *Nanosystems: Molecular Machinery, Manufacturing & Computation* by K. Eric Drexler. Copyright © 1992 by Wiley Publishing, Inc. All rights reserved. Reproduced here by permission of the publisher.

[2] R. E. Smalley, Scientific American **285**, Number 3, 76 (2001).

[3] L. T. Scott, M. M. Boorum, B. J. McMahon, S. Hagen, M. Mack, J. Blank, H.Wagner, and A.de Meijere, Science **295**, 1500 (2002).

[4] R. C. Merkle, Nanotechnology **4**, 86 (1993).

[5] K. E. Drexler, *Engines of Creation*, (Anchor Books, New York, 1990) p. 172.

[6] Bill Joy, "Why the Future Does not Need us.", Wired magazine, April, 2000.

[7] F. Fukuyama, *Our Posthuman Future*, (Farrar, Straus and Giroux, New York, 2002) pp. 7, 8.

[8] Courtesy Cornelia Büchen-Osmond, Earth Institute, Columbia University.

[9] S. Albin, "Plague Patient Slowly Recovering", New York Times, January 17, 2003.

[10] D. Grady and G. Kolata, "Gene Therapy Used to Treat Patients with Parkinson's", New York Times, August 19, 2003.

[11] J. Cello, A. V. Paul, and E. Wimmer, Science **297**, 1016 (2002).

[12] R. Kurzweil, *The Age of Spiritual Machines* (Penguin, New York, 1999).

[13] R. A. Brooks, *Flesh and Machines: How Robots Will Change Us* (Vintage Books, New York, 2003) p. 200, 235, 236.

[14] "Futures of Artificial Life: Researchers Involved in Synthetic Biology Need to Take Steps to Engage More With the Public", Nature **431**, 613 (2004).

[15] P. Ball, "Starting from Scratch: Genetic Engineering is Old Hat. Biologists Are Now Synthesizing Genomes, Altering the Genetic Code, and Contemplating New Life Forms", Nature **431**, 624 (2004).

[16] A. Pollock, "Custom-Made Microbes, At Your Service: The Rise of Synthetic Biology", New York Times, January 17, 2006.

[17] G. Stix, "Owning the Stuff of Life", Scientific American **294**, Number 2, p. 76 (2006).

Credits

IBM Research credit for Figures 3.9 and 7.3

Eastman Kodak Research for Figure 6.6

R. M. Eisberg and R. Resnick, *Quantum Physics of Atoms, Molecules, Solids, Nuclei, and Particles*, (2nd Edition) (Wiley, New York, 1985) for Figures 4.8, 5.1, 5.8, 5.9 and 5.10

K. E. Drexler, *Nanosystems* (Wiley, New York, 1992) for Figures 5.3, 6.1 and 10.1

Charles Kittel, *Introduction to Solid State Physics* (Sixth Edition) (Wiley, New York, 1986), for Table 5.3, Figures 5.20 and 5.21

K. S. Krane, *Modern Physics* (2nd Edition) (Wiley, New York, 1995), for Figure 5.2.

F. J. Pilar, *Elementary Quantum Chemistry*, (Dover, New York., 2001), for Table 4.1 and Figures 4.9, 4.10 and 4.11

Glossary of Abbreviations

A	adenosine; one of the four bases in RNA; one of the four bases in the DNA double helix (joins with T)
ABM	Anti-Ballistic Missile (treaty); suggested as a prototype for limits on robots
ADC	analog-to-digital converter
AFM	atomic force microscope
amu	atomic mass unit (u); defined as 1/12 of mass of carbon-12 (^{12}C): $u = 1.661 \times 10^{-27}$ kg
ATP	adenosine triphosphate; biological energy source, leading to ADP
bar	one atmosphere, 101.3 kPa
BCS	Bardeen, Cooper, Schrieffer (theory); basic accepted theory of superconductivity
BOX	buried oxide layer; in Si, used to reduce capacitances in FET devices (*see* SOI)
C	cytosine; one of the four bases in RNA: one of the four bases in the DNA double helix (joins with G)
CIP	current in plane; geometry for spin valve type of GMR magnetic field sensor
CMOL	CMOS / nanowire / molecular hybrid; proposed computer logic
CMOS	complementary metal oxide semiconductor; computer logic
CPP	current perpendicular to plane; geometry in TMR (tunnel valve) magnetic sensor
CPU	central processing unit; in computer
CVD	chemical vapor deposition; a rapid means of depositing layers in semiconductor
D	debye; unit of electric dipole moment: $1\ D = 3.3 \times 10^{-30}$ C m
DMF	dimethylformamide; a polar molecule
DNA	deoxyribonucleic acid; polymer, forms single and double helices; latter hydrogen-bonded by bases C,G,A,T, encoding information for assembly of proteins
DNT	dinitrotoluene; an explosive ingredient; a polar molecule, potentially detectable

Nanophysics and Nanotechnology: An Introduction to Modern Concepts in Nanoscience. Second Edition.
Edward L. Wolf

ISBN: 3-527-40651-4

DOS	density of states; usually stated per unit energy per unit volume, for electrons
DRAM	dynamic random access memory
dsDNA	double-strand DNA
E_F	Fermi energy
E_g	energy gap of a semiconductor, typically in eV
ESR	electron spin resonance
eV	electronvolt; 1 eV = 1.6×10^{-19} J
fcc	face-centered cubic
FET	field effect transistor
fm	femtometer; 10^{-15} m, size scale of the atomic nucleus
FQHE	fractional quantum Hall effect
G	guanine; one of the four bases in RNA; one of the four bases in the DNA double helix (joins with C)
GMR	giant magnetoresistance; basis of "spin valve" CIP hard drive *B*-field sensor
hcp	hexagonal close packed
HEMT	high electron mobility transistor
HTS	high-temperature superconductor
JJ	Josephson tunnel junction
laser	light amplification by stimulated emission of radiation
maJ	milliattojoule; possible energy unit for small systems
MBE	molecular beam epitaxy; a method for depositing atomically perfect crystal layers
MEMS	micro-electromechanical system
meV	millielectronvolts; thousandth of an electronvolt
MeV	megaelectronvolts; million electronvolts
MFM	magnetic force microscope
MOSFET	metal oxide semiconductor field effect transistor; basic switch in electronics
MRAM	magnetic random access memory
MRFM	magnetic resonance force microscope
MRI	magnetic resonance imaging; use of proton (spin) magnetic resonance to map locations of water molecules in living specimens
MWNT	multi-wall (carbon) nanotube
NEMS	nano-electromechanical system
NMR	nuclear magnetic resonance
NW	nanowire
PMMA	polymethylmethacrylate; used as a photo-resist in silicon technology
pn junction	junction between p (positively doped) and n (negatively doped) semiconductors; rectifier, element of transistor, and voltage-variable capacitor
Py	Permalloy; a high-permeability Ni–Fe ferromagnet
PZT	lead zirconate titanate; piezoelectric used in sonar, SPM and SBSL transducers

QCA	quantum cellular automata
QD	quantum dot; a three-dimensionally small object, "artificial atom"
QED	quantum electrodynamics; interaction of radiation with matter, leading, for example, to small change in g-factor of electron spin from 2.0 to 2.0023
QHE	quantum Hall effect
QPC	quantum point contact
radar	radio detection and ranging
RAM	random access memory
recA	protein of prototypical bacterium *E. coli* used in DNA assembly of nano-structures
RFSET	radio-frequency single electron transistor
RNA	ribonucleic acid; polymer, similar to DNA, but uses base U instead of base T; evolutionary precursor of DNA, retains central roles in protein synthesis
RNAP	RNA polymerase; enzyme that transcribes DNA template into messenger RNA
RSFQ	rapid single flux quantum; form of superconducting computer logic
RTD	resonant tunneling diode; plus related forms (e.g., TBRTD) of transistor and logic
SBSL	single bubble sonoluminescence
SEM	scanning electron microscope
SET	single electron transistor
SFQ	single flux quantum
SHO	simple harmonic oscillator
SOI	silicon on insulator; sometimes implemented on single-crystal Si by implanting a deep layer of oxygen, followed by annealing to produce quartz insulator (*see* BOX)
sonar	sound navigation and ranging; in modern forms uses piezoelectric transducers for sound generation underwater
SPM	scanning probe microscope
SQUID	superconducting quantum interference detector
ssDNA	single-strand DNA
STM	scanning tunneling microscope
SWNT	single-wall (carbon) nanotube
T	thymine; not present in RNA; one of the four bases in the DNA double helix (joins with A)
TBRTD	triple-barrier resonant tunneling diode (*see* RTD)
TEM	transmission electron microscope
TMR	tunnel magnetoresistance; basis of "tunnel valve" CPP magnetic field sensor
TPa	terapascal; $1\ \text{TPa} = 10^{12}\ \text{N/m}^2$; possible value of Young's modulus
u	atomic mass unit (amu); defined as 1/12 of mass of carbon-12 (^{12}C): $1\ \text{u} = 1.661 \times 10^{-27}$ kg

U	uracil; one of the four bases (replacing T) in RNA (joins with A); not present in DNA
2DEG	two-dimensional electron gas

Exercises

Exercises – Chapter 1

1. Referring to Figure 1.2: if there are 10 million transistors uniformly distributed on a one centimeter square silicon chip, what is the linear size of each unit?
2. A contemporary computer chip dissipates 54 Watts on an area of one centimeter square. Assuming that transistor elements in succeeding computer generations require constant power independent of their size (a hypothetical assumption), estimate the power that will be needed for a one centimeter square silicon chip in 5 years. Base your estimate on the Moore's Law trend of doubling the transistor count every 1.5 years. (The industry is confident of continuing this trend.)
3. Extrapolate the line in Figure 1.2, to estimate in which year the size of the transistor cell will be 10 nm.
4. In Figure 1.1, the vibrational motions of the single-crystal-silicon bars are transverse (vertical as seen in the picture) and the lowest frequency vibration corresponds to an anti-node at the middle, with nodes at each end of the bar. The resonances occur when $L = n\lambda/2$, $n = 1,2,3\ldots$ If the fundamental frequency of the 2 micrometer (uppermost) bar is 0.4 GHz, what frequency does that bar have when oscillating in its 2nd harmonic? How many nodes will occur across the bar in that motion?
5. Regarding Figure 1.1, the supporting article states that the vibrations of the bars are generated with electromagnetic radiation. If the radiation used to excite the shortest (2 micrometer) bar is tuned to the bar's fundamental frequency of 0.4 GHz, what is the vacuum wavelength of the radiation?
6. Continuing from Exercise 5, suggest possible means of detecting the motion of a bar in order to confirm a resonance. [Hints: look in a transmission electron microscope (TEM), or Scanning Electron Microscope (SEM) for blurring of the image: look for power absorption from the source....]

Nanophysics and Nanotechnology: An Introduction to Modern Concepts in Nanoscience. Second Edition.
Edward L. Wolf

ISBN: 3-527-40651-4

Exercises – Chapter 2

1. It is stated that the linear vibration frequency of the CO molecule is 64.2 THz. Using the masses of C and O as 12 and 16, respectively (in units of $u = 1.66 \times 10^{-27}$ kg), can you show that this vibration frequency is consistent with an effective spring constant K of 1860 N/m? [Hint: in a case like this the frequency becomes $\omega = 2\pi f = (K/\mu)^{1/2}$, with effective mass $\mu = m_1 m_2/(m_1 + m_2)$.] What is the wavelength of radiation matching this frequency?
2. Treating the CO molecule of Exercise 1 as a classical oscillator in equilibrium with a temperature 300 K, estimate the amplitude of its thermal vibration.
3. Show that the spring constant K' of one piece of a spring cut in half is $K' = 2K$.
4. A small particle of radius 10 micrometers and density 2000 Kg/m^3 falls in air under the action of gravity. If the viscosity of air is 1.8×10^{-5} Pa·s, show that the terminal velocity is about 23 mm/s.
5. For the particle considered in Exercise 4, find the diffusion length during a time 1 s, in air, at 300 K.

Exercises – Chapter 3

1. In the Millikan "oil drop experiment", find the electric field in V/m needed to arrest the fall of a 10 micrometer radius oil particle with net charge $4e$. Take the density of the oil drop as 1000 Kg/m^3; the acceleration of gravity, g, as 9.8 m/s^2; and neglect the buoyant effect of the air in the chamber on the motion of the droplet.
2. An FM radio station transmits 1000 Watts at 96.3 MHz. How many photons per second does this correspond to? What is the wavelength? What is the significance of the wavelength, if any, from the photon point of view?
3. A photon's energy can be expressed as pc, where p is its momentum and c is the speed of light. Light carries energy and momentum but has zero mass! Calculate the force exerted by a 1000 Watt beam of light on a perfectly absorbing surface. Why is this force doubled if the surface, instead of absorbing, perfectly reflects all of the light?
4. It is found that short-wavelength light falling on a certain metal, causes emission (photoemission) of electrons of maximum kinetic energy, K, of 0.3 eV. If the work function, ϕ, of the metal in question is 4 eV, what is the wavelength of the light?
5. Explain, in the context of Exercise 4, how a value for Planck's constant h can be found as the slope of a plot of K vs. $f = c/\lambda$, where the light frequency f is assumed to be varied. Such experiments give the same value of Planck's constant h as was derived by Planck by fitting the spectrum of glowing light from a heated body. Einstein was awarded a Nobel Prize in part for his (1905) analysis of this *photoelectric effect*.

Exercises – Chapter 4

1. Using the Uncertainty Principle, estimate the minimum velocity of a bacterium (modeled as a cube of side 1 micrometer, and having the density of water), known to be located with uncertainty 0.1nm at $x = 0$, in vacuum and at $T = 0$.
2. Explain why the preceding question makes no sense if the bacterium is floating in water at 300 K. Explain, briefly, what is meant by Brownian motion.
3. Find the minimum speed of a C_{60} molecule in vacuum, if it is known to be located precisely at $x,y,z = 0$ plus or minus 0.01 nm in each direction.
4. A Buckyball C_{60} molecule of mass 1.195×10^{-24} Kg is confined to a one-dimensional box of length $L = 100$ nm. What is the energy of the $n = 1$ state? What quantum number n would be needed if the kinetic energy is 0.025 eV (appropriate to room temperature)?
5. A particle is in the $n = 7$ state of a one-dimensional infinite square well potential along the x-axis: $V = 0$ for $0 < x < L = 0.1$ nm; $V =$ infinity, elsewhere. What is the exact probability at $T = 0$ of finding the particle in the range $0.0143 \text{ nm} < x < 0.0286$ nm? (Answer is 1/7, by inspection of the plot of $P(x)$.)
6. Consider a hypothetical semiconductor with bandgap 1 eV, with relative electron mass 0.05, and relative hole mass 0.5. In a cube-shaped quantum dot of this material with $L = 3$ nm, find the energy of transition of an electron from the (211) electron state to the (111) hole state. (This transition energy must include the bandgap energy, and the additional energy is referred to as the "blue shift" of the fluorescent emission by the "quantum size effect".) Note that the "blue shift" can be tuned by adjusting the particle size, L.

Exercises – Chapter 5

1. Make an *estimate* of the vibrational frequency of the H_2 molecule, from the lower curve in Figure 5.1. (You can do this by approximating the minimum in the curve as $1/2\, K(r-r_o)^2$, where r is the interatomic spacing, and K is an effective spring constant, in eV/(Angstrom)^2. Estimate K from Figure 5.1; then adapt the usual formula for the frequency of a mass on a spring).
2. Use the result from Exercise 5.1 to find the "zero point energy" of di-hydrogen treated as a linear oscillator. Estimate the "zero-point motion" of this oscillator, in nm, and as a fraction of the equilibrium spacing. Compare this answer to one obtained using the uncertainty principle.
3. At $T = 300$ K, what is the approximate value of the oscillator quantum number, n, for the di-hydrogen vibrations, expected from $kT/2 = (n + 1/2)h\nu$? (At 300 K, an approximate value for kT is 0.025 eV.)
4. Explain, in a paragraph or two, the connections between the covalent bond energy of symmetric molecules, such as di-hydrogen; the notion of anti-symmetry under exchange for fermi particles, such as electrons; and the phenomenon of ferromagnetism (a basis for hard disk data storage).

5. Explain (a) why *mean free paths* of electrons and holes in semiconductors can be much larger than the spacing between the atoms in these materials. (b) Explain, in light of the nanophysical Kronig–Penney model, why the *mean free paths* of electrons and holes in exquisitely ordered solids (such as the semiconductor industry's 6 inch diameter single crystal "boules" of silicon) are not *infinitely* large (i.e., limited by the size, L, of the samples). Consider the effects of temperature, T, the impurity content N_i; and possible isotopic mass differences among the silicon atoms.
6. Why cannot a filled band provide electrical conductivity?
7. Why can the effective mass in a solid differ from the mass of the electron in vacuum?
8. Why does a hole have a charge of $+e$, and a mass equal to the mass of the valence band electron that moves into the vacant site?
9. How does Bragg reflection ($n\lambda = d\sin\theta$) influence the motion of carriers in a (one-dimensional) semiconductor with lattice constant a? Why does this lead to the answer that the "zone-boundary" is at $k = \pi/a$?
10. Explain why the group velocity of carriers is zero at "the zone boundary", $k = \pi/a$.
11. Explain the difference between the two (standing wave) wavefunctions valid at $k = \pi/a$. Explain, qualitatively, the difference in their electrostatic energies, and why this energy difference is precisely the "band gap" into the next higher band.
12. Estimate the lifetime τ for the H (or D) atom to lose its electron by field ionization at (a) E = 25 V/nm and (b) E = 2.5 V/nm, using a one-dimensional tunneling model.
Hints: The ground state with $E = 0$ is spherically symmetric, depending only on r. With an electric field $E_x = E$, the situation approximately depends only on x:

electron potential energy $U(x) = -k_C e^2/x - eEx$,

where $k_C = 9 \times 10^9$, with E in V/m. Assume that the ground-state energy is unchanged by E, so

$E = -E_0 = -13.6$ eV.

The electron barrier potential energy is $U(x) - E_0$. For $E = 0$, the electron is assumed to have energy $U = -E_0$ at $x = a_o = 0.053$ nm, the Bohr radius. When $E > 0$, there is a second location, near $x_2 = E_0/eE$, where U is again E_0. The problem is to find the time τ to tunnel from $x = a_o$ to $x = x_2 = E_0/eE$ through the potential barrier $U(x) - E_0$. For small E, the maximum height of the barrier seen by the electron is $\approx E_0 = 13.6$ eV, and the width t of the tunnel barrier is essentially $\Delta x = t = x_2 = E_0/eE$. If the barrier potential $U(x)$ were V_B (independent of x), the tunneling transmission probability T^2 would be given by (4.57) and (4.64), retaining only the exponential part of (4.64). Thus

$T^2 = \exp[-2(2mV_B)^{1/2}\Delta x/\hbar]$,

with barrier width Δx and $\hbar = h/2\pi$, with h Planck's constant.
[A more accurate conventional approach would replace $(V_B)^{1/2}\Delta x$ by $\int[U(x) - E_0]^{1/2}\,dx$.]

To simplify, consider the *average value* of the barrier. Approximate the barrier function $U(x) - E_0$ by half its maximum value, $\frac{1}{2}V_{Bmax}$, over the width Δx. The escape rate, $f_{escape} = 1/\tau$, where τ is the desired ionization lifetime, is $T^2 f_{approach}$. The orbital frequency of the electron will be taken from Bohrs's rule for the angular momentum $L = mvr = h/2\pi$, equation (4.2), so

$f_{approach} = h/[(2\pi)^2 m a_o{}^2] = 6.5 \times 10^{15}\ s^{-1}$.

The working formula $f_{escape} = 1/\tau$ is then

$f_{escape} = f_{approach}\ \exp[-2(mV_{Bmax})^{1/2}\Delta x/\hbar]$.

So as remaining steps for the reader:
(A) Solve the quadratic $-E_0 = -ke^2/x - eEx$, show that $\Delta x = [(E_0/e)^2 - 4keE]/E$. (This limits the applicability of the approximation to $E < 32.1$ V/nm, where $\Delta x = 0$. This alone is a rough criterion for the onset of rapid field ionization.)
(B) Find dU/dx and set it to zero, to show that the position x' of the maximum in $U(x)$ occurs at $x' = (ke/E)^{1/2}$. Insert this value x' into $U(x)$ to show that its maximum value is $U(x') = -2e(keE)^{1/2}$. Then show that $V_{Bmax} = E_0 - 2e(keE)^{1/2}$.
(C) Evaluate f_{escape} from the working formula, taking $E = 25$ V/nm. (The exponential should be about exp(−2.35). The lifetime is about 1.7×10^{-15} s.
(D) Evaluate f_{escape} from the working formula for $E = 2.5$ V/nm. (The exponential should be exp(−118.8). The lifetime is about 6×10^{33} s or about 1.9×10^{26} y. (The famous result of Oppenheimer is a lifetime of $(10^{10})^{10}$ s for $E = 1000$ V/m.)

13. Estimate the ion current in the neutron generation apparatus using a positively charged tip as described in relation to Figure 5.23. Assume that the radius r around the tip center leading to 100% ionization of the deuterium is 600 nm. The assumption is that all D_2 molecules which fall on the front half of this surface (of area $2\pi r^2$) contribute 2e to the ion current. The rate of molecules crossing a surface in a dilute gas is $nv/4$ (molecules per unit area per unit time), where n is the number per unit volume of molecules whose rms speed is Iv.
(A) In a gas of deuterium molecules at temperature $T = 270$ K, show that the rms speed is 1294 m/s. The deuterium molecule has a mass of 4 amu.
(B) The gas is said to be at a pressure of 0.7 Pa at $T = 270$ K. Using the ideal gas law $pV = RT$ (for one mole, corresponding to Avogadro's number of molecules), show that the number density n of molecules is $1.9 \times 10^{20}\ m^{-3}$. On this basis, show that the ion current would be about 44.5 nA (the observed value is 4 nA).
(C) Verify that the mean free path of the D ion exceeds the dimension of the container, so that straight-line trajectories can be assumed. A formula for

the mean free path is $\lambda = 1/ns$, where the cross-section can be taken as $s = \pi\rho^2$, and take the radius ρ of the D_2 molecule as 0.037 nm (look at Figure 5.1).

14. Under adiabatic compression of an ideal gas bubble of initial radius 0.1 mm at 1 bar (one atmosphere = 101.3 kPa) and 300 K to final radii (A) r_{min} = 14.1 μm and (B) r_{min} = 10 nm, find the resulting final pressures, expressed in atmospheres (bar). See text related to Figure 5.25.
15. For adiabatic compression of an ideal gas, find expressions for the pressure as a function of the temperature, volume and radius r of the bubble. Follow the method used in the text in deriving the relation $T \propto (1/V)\gamma^{-1}$.
16. The classical equation of Laplace relates the pressure difference across the surface of a spherical bubble of radius r in a liquid as

$$\Delta p = 2\gamma/r,$$

where γ is the surface tension, which is 0.0728 N/m for water at 25 °C. Evaluate this equilibrium pressure difference in atmospheres for water bubbles of radii 0.1 mm, 1 μm and 10 nm.

Exercises – Chapter 6

1. The basic measure of thermal agitation energy is $1/2\,kT$ per degree of freedom, where k is Boltzmann's constant. How does thermal energy (at 400 K) compare with the energy differences among the computer-modeled simulations of octane presented in Figure 6.1?
2. In Figure 6.1, 10^{11} Hz is a characteristic rate at which the octane molecule switches from one shape (conformation) to another. For comparison, find the lowest longitudinal vibration frequency, using the methods of Chapter 2, modeling octane as a chain of eight carbon atoms (mass 12) connected by springs of strength $K = 440$ N/m and length 0.15 nm.
3. The variety of different conformations of octane, exhibited in Figure 6.1, seem incompatible with using octane to carry a single (small-diameter) tip, which is a required function of the proposed "molecular assembler". With regard to stability, give reasons why acetylene C_2H_2 (a) and a small-diameter carbon nanotube (b), should be superior.
4. Consider the FET device shown in Figure 6.4. Why is the forward current in this FET so large? How could such FET devices be assembled at a density of 10^7 /cm^2 on a silicon chip? (If you have a good answer, have your statement notarized and hire a patent lawyer!)
5. How fast can the FET device shown in Figure 6.4 respond to a gate voltage turned on at time zero? It is said that the electron transport along a nanotube is "ballistic", which means that, if the force is eE, then the velocity is eEt/m, and the distance covered in time t is $1/2(eE/m)t^2$. For the device geometry shown, where $E = V/L$, set $V = 1$ Volt and find: (a) the corresponding value of t. Secondly, (b) for the device geometry shown, convert this result into an effective mobility μ, such that $v = \mu E = \mu V/L$, and $t = L^2/\mu V$. Note that the mo-

bility μ is conventionally given in units of cm^2/Vs. Compare this mobility to values listed in Table 5.1.

6. Consider the magnetotactic bacterium shown in Figure 6.7. (a) Calculate the torque exerted on the bacterium in an earth's magnetic field of 1 Gauss (10^{-4} T) if the bacterium is oriented at 90°. (b) Explain why a strain of magnetotactic bacteria found in the northern hemisphere, if transported to an ocean in the southern hemisphere, would surely die out.
7. In the STM study of ordered arrays of C_{60} shown in Figure 6.9 it was observed that the molecules were locked into identical orientations at 5 K, but were apparently rotating at 77 K. (a) From this information, make an estimate of the interaction energy between nearest-neighbor C_{60} molecules. Is this energy in the right range to be a van der Waals interaction? (b) Compare this temperature information to the melting point of a "C_{60} molecular solid" which you can estimate from Figure 5.2.

Exercises – Chapter 7

1. Compare the conventional, state-of-the-art, spatial resolution for modern photolithography, presented as 180 nm, to the resolution, in principle, that is available at wavelength 248 nm, see Figure 7.3.
2. Explain why e-beam writing is not competitive as a production process for writing wiring on Pentium chips.
3. What features of Figure 7.3 are inconsistent with the stated exposure wavelength of 248 nm?
4. Compare e-beam writing to conventional lithography with regard to *resolution* and *throughput*.
5. It is reported in Consumer Reports (October 2003, p. 8) that the "Segway Human Transporter" has five independent *electronic gyroscope sensors*, with, presumably, a corresponding set of servo systems. *Write an essay on how such rotation sensors might be fabricated (clue, one method might be based on Figure 7.1).* In principle, how would the performance of such rotation sensors be expected to change with a uniform reduction in scale, *xyz* of $\times$ 10?
6. What are the chief advantages of the superconductive RSFQ computing technology, and what are the chief disadvantages of the same technology?
7. Why is the spatial resolution of the STM better than that of the AFM?
8. Why can't a single tip in the STM/AFM technology possibly perform operations at a rate greater then 1 GHz?
9. Why cannot a single tip in the STM/AFM technology perform operations beyond *observing, nudging,* and *exciting* an atom (or molecule) in its view? (Explain what is meant by "nudging", and why it is easier for atoms, like Xe, or molecules, like benzene, with a large number of electrons.) Other desirable, but unavailable, functions for a single tip would be picking up, orienting, and depositing, atoms or molecules. *Why are these functions unavailable?*

Exercises – Chapter 8

1. With reference to the initial discussion of qubits in Chapter 8, explain in words why several independent measurements of identically prepared quantum states are needed to determine the coefficients a and b. Can a and b ever be precisely known in such a case?
2. Verify the size of effective magnetic field B^* needed to produce an energy difference between (electron) spin up and spin down of about twice 4.74 eV, as suggested in Figure 5.1.
3. Make an argument based on symmetry to explain why the average magnetic dipole–dipole interaction for the proton spin and the ground-state electron spin in hydrogen is zero. If this is so, what is the origin of the 21.1 cm line emission observed from hydrogen in outer space?
4. With regard to Figure 8.2, explain, for ferromagnetic hcp-Co, the statement that the "density of spin-down states at E_F exceeds the spin-up density", while the majority spins are "up".
5. Explain why the DNA strands seen in Figure 8.17, deposited initially in a drop of solvent directly over the trench in the Si, eventually, after evaporation of the solvent, cross the trench in straight lines which are accurately perpendicular to the trench.

Exercises – Chapter 9

1. Evaluate the capacitance of a nanotube of radius 0.5 nm and length 1 m, assuming a coaxial counter-electrode of radius 250 nm (see text near Figure 9.7). [Answer: 36 pF/m]
2. Using the result from above, and the stated capacitance, 10 nF, for the chemical sensing device of Figure 9.7, estimate the total length of the nanotube network forming the chemically sensitive electrode. [Answer: 279 m]
3. Published data have shown roughly 0.1 μA as an observed source–drain current in carbon nanotube-based FET, at limits of the gate voltage range. Estimate the current density, in A/cm^2, assuming the nanotube of diameter 1.4 nm is (A) a solid cylinder and (B) a cylindrical shell of diameter 1.4 nm and shell thickness 0.1 nm.
4. In connection with Figures 9.9 and 9.10, estimate the electrostatic attractive force in newtons between two crossing (10,10) single-wall nanotubes, assuming the voltages are +5 V and –5 V, and the tubes are 20 nm long. Take the initial wall-to-wall spacing s to be 1.2 nm. The diameter of the (n,m) nanotube is given as $d_{\text{tube}} = a\pi^{-1}(nn + mm + nm)^{1/2}$ where $a = 0.249$ nm.
5. Assume that a (10,10) single-wall carbon nanotube as in Exercise 9.4 (see Figure 9.9) of length $L = 20$ nm is clamped at each end. Estimate the transverse vibrational frequency of one such section.
   Hints: Use the formulas in Section 2.1, applicable to an elastic beam (assume Young's modulus is 1 TPa) clamped at each end. Estimate the mass per unit

length of the (10,10) single-wall nanotube from the mass per carbon atom, $12 \times 1.66 \times 10^{-27}$ kg, the C–C bond length of 0.142 nm, and the tube's surface area $\pi d_{tube} L$ for length L. (A published estimate for this frequency is about 100 GHz.)

6. Continuing from Exercise 9.5, the formulas in Section 2.1 give an approach to estimate the spring constant K for the (10,10) nanotube. Using this estimate, find the force needed to displace the tube by 1 nm. Compare this force to the electrostatic force estimated in Exercise 9.4.

Exercises – Chapter 10

1. Estimate the diameters of the proposed molecules in Figure 10.1.
2. Provide a critique of the suggested size of the "molecular assembler" as one billion atoms (it may be on the small side?) What would be the expected radius for a spherical assembly of one billion atoms, in nm? Can you estimate the number of atoms in a typical enzyme, as a comparison?
3. It has been suggested that the time for a typical bacterium to grow is about an hour. Suppose a rectangular (cube) bacterium has dimension $L = 1$ micrometer, the density of water, and is made of carbon 12. Roughly how many atoms are there in this bacterium? Calculate the rate of growth in atoms per second if this forms in one hour. Compare your answer to the performance of the IBM Millipede AFM, discussed in Chapter 7.
4. The example of the magnetotactic bacterium (see Chapter 6) proves that some robust inorganic solids, eg. Fe_3O_4, can be grown at room temperature from (ionic) solution. Compare this case with the case of diamond (and the proposed diamondoid inorganic molecules), which require high temperatures/pressures to form (and apparently cannot be grown from solution). What is the crucial difference?
5. What is the most significant difference between a virus and a living organism? (Can a virus live forever?)
6. Poll your colleagues to ask if they think that "computers" (or artificially intelligent robots) will ever be able to establish separate identities, consciousness, and other human qualities. If so, how threatening does this prospect seem?

Some Useful Constants

Avogadro's number	N_A	6.022×10^{23} particles/mole
Boltzmann's constant	k_B	1.381×10^{-23} J/K
Ideal gas constant	$R = N_A k_B$	8.315 J/(mol K)
Fundamental charge	e	1.602×10^{-19} C
Mass of electron	m_e	9.109×10^{-31} kg (= 511 keV/c^2)
Mass of proton	m_p	1.672×10^{-27} kg (= 938.3 MeV/c^2)
Planck's constant	h	6.63×10^{-34} J s (= 4.136×10^{-15} eV s)
	$\hbar = h/2\pi$	1.055×10^{-34} J s (= 6.58×10^{-16} eV s)
Bohr magneton	$\mu_B = e\hbar/2m_e$	9.274×10^{-24} J/T (= 5.79×10^{-5} eV/T)
Coulomb's constant	$k_C = 1/(4\pi\varepsilon_o)$	8.988×10^9 Nm^2/C^2
Permittivity of space	μ_o	$4\pi \times 10^{-7}$ N/A^2
Speed of light	c	2.998×10^8 m/s (= 0.2998 mm/ps)
Photon energy	$hc/\lambda = hf$	1240 (eV nm)/nm
Hydrogen atom binding energy	$k_C e^2/2a_o$	13.609 eV
Bohr radius	a_o	0.0529 nm
Conductance quantum= $(R_{\text{Klitzing}})^{-1}$	$e^2/h = 2\alpha/\mu_o c$	1/(25.8 kΩ)
Flux quantum, superconducting	$hc/2e$	2.07×10^{-15} Wb= 2.07×10^{-7} Oe cm^2
Electronvolt	eV	1.602×10^{-19} J (= 23.06 kcal/mol)
Fine structure constant	$\alpha = \mu_o c e^2/2h$	1/137.036
Josephson frequency/voltage ratio	$2e/h$	4.836×10^{14} Hz/V
		483.6 MHz/μV
Polarizability of hydrogen orbit	$\alpha_H = p/E = 4.5a_o^3/k_C$	7.40×10^{-41} C^2 m^2/J

Nanophysics and Nanotechnology: An Introduction to Modern Concepts in Nanoscience. Second Edition.
Edward L. Wolf

ISBN: 3-527-40651-4

Index

Nanophysics and Nanotechnology: An Introduction to Modern Concepts in Nanoscience. Second Edition.
Edward L. Wolf

ISBN: 3-527-40651-4

f

g

h

n

q

t